Quantum Mechanics

Franz Schwabl

Quantum Mechanics

Fourth Edition
With 123 Figures, 16 Tables,
Numerous Worked Examples and 127 Problems

 Springer

Professor Dr. Franz Schwabl

Physik-Department
Technische Universität München
James-Franck-Strasse 2
85748 Garching, Germany
E-mail: schwabl@ph.tum.de

The first edition, 1992, was translated by Dr. Ronald Kates

Title of the original German edition: *Quantenmechanik* 7th edition

© Springer-Verlag Berlin Heidelberg 2010

ISBN   978-3-642-09107-0   e-ISBN   978-3-540-71933-5

Springer is a part of Springer Science+Business Media

springer.com

© Springer-Verlag Berlin Heidelberg 2010

Cover design: eStudio Calamar S.L., F. Steinen-Broo, Pau/Girona, Spain

Preface to the Fourth Edition

In this latest edition new material has been added, which includes many additional clarifying remarks to some of the more advanced chapters. The design of many figures has been reworked to enhance the didactic appeal of the book. However, in the course of these changes, I have attempted to keep intact the underlying compact nature of the book.

I am grateful to many colleagues for their help with this substantial revision. Special thanks go to Uwe Täuber and Roger Hilton for discussions, comments and many constructive suggestions on this new edition. Some of the figures which were of a purely qualitative nature have been improved by Robert Seyrkammer in now being computer-generated. I am very obliged to Andrej Vilfan for redoing and checking the computation of some of the scientifically more demanding figures. I am also very grateful to Ms Ulrike Ollinger who undertook the graphical design of the diagrams. It is my pleasure to thank Dr. Thorsten Schneider and Mrs Jacqueline Lenz of Springer for their excellent co-operation, as well as the LE-TEX setting team for their careful incorporation of the amendments for this new edition. Finally, I should like to thank all colleagues and students who, over the years, have made suggestions to improve the usefulness of this book.

Munich, August 2007   *F. Schwabl*

Preface to the First Edition

This is a textbook on quantum mechanics. In an introductory chapter, the basic postulates are established, beginning with the historical development, by the analysis of an interference experiment. From then on the organization is purely deductive. In addition to the basic ideas and numerous applications, new aspects of quantum mechanics and their experimental tests are presented. In the text, emphasis is placed on a concise, yet self-contained, presentation. The comprehensibility is guaranteed by giving all mathematical steps and by carrying out the intermediate calculations completely and thoroughly.

The book treats nonrelativistic quantum mechanics without second quantization, except for an elementary treatment of the quantization of the radiation field in the context of optical transitions. Aside from the essential core of quantum mechanics, within which scattering theory, time-dependent phenomena, and the density matrix are thoroughly discussed, the book presents the theory of measurement and the Bell inequality. The penultimate chapter is devoted to supersymmetric quantum mechanics, a topic which to date has only been accessible in the research literature.

For didactic reasons, we begin with wave mechanics; from Chap. 8 on we introduce the Dirac notation. Intermediate calculations and remarks not essential for comprehension are presented in small print. Only in the somewhat more advanced sections are references given, which even there, are not intended to be complete, but rather to stimulate further reading. Problems at the end of the chapters are intended to consolidate the student's knowledge.

The book is recommended to students of physics and related areas with some knowledge of mechanics and classical electrodynamics, and we hope it will augment teaching material already available.

This book came about as the result of lectures on quantum mechanics given by the author since 1973 at the University of Linz and the Technical University of Munich. Some parts of the original rough draft, figures, and tables were completed with the help of R. Alkofer, E. Frey and H.-T. Janka. Careful reading of the proofs by Chr. Baumgärtel, R. Eckl, N. Knoblauch, J. Krumrey and W. Rossmann-Bloeck ensured the factual accuracy of the translation. W. Gasser read the entire manuscript and made useful suggestions about many of the chapters of the book. Here, I would like to express my sincere gratitude to them, and to all my other colleagues who gave important assistance in producing this book, as well as to the publisher.

Munich, June 1991   *F. Schwabl*

Table of Contents

1. Historical and Experimental Foundations

1.1 Introduction and Overview

In spite of the multitude of phenomena described by classical mechanics and electrodynamics, a large group of natural phenomena remains unexplained by classical physics. It is possible to find examples in various branches of physics, for example, in the *physics of atomic shells*, which provide a foundation for the structure of electron shells of atoms and for the occurrence of discrete energy levels and of homopolar and Van der Waals bonding. *The physics of macroscopic bodies* (solids, liquids, and gases) is not able to give – on the basis of classical mechanics – consistent explanations for the structure and stability of condensed matter, for the energy of cohesion of solids, for electrical and thermal conductivity, specific heat of molecular gases and solids at low temperatures, and for phenomena such as superconductivity, ferromagnetism, superfluidity, quantum crystals, and neutron stars. Nuclear physics and elementary particle physics require absolutely new theoretical foundations in order to describe the structure of atomic nuclei, nuclear spectra, nuclear reactions (interaction of particles with nuclei, nuclear fission, and nuclear fusion), and the stability of nuclei, and similarly in order to make predictions concerning the size and structure of elementary particles, their mechanical and electromagnetic properties (mass, angular momentum (spin), charge, magnetic moment, isospin), and their interactions (scattering, decay, and production). Even in *electrodynamics and optics* there are effects which cannot be understood classically, for example, blackbody radiation and the photoelectric effect.

All of these phenomena can be treated by quantum theoretical methods. (An overview of the elements of quantum theory is given in Table 1.1.) This book is concerned with the nonrelativistic quantum theory of stable particles, described by the Schrödinger equation.

First, a short summary of the essential concepts of classical physics is given, before their limitations are discussed more thoroughly in Sect. 1.2.

At the end of the nineteenth century, physics consisted of classical mechanics, which was extended in 1905 by Albert Einstein's theory of relativity, together with electrodynamics.

Classical mechanics, based on the Newtonian axioms (*lex secunda*, 1687), permits the description of the dynamics of point masses, e.g., planetary mo-

Table 1.1. The elements of quantum theory

	Nonrelativistic	Relativistic
Quantum theory of stable particles	Schrödinger equation	Dirac equation (for fermions)
Quantum theory of creation and annihilation processes	Nonrelativistic field theory	Relativistic field theory

tion, the motion of a rigid body, and the elastic properties of solids, and it contains hydrodynamics and acoustics. Electrodynamics is the theory of electric and magnetic fields in a vacuum, and, if the material constants ε, μ, σ are known, in condensed matter as well. In classical mechanics, the state of a particle is characterized by specifying the position $\boldsymbol{x}(t)$ and the momentum $\boldsymbol{p}(t)$, and it seems quite obvious to us from our daily experience that the simultaneous specification of these two quantities is possible to arbitrary accuracy. Microscopically, as we shall see later, position and momentum cannot simultaneously be specified to arbitrary accuracy. If we designate the uncertainty of their components in one dimension by Δx and Δp, then the relation $\Delta x \Delta p \geq \hbar/2$ must always hold, where $\hbar = 1.0545 \times 10^{-27}$ erg s is the Planck quantum of action[1]. *Classical particles* are thus characterized by position and velocity and represent a spatially bounded "clump of matter".

On the other hand, *electromagnetic waves*, which are described by the potentials $\boldsymbol{A}(\boldsymbol{x},t)$ and $\Phi(\boldsymbol{x},t)$ or by the fields $\boldsymbol{E}(\boldsymbol{x},t)$ and $\boldsymbol{B}(\boldsymbol{x},t)$, are spatially extended, for example, plane waves $\exp\{i(\boldsymbol{k} \cdot \boldsymbol{x} - \omega t)\}$ or spherical waves $(1/r) \exp\{i(kr - \omega t)\}$. Corresponding to the energy and momentum density of the wave, the energy and momentum are distributed over a spatially extended region.

In the following, using examples of historical significance, we would like to gain some insight into two of the main sources of the empirical necessity for a new theoretical basis: (i) on the one hand, the impossibility of separating the particle and wave picture in the microscopic domain; and (ii) the appearance of discrete states in the atomic domain, which forms the point of departure for the Bohr model of the atom.

---

[1] 1 erg $= 10^{-7}$ J

1.2 Historically Fundamental Experiments and Insights

At the end of the nineteenth and the beginning of the twentieth century, the inadequacy of classical physics became increasingly evident due to various empirical facts. This will be illustrated by a few experiments.

1.2.1 Particle Properties of Electromagnetic Waves

1.2.1.1 Black-Body Radiation

Let us consider a cavity at temperature T in radiation equilibrium (Fig. 1.1).

The volume of the cavity is $V = L^3$, the energy density (energy per unit volume and frequency) $u(\omega)$. Here $u(\omega)d\omega$ expresses the energy per unit volume in the interval $[\omega, \omega + d\omega]$. Classically, the situation is described by the Rayleigh–Jeans law

$$u(\omega) = \frac{k_B T}{\pi^2 c^3}\omega^2 \quad . \tag{1.1}$$

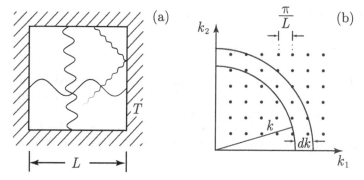

Fig. 1.1a,b. Black-body radiation. **(a)** The radiation field. **(b)** k-space: 1 point per volume $(\pi/L)^3$

One can easily make this plausible by considering standing plane waves in a cavity with reflecting metal walls. The components of the electric field are

$$E_1(x) \sim \cos k_1 x_1 \sin k_2 x_2 \sin k_3 x_3 \dots \quad \text{with} \quad k = \frac{\pi}{L}(n_1, n_2, n_3) \quad ,$$

$$\vdots \qquad\qquad n_i = 1, 2, 3, \dots \quad .$$

The number of waves in the interval $[\omega, \omega + d\omega]$ is, considering the vacuum dispersion relation $\omega = ck$, equal to the number dN of wave vector points in

$1/8$ of the spherical shell[2] $[k, k + dk]$, that is

$$dN = \frac{1}{8} \frac{\text{Volume of the } k\text{-space spherical shell}}{k\text{-space volume per point}}$$
$$= \frac{4\pi k^2 dk}{8 \,(\pi/L)^3} = \frac{L^3}{2\pi^2 c^3} \omega^2 d\omega \quad .$$

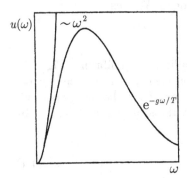

Fig. 1.2. The Rayleigh–Jeans law and the Planck radiation law

Furthermore, since the energy of an oscillator is $k_B T$ (where the Boltzmann constant is $k_B = 1.3806 \times 10^{-16}$ erg/K), one obtains because of the two directions of polarization

$$u(\omega)d\omega = 2 \frac{L^3}{2\pi^2 c^3} \omega^2 d\omega \frac{k_B T}{L^3} = \frac{k_B T}{\pi^2 c^3} \omega^2 d\omega \quad ,$$

i.e., Eqn. (1.1). However, because of $\int_0^\infty u(\omega)d\omega = \infty$, this classical result leads to the so-called "ultraviolet catastrophe", i.e., the cavity would have to possess an infinite amount of energy (Fig. 1.2).

Although experiments at low frequencies were consistent with the Rayleigh–Jeans formula, Wien found empirically the following behavior at high frequencies:

$$u(\omega) \overset{\omega \to \infty}{\longrightarrow} A\omega^3 e^{-g\omega/T} \quad (A, g = \text{const}) \quad .$$

Then in 1900, Max Planck discovered (on the basis of profound thermodynamical considerations, he interpolated the second derivative of the entropy between the Rayleigh–Jeans and Wien limits) an interpolation formula (the Planck radiation law):

$$u(\omega) = \frac{\hbar}{\pi^2 c^3} \frac{\omega^3}{\exp\{\hbar\omega/k_B T\} - 1} \quad , \quad \hbar = 1.0545 \times 10^{-27} \, \text{erg s} . \qquad (1.2)$$

---

[2] Remark: The factor $1/8$ arises because the k_i-values of the standing wave are positive. One obtains the same result for dN in the case of periodic boundary conditions with $\exp\{i\boldsymbol{k} \cdot \boldsymbol{x}\}$ and $\boldsymbol{k} = (n_1, n_2, n_3)2\pi/L$ and $n_i = 0, \pm 1, \pm 2, \ldots$.

He also succeeded in deriving this radiation law on the basis of the hypothesis that energy is emitted from the walls into radiation only in multiples of $\hbar\omega$, that is $E_n = n\,\hbar\omega$.

This is clear evidence for the *quantization of radiation energy.*

1.2.1.2 The Photoelectric Effect

If light of frequency ω (in the ultraviolet; in the case of alkali metals in the visible as well) shines upon a metal foil or surface (Hertz 1887, Lenard), one observes that electrons with a maximal kinetic energy of

$$E_e = \frac{mv_e^2}{2} = \hbar\omega - W \quad (W = \text{work function})$$

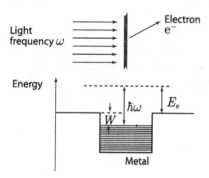

Fig. 1.3. The photoelectric effect

are emitted (Fig. 1.3). This led Albert Einstein in 1905 to the hypothesis that light consists of photons, quanta of energy $\hbar\omega$. According to this hypothesis, an electron that is bound in the metal can only be dislodged by an incident photon if its energy exceeds the energy of the work function W.

In classical electrodynamics, the energy density of light in vacuum is given by $(1/8\pi)(E^2 + H^2)$ (proportional to the intensity) and the energy flux density by $\boldsymbol{S} = (c/4\pi)\,\boldsymbol{E} \times \boldsymbol{H}$. Thus, one would *expect classically* at small intensities that only after a certain time would enough energy be transmitted in order to cause electron emission. Also, there should not be a minimum light frequency for the occurrence of the photoelectric effect. However, what one actually observes, even in the case of low radiation intensity, is the *immediate* onset of electron emission, albeit in small numbers (Meyer and Gerlach), and no emission occurs if the frequency of the light is lowered below $W/\hbar$, consistent with the quantum mechanical picture. Table 1.2 shows a few examples of real work functions.

We thus arrive at the following hypothesis: Light consists of photons of energy $E = \hbar\omega$, with velocity c and propagation direction parallel to the electromagnetic wave number vector $\boldsymbol{k}$ (reason: light flash of wave number $\boldsymbol{k}$).

Table 1.2. Examples of real work functions

Element	W	Ta	Ni	Ag	Cs	Pt
W in eV	4.5	4.2	4.6	4.8	1.8	5.3

$1\,\mathrm{eV} \triangleq \lambda = 1.24 \times 10^{-4}\,\mathrm{cm} \triangleq 1.6 \times 10^{-12}\,\mathrm{erg}$
$4\,\mathrm{eV} \triangleq \lambda = 3.1 \times 10^{-5}\,\mathrm{cm}$, i.e. ultraviolet

With this we can already make a statement about the momentum and mass of the photon.

From relativity theory, one knows that

$$E = \sqrt{p^2c^2 + m^2c^4} \quad , \quad v = \frac{\partial E}{\partial p} = \frac{pc^2}{\sqrt{p^2c^2 + m^2c^4}} \quad . \tag{1.3}$$

Since $|v| = c$, it follows from (1.3) that $m = 0$ and thus $E = pc$. If we compare this with $E = \hbar\omega = \hbar ck$ (electromagnetic waves: $\omega = ck$), then $p = \hbar k$ results. Because $\boldsymbol{p}$ and $\boldsymbol{k}$ are parallel, it also follows that $\boldsymbol{p} = \hbar\boldsymbol{k}$. Thus

$$\left.\begin{array}{l} E = \hbar\omega \\ \boldsymbol{p} = \hbar\boldsymbol{k} \end{array}\right\} \quad \text{four-vector } p^\mu : \quad \begin{pmatrix} E/c \\ \boldsymbol{p} \end{pmatrix} = \hbar\begin{pmatrix} k \\ \boldsymbol{k} \end{pmatrix} \quad . \tag{1.4}$$

1.2.1.3 The Compton Effect[3]

Suppose that X-rays strike an electron (Fig. 1.4), which for the present purposes can be considered as free and at rest. In an elastic collision between an electron and a photon, the four-momentum (energy and momentum) remains conserved. Therefore,

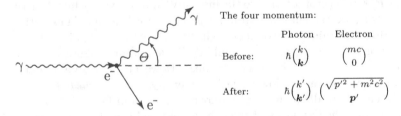

Fig. 1.4. Collision of a photon γ and an electron e^-

[3] A.H. Compton, A. Simon: Phys. Rev. **25**, 306 (1925)

$$\hbar \begin{pmatrix} k \\ \mathbf{k} \end{pmatrix} + \begin{pmatrix} mc \\ 0 \end{pmatrix} = \hbar \begin{pmatrix} k' \\ \mathbf{k}' \end{pmatrix} + \begin{pmatrix} \sqrt{p'^2 + m^2 c^2} \\ \mathbf{p}' \end{pmatrix} . \tag{1.5}$$

If we bring the four-momentum of the photon after the collision over to the left side of (1.5) and construct the four-vector scalar product ($v^\mu q_\mu \equiv v^0 q^0 - \mathbf{v} \cdot \mathbf{q}$ = product of the timelike components v^0, q^0 minus the scalar product of the spacelike ones) of each side with itself, then since $p^\mu p_\mu = p'^\mu p'_\mu = m^2 c^2$, $k^\mu k_\mu = k'^\mu k'_\mu = 0$:

$$m^2 c^2 + 2\hbar (k - k') mc - 2\hbar^2 (kk' - \mathbf{k} \cdot \mathbf{k}') = m^2 c^2 \quad ,$$

$$k - k' = \frac{\hbar}{mc} kk' (1 - \cos \Theta) \quad .$$

Because of $k = 2\pi/\lambda$ one obtains for the change of wavelength

$$\lambda' - \lambda = \frac{4\pi\hbar}{mc} \sin^2 \frac{\Theta}{2} = 4\pi\lambda_c \sin^2 \frac{\Theta}{2} \quad , \tag{1.6}$$

where $\lambda_c = \hbar/m_e c = 3.86 \times 10^{-11}$ cm is the Compton wavelength of the electron ($m_e = 0.91 \times 10^{-27}$ g, $c = 2.99 \times 10^{10}$ cm s^{-1}). For the scattering of X-rays from electrons in carbon, for example, one finds the intensity distribution of Fig. 1.5.

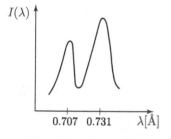

0.707 Å: unscattered photons
0.731 Å: scattered photons
The collision of a photon with an electron leads to an energy loss, i.e., to an increase in the wavelength.

Fig. 1.5. Intensity distribution for scattering of X-rays from carbon

The experiments just described reveal clearly the *particle character* of light. On the other hand, it is certain that light also possesses *wave properties*, which appear for example in interference and diffraction phenomena.

Now, a *duality* similar to that which we found for light waves also exists for the conventional particles of classical physics.

1.2.2 Wave Properties of Particles, Diffraction of Matter Waves

Davisson and Germer (1927), Thomson (1928), and Rupp (1928) performed experiments with electrons in this connection; Stern did similar experiments with helium. If a matter beam strikes a grid (a crystal lattice in the case of electrons, because of their small wavelength), interference phenomena result which are well known from the optics of visible light. Empirically one obtains in this way for nonrelativistic electrons (kinetic energy $E_{kin} = p^2/2m$)

$$\lambda = \frac{2\pi\hbar}{p} = \frac{2\pi\hbar c}{\sqrt{2mc^2(p^2/2m)}} = \frac{12.2 \text{ Å}}{\sqrt{E_{kin}(\text{eV})}} \quad . \tag{1.7}$$

This experimental finding is in exact agreement with the hypothesis made by de Broglie in 1923 that a particle with a total energy E and momentum p is to be assigned a frequency $\omega = E/\hbar$ and a wavelength $\lambda = 2\pi\hbar/p$. The physical interpretation of this wave will have to be clarified later (see Sect. 2.1). On the other hand, it is evident on the basis of the following phenomena that in the microscopic domain the particle concept also makes sense:

- Ionization tracks in the Wilson chamber: The electrons that enter the chamber, which is filled with supersaturated water vapor, ionize the gas atoms along their paths. These ions act as condensation seeds and lead to the formation of small water droplets as the water vapor expands and thus cools.
- Scattering and collision experiments between microscopic particles.
- The Millikan experiment: Quantization of electric charge in units of the elementary charge $e_0 = 1.6021 \times 10^{-19}$ C $= 4.803 \times 10^{-10}$ esu.
- The discrete structure of solids.

1.2.3 Discrete States

1.2.3.1 Discrete Energy Levels

The state of affairs will be presented by means of a short summary of the recent history of atomic theory.

Thomson's model of the atom assumed that an atom consists of an extended, continuous, positive charge distribution containing most of the mass, in which the electrons are embedded.[4] Geiger, and Geiger and Marsden (1908) found backward and perpendicular scattering in their experiments, in which alpha particles scattered off silver and gold. Rutherford immediately realized that this was inconsistent with Thomson's picture and presented his model of the atom in 1911, according to which the electrons orbit like planets about a positively charged nucleus of very small radius, which carries nearly the

---

[4] By means of P. Lenard's experiments (1903) – cathode rays, the Lenard window – it was demonstrated that atoms contained negatively charged $(-e_0)$ particles – electrons – about 2 000 times lighter than the atoms themselves. Thomson's model of the atom (J.J. Thomson, 1857–1940) was important because it attempted to explain the structure of the atom on the basis of electrodynamics; according to his theory, the electrons were supposed to undergo harmonic oscillations in the electrostatic potential of the positively charged sphere. However, it was only possible to explain a single spectral line, rather than a whole spectrum.

entire mass of the atom. Rutherford's theory of scattering on a point nucleus was confirmed in detail by Geiger and Marsden. It was an especially fortunate circumstance (Sects. 18.5, 18.10) for progress in atomic physics that the classical Rutherford formula is identical with the quantum mechanical one, but it is impossible to overlook the difficulties of Rutherford's model of the atom. The orbit of the electron on a curved path represents an accelerated motion, so that the electrons should constantly radiate energy away like a Hertz dipole and spiral into the nucleus. The orbital frequency would vary continuously, and one would expect a continuous emission spectrum. However, in fact experiments reveal discrete emission lines, whose frequencies, as in the case of the hydrogen atom, obey the generalized Balmer formula

$$\hbar\omega = \mathrm{Ry}\left(\frac{1}{n^2} - \frac{1}{m^2}\right)$$

(Ry is the Rydberg constant, n and m are natural numbers). This result represents a special case of the Rydberg–Ritz combination principle, according to which the frequencies can be expressed as differences of spectral terms.

In 1913, Bohr introduced his famous quantization condition. He postulated as stationary states the orbits which fulfill the condition $\oint p\,dq = 2\pi\hbar n$.[5] This was enough to explain the Balmer formula for circular orbits. While up to this time atomic physics was based exclusively on experimental findings whose partial explanation by the Bohr rules was quite arbitrary and unsatisfactory – the Bohr theory did not even handle the helium atom properly – Heisenberg (matrix mechanics 1925, uncertainty relation 1927) and Schrödinger (wave mechanics 1926) laid the appropriate axiomatic groundwork with their equivalent formulations for quantum mechanics and thus for a satisfactory theory of the atomic domain.

Aside from the existence of discrete atomic emission and absorption spectra, an experiment by J. Franck and G. Hertz in 1913 also shows quite clearly the presence of discrete energy levels in atoms.

In an experimental setup shown schematically in Fig. 1.6, electrons emitted from the cathode are accelerated in the electric field between cathode and grid and must then penetrate a small counterpotential before reaching the anode. The tube is filled with mercury vapor. If the potential difference between C and G is increased, then at first the current I rises. However, as soon as the kinetic energy of the electrons at the grid is large enough to knock

---

[5] More precisely, the Bohr theory consists of three elements: (i) There exist stationary states, i.e., orbits which are constant in time, in which no energy is radiated. (ii) The quantization condition: Stationary states are chosen from among those which are possible according to Newtonian mechanics on the basis of the Ehrenfest adiabatic hypothesis , according to which adiabatically invariant quantities – that is, those which remain invariant under a slow change in the parameters of the system – are to be quantized. (iii) Bohr's frequency condition: In an atomic transition from a stationary state with energy E_1 to one with energy E_2, the frequency of the emitted light is $(E_1 - E_2)/\hbar$.

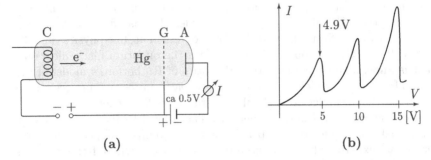

Fig. 1.6a,b. The Franck–Hertz effect. (**a**) Experimental setup: C cathode, G grid, A anode. (**b**) Current I versus the voltage V: Fall-off, if electrons can excite Hg before G, half-way and before G, etc.

mercury atoms into the first excited state during a collision, they lose their kinetic energy for the most part and, because of the negative countervoltage, no longer reach the anode. This happens for the first time at a voltage of about 5 V. At 10 V, the excitation process occurs at half the distance between cathode and grid and again at the grid, etc. Only well defined electron energies can be absorbed by the mercury atoms, and the frequency of the radiated light corresponds to this energy.

1.2.3.2 Quantization of Angular Momentum (Space Quantization)

In 1922, Stern and Gerlach shot a beam of paramagnetic atoms into a strongly inhomogeneous magnetic field and observed the ensuing deflections (Fig. 1.7). According to electrodynamics, the force acting on a magnetic moment $\boldsymbol{\mu}$ under such conditions is given by

$$\boldsymbol{F} = \boldsymbol{\nabla}(\boldsymbol{\mu} \cdot \boldsymbol{B}) \quad . \tag{1.8}$$

Here $B_z \gg B_x, B_y$, and hence the magnetic moment precesses about the z-direction and $\boldsymbol{\mu} \cdot \boldsymbol{B} \cong \mu_z B_z$. Now, the x- and y-dependence of B_z can be neglected in comparison to the z-dependence, so that

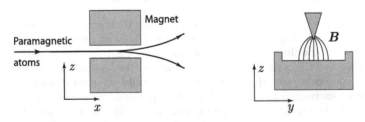

Fig. 1.7. The Stern–Gerlach experiment

$$\boldsymbol{F} = \mu_z \frac{\partial B_z}{\partial z} \boldsymbol{e}_z \quad , \tag{1.9}$$

where $\boldsymbol{e}_z$ is a unit vector in the z-direction.

The deflection thus turns out to be proportional to the z-component of the magnetic moment. Since classically μ_z varies continuously, one would expect the beam to fan out within a broad range. However, what one actually finds experimentally is a discrete number of beams, two in the case of hydrogen. Apparently, only a few orientations of the magnetic moment $\boldsymbol{\mu}$ with respect to the field direction are allowed. Thus, the Stern–Gerlach experiment gives evidence for the existence of spin.

2. The Wave Function and the Schrödinger Equation

2.1 The Wave Function and Its Probability Interpretation

According to the considerations of Sect. 1.2.2 in connection with electron diffraction, electrons also have wavelike properties; let this wave be $\psi(\boldsymbol{x}, t)$. For *free* electrons of momentum $\boldsymbol{p}$ and energy $E = \boldsymbol{p}^2/2m$, in accordance with diffraction experiments, one can consider these to be free plane waves, i.e., ψ takes the form

$$\psi(\boldsymbol{x}, t) = C\, e^{i(\boldsymbol{k} \cdot \boldsymbol{x} - \omega t)} \quad \text{with} \quad \omega = E/\hbar \quad , \quad \boldsymbol{k} = \boldsymbol{p}/\hbar \quad . \tag{2.1}$$

Now let us consider the question of the physical significance of the *wave function*. For this we shall consider an idealized diffraction experiment ("thought experiment").

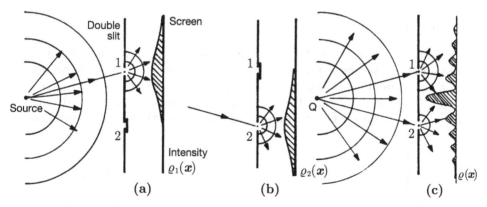

Fig. 2.1a–c. Diffraction at the double slit (**a**) with slit 1 open, (**b**) with slit 2 open, (**c**) both slits open

Suppose electrons are projected onto a screen through a double slit (Fig. 2.1). A photographic plate (or counter) in the plane of the screen behind the double slit provides information on the image created by the incident electrons. Suppose first that one or the other of the slits is closed. One then obtains the distributions $\varrho_1(\boldsymbol{x})$ and $\varrho_2(\boldsymbol{x})$, respectively, on the screen (Fig. 2.1a,b). If both slits are open, an interference pattern is created (Fig. 2.1c) with an amplification of the intensity where the path length difference Δl between the slits is an integral multiple of the electron wavelength λ, that is, $\Delta l = n\lambda$. Because of the interference, one has for the intensities $\varrho(\boldsymbol{x}) \neq \varrho_1(\boldsymbol{x}) + \varrho_2(\boldsymbol{x})$. We are familiar with such interference phenomena with just such screen patterns in the optics of light and also in water waves. If a cylindrical electromagnetic wave goes out from slit 1 with electric field $\boldsymbol{E}_1(\boldsymbol{x}, t)$, and one from slit 2 with electric field $\boldsymbol{E}_2(\boldsymbol{x}, t)$, one gets the following for the above experimental setup:

If only slit 1 is open, one has the intensity distribution $I_1(\boldsymbol{x}) = |\boldsymbol{E}_1(\boldsymbol{x}, t)|^2$ on the screen, whereas if only slit 2 is open, one gets $I_2(\boldsymbol{x}) = |\boldsymbol{E}_2(\boldsymbol{x}, t)|^2$. Here we have assumed that $\boldsymbol{E}_j(\boldsymbol{x}, t) \propto \exp\{-\mathrm{i}\omega t\}$, which is equivalent to time-averaging the intensities of real fields, up to a factor of 2. If both slits are open, one must superimpose the waves, and one obtains

$$\boldsymbol{E}(\boldsymbol{x}, t) = \boldsymbol{E}_1(\boldsymbol{x}, t) + \boldsymbol{E}_2(\boldsymbol{x}, t) \quad ,$$
$$I = |\boldsymbol{E}(\boldsymbol{x}, t)|^2 = I_1 + I_2 + 2\,\mathrm{Re}\,(\boldsymbol{E}_1^* \cdot \boldsymbol{E}_2) \quad .$$

The third term in the total intensity represents the so-called interference term.

Comparison with our electron experiment allows the following conclusion:

Hypothesis. The wave function $\psi(\boldsymbol{x}, t)$ gives the probability distribution

$$\varrho(\boldsymbol{x}, t) = |\psi(\boldsymbol{x}, t)|^2 \tag{2.2}$$

that an electron occupies the position $\boldsymbol{x}$. Thus, $\varrho(\boldsymbol{x}, t)\,d^3x$ is the probability of finding the electron at the location $\boldsymbol{x}$ in the volume element d^3x. According to this picture, the electron waves $\psi_1(\boldsymbol{x}, t)$ and $\psi_2(\boldsymbol{x}, t)$, which cause screen darkening $\varrho_1(\boldsymbol{x}, t) = |\psi_1(\boldsymbol{x}, t)|^2$ and $\varrho_2(\boldsymbol{x}, t) = |\psi_2(\boldsymbol{x}, t)|^2$, are emitted from slits 1 and 2, respectively. If both slits are open, then there is a superposition

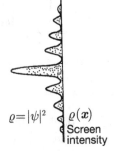

$\varrho = |\psi|^2$ $\varrho(\boldsymbol{x})$
Screen
intensity

Fig. 2.2. An interference pattern and its probability interpretation: Each electron makes a localized impact on the screen. The interference pattern becomes visible after the impact of many electrons with the same wave function $\psi(\boldsymbol{x}, t)$

of the wave functions $\psi_1(x,t) + \psi_2(x,t)$, and the darkening is proportional to $|\psi_1 + \psi_2|^2$ (Fig. 2.2). Two important remarks:

(i)   Each electron makes a local impact, and the darkening of the photographic plate by a single electron is not smeared out. $\varrho(x,t)$ is not the charge distribution of the electron, but rather gives the probability density for measuring the particle at the position x at the time t.

(ii)   This probability distribution does not occur by interference of many simultaneously incoming electrons, but rather one obtains the same interference pattern if each electron enters separately, i.e., even for a very low intensity source. The wave function thus applies to every electron and describes the state of a single electron.

We shall try to construct a theory that provides the wave function $\psi(x,t)$ and thus a statistical description for the results of experiments. This theory should reduce to classical mechanics in the limit of macroscopic objects.

2.2 The Schrödinger Equation for Free Particles

The equation of motion for $\psi(x,t)$ should satisfy the following basic demands:

(i)   It should be a first order differential equation in time so that $\psi(x,t)$ will be determined by the initial distribution $\psi(x,0)$.

(ii)   It must be linear in ψ in order for the principle of superposition to hold, i.e., linear combinations of solutions are again solutions, and thus interference effects such as those of optics occur. (These follow in the same way from the linearity of the Maxwell equations.) For the same reason, the constants in the equation may not contain any quantities that depend on the particular state of the particle such as its energy or momentum.

(iii) It should be homogeneous, so that

$$\int d^3x |\psi(x,t)|^2 = 1 \tag{2.3}$$

is satisfied at all times, because the total probability of finding the particle somewhere in space is 1 (normalization).

Remark: If an inhomogeneity q in the equation were to occur, e.g.,

$$\frac{\partial}{\partial t}\psi(x,t) = D\psi(x,t) + q \quad ,$$

then one would have

$$\frac{d}{dt}\int d^3x |\psi(x,t)|^2 = \int d^3x(\dot{\psi}\psi^* + \psi\dot{\psi}^*)$$
$$= \int d^3x((D\psi)\psi^* + \psi(D\psi)^*) + \int d^3x(q\psi^* + \psi q^*) \quad .$$

If D is the differential operator of the Schrödinger equation, then by Gauss's integral theorem (j is current density; see (2.58)–(2.60))

$$= -\int_S da \cdot j + \int d^3x\, 2\,\mathrm{Re}\{q\psi^*\} \quad .$$

The first term is 0, if ψ decreases sufficiently rapidly, e.g., $\psi \in L^2$, but the second term is in general nonvanishing.

(iv) Finally, plane waves

$$\psi(\boldsymbol{x},t) = C\,\exp\left\{ i\left(\boldsymbol{p} \cdot \boldsymbol{x} - \frac{p^2}{2m}t \right) \Big/ \hbar \right\}$$

should be solutions of the equations. For plane waves,

$$\frac{\partial}{\partial t}\psi(\boldsymbol{x},t) = -\frac{i}{\hbar}\frac{p^2}{2m}\psi(\boldsymbol{x},t) = \frac{i}{\hbar}\frac{\hbar^2}{2m}\boldsymbol{\nabla}^2\psi(\boldsymbol{x},t) \quad .$$

From postulates (i–iv) we thus obtain

$$i\hbar\frac{\partial}{\partial t}\psi(\boldsymbol{x},t) = -\frac{\hbar^2}{2m}\boldsymbol{\nabla}^2\psi(\boldsymbol{x},t) \quad . \tag{2.4}$$

This is the *time dependent Schrödinger equation* for *free particles*.

2.3 Superposition of Plane Waves

The plane waves

$$\psi(\boldsymbol{x},t) = C\,\exp\left\{ \frac{i}{\hbar}\left(\boldsymbol{p} \cdot \boldsymbol{x} - \frac{p^2}{2m}t \right) \right\}$$

have a spatially homogeneous probability density $|\psi(\boldsymbol{x},t)|^2 = C^2$. If we imagine that the particle is enclosed within a box of volume V, then the normalization condition $\int_V d^3x\, C^2 = 1$ for C gives the value $C = 1/\sqrt{V}$.

Localized states, that is, states with spatially concentrated extension, are obtained by superposition of plane waves:[1]

$$\psi(\boldsymbol{x},t) = \underbrace{\int \frac{d^3p}{(2\pi\hbar)^3}\varphi(\boldsymbol{p})\exp\left\{ \frac{i}{\hbar}\left(\boldsymbol{p} \cdot \boldsymbol{x} - \frac{p^2}{2m}t \right) \right\}}_{} \quad . \tag{2.5}$$

(three-dimensional wave packet)

The relationship is especially simple for a *one-dimensional Gaussian wave packet* , i.e.,

$$\varphi(p) = A\,\exp\{-(p-p_0)^2 d^2/\hbar^2\} \quad . \tag{2.6}$$

---

[1] We sometimes leave out the limits of integration, as in (2.5). In this case, these are always $-\infty$ and $+\infty$.

(The generalization to three dimensions is trivial, because the three-dimensional Gaussian wave packet $\exp\{-(\boldsymbol{p}-\boldsymbol{p}_0)^2 d^2/\hbar^2\}$ factorizes into three one-dimensional Gaussians.) In order to calculate (2.5) we temporarily introduce the following abbreviations:

$$
a = \frac{d^2}{\hbar^2} + \mathrm{i}\frac{t}{2m\hbar} \quad , \quad b = \frac{d^2 p_0}{\hbar^2} + \mathrm{i}\frac{x}{2\hbar} \quad , \quad c = \frac{d^2 p_0^2}{\hbar^2} \quad , \tag{2.7}
$$

by means of which (2.5) and (2.6) result in

$$
\begin{aligned}
\psi(x,t) &= \frac{A}{2\pi\hbar} \int dp \, \exp\left\{ -a\left(p - \frac{b}{a}\right)^2 + \frac{b^2}{a} - c \right\} \\
&= \frac{A}{2\pi\hbar}\sqrt{\frac{\pi}{a}} \exp\left\{ \frac{b^2}{a} - c \right\} \quad ,
\end{aligned} \tag{2.8}
$$

where we make use of the well known Gaussian integral

$$
\int_{-\infty}^{+\infty} dx \, \mathrm{e}^{-\alpha x^2} = \sqrt{\frac{\pi}{\alpha}} \quad . \tag{2.9}
$$

In the following we will primarily be interested in the probability density

$$
|\psi(\boldsymbol{x},t)|^2 = \left(\frac{A}{2\pi\hbar}\right)^2 \frac{\pi}{|a|} \exp\left\{ 2\,\mathrm{Re}\left\{\frac{b^2 - ac}{a}\right\} \right\} \quad . \tag{2.10}
$$

The exponent in (2.10) becomes

$$
2\,\mathrm{Re}\left\{(b^2 - ac)a^*\right\}/|a|^2 = -(x - vt)^2/2d^2(1 + \Delta^2) \quad , \tag{2.11}
$$

with

$$
v = \frac{p_0}{m} \text{ and } \Delta \equiv \Delta(t) = \frac{t\hbar}{2md^2} \quad . \tag{2.12}
$$

Now, using (2.7), (2.9), and (2.11), we can fix the normalization factor A such that $\int dx |\psi(\boldsymbol{x},t)|^2 = 1$, with the result

$$
A = \sqrt[4]{8\pi d^2} \quad . \tag{2.13}
$$

Thus, we finally obtain the complete result

$$
|\psi(x,t)|^2 = \frac{1}{d\sqrt{2\pi(1 + \Delta^2)}} \exp\left\{ -\frac{(x - vt)^2}{2d^2(1 + \Delta^2)} \right\} \quad , \tag{2.14}
$$

i.e., a Gaussian distribution in configuration space as well. The maximum of the wave packet moves with the group velocity $v = p_0/m = \partial E/\partial p|_{p_0}$ like a classical particle, whereas the individual superimposed plane waves have the phase velocities $v_{\mathrm{ph}} = E_p/p = p/2m$. The quantity Δ increases with time t. This means that the function $|\psi|^2$ gets flatter or "spreads" as time goes on, and thus its degree of localization is reduced.

We are also interested in the average value and the root-mean-square deviation of position for the present probability density (2.14). The expectation value of the position is calculated as

$$\langle x \rangle = \int_{-\infty}^{+\infty} |\psi(x,t)|^2 x \, dx$$

$$= \int_{-\infty}^{+\infty} dx |\psi(x,t)|^2 (x - vt) + \int_{-\infty}^{+\infty} dx |\psi(x,t)|^2 vt = vt \quad .$$

The first integral vanishes, since $|\psi(x,t)|^2$ is an even function of $(x - vt)$. For the mean-square deviation , one obtains

$$(\Delta x)^2 = \langle (x - \langle x \rangle)^2 \rangle$$

$$= \frac{\int_{-\infty}^{+\infty} dx |\psi(x,t)|^2 (x - vt)^2}{\int_{-\infty}^{+\infty} dx |\psi(x,t)|^2} = d^2(1 + \Delta^2) \quad .$$

Here we make use of (2.9) and its derivative with respect to α:

$$\int_{-\infty}^{+\infty} dx \, x^2 \, e^{-\alpha x^2} = \sqrt{\pi}/2\alpha^{3/2} \quad .$$

Thus,

position expectation value: $\langle x \rangle = vt$, $\qquad\qquad$ (2.15)

position uncertainty: $\Delta x = d\sqrt{1 + \Delta^2}$. $\qquad\qquad$ (2.16)

In order to illustrate these results, we consider two examples.

(i)   Let the particle being described by a Gaussian wave packet be a macroscopic body of mass $m = N m_{\mathrm{p}} \cong 10^{23} \times 10^{-24} \, \mathrm{g} = 10^{-1} \, \mathrm{g}$. In this case, one thus finds $\Delta = t\hbar/2md^2 \approx 10^{-26} t/d^2$ (t and d in cgs-units, Δ dimensionless). Such a body with initial positional uncertainty $\Delta x = d = 10^{-8} \, \mathrm{cm}$ does not have $\Delta = 1$ until 10^{10} s and thus has the width $\Delta x = \sqrt{2} \, d$. This value is quite irrelevant in comparison to the extension of a macroscopic body.

(ii)   On the other hand, for an α-particle, one gets

$$\Delta = (10^{-27}/2 \times 4 \times 1.6 \times 10^{-24}) \frac{t}{d^2} \cong 10^{-4} \frac{t}{d^2} \quad .$$

With $\Delta x = d = 10^{-11} \, \mathrm{cm}$ at the time $t = 0$ one finds $\Delta = 1$ for $t \approx 10^{-18}$ s. Although this time is very short, whether or not the spreading is significant depends entirely on the problem. For example, an α-particle with speed $v = c/30$ traverses a distance 10^{-9} cm during this time, which is much larger than a nuclear radius ($\approx 10^{-12}$ cm). However, this implies that during the collision with a nucleus the trajectory can be described classically!

The time evolution of a Gaussian wave packet is sketched in Fig. 2.3.

$|\psi(x,t)|^2$

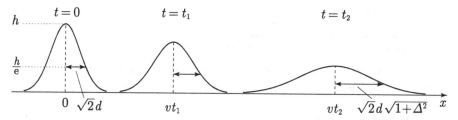

Fig. 2.3. Motion and spreading of a Gaussian wave packet. The "width" of the probability density grows with time

2.4 The Probability Distribution for a Measurement of Momentum

Now we consider the question of what probability density describes the realization of particular values of momentum. In position space the probability of finding a particle at the position x in the volume d^3x was given by $\varrho(x,t)d^3x = |\psi(x,t)|^2 d^3x$. Correspondingly, let the probability of finding the particle with momentum p in d^3p be represented by $W(p,t)d^3p$. Here, the total probability is also normalized to 1:

$$\int d^3p\, W(p,t) = 1 \quad . \tag{2.17}$$

If one expresses in analogy to (2.5) $\psi(x,t)$ in terms of its Fourier transform (see Appendix A) $\varphi(p,t)$, that is

$$\psi(x,t) = \int \frac{d^3p}{(2\pi\hbar)^3}\varphi(p,t)\,e^{i\,p\cdot x/\hbar} \quad ,$$

one then gets

$$\int d^3x |\psi(x,t)|^2$$
$$= \int d^3x \int \frac{d^3p}{(2\pi\hbar)^6} \int d^3p'\, \exp\left\{\frac{i}{\hbar}(p-p')\cdot x\right\}\varphi(p,t)\varphi^*(p',t)$$
$$= \int d^3p \int \frac{d^3p'}{(2\pi\hbar)^3}\delta^{(3)}(p-p')\varphi(p,t)\varphi^*(p',t) \quad , \tag{2.18}$$

because of

$$\int d^3x \exp\left\{\frac{i}{\hbar}(p-p')\cdot x\right\} = (2\pi\hbar)^3\delta^{(3)}(p-p') \quad .$$

Consequently,

$$\int d^3x |\psi(\boldsymbol{x},t)|^2 = \int d^3p \frac{1}{(2\pi\hbar)^3} |\varphi(\boldsymbol{p},t)|^2 \tag{2.19}$$

$\underbrace{\phantom{\int d^3x |\psi(\boldsymbol{x},t)|^2 = \int d^3p \frac{1}{(2\pi\hbar)^3} |\varphi(\boldsymbol{p},t)|^2}}$

(Parseval's theorem of Fourier transforms)

results from (2.18). This suggests for the probability density in momentum space the following definition:

$$W(\boldsymbol{p},t) = \frac{1}{(2\pi\hbar)^3} |\varphi(\boldsymbol{p},t)|^2 \quad . \tag{2.20}$$

This is consistent with the idea that for a plane wave with momentum $\boldsymbol{p}_0$ the Fourier transform $\varphi(\boldsymbol{p},t)$ differs from zero only for $\boldsymbol{p} = \boldsymbol{p}_0$.

Let us now return to the Gaussian wave packet in one dimension ((2.5), specialized to one dimension, and (2.6)). For this special case, one obtains the probability density

$$W(p,t) = \frac{1}{2\pi\hbar} |\varphi(p)|^2 = \sqrt{\frac{2}{\pi}} \frac{d}{\hbar} \exp\left\{-2(p - p_0)^2 d^2/\hbar^2\right\} \quad . \tag{2.21}$$

This is time independent, since we are considering free particles. With (2.21) the expectation value of the momentum is calculated as

$$\langle p \rangle = \int dp\, W(p,t)p = \int dp\, W(p,t)(p - p_0) + \int dp\, W(p,t)p_0 = p_0 \quad ,$$

and the corresponding mean-square fluctuation is

$$(\Delta p)^2 = \langle (p - p_0)^2 \rangle = \int dp\, W(p,t)(p - p_0)^2 = \left(\frac{\hbar}{2d}\right)^2 \quad .$$

Thus:

momentum expectation value: $\langle p \rangle = p_0$, $\tag{2.22}$

momentum uncertainty: $\Delta p = \hbar/2d$. $\tag{2.23}$

Together with (2.16), this leads to

$$\Delta x\, \Delta p = \frac{\hbar}{2}\sqrt{1 + \Delta^2} \quad . \tag{2.24}$$

Equation (2.24) represents a special case of the general uncertainty relation

$$\Delta x\, \Delta p \geq \hbar/2 \quad .$$

In the present context, it enters as a property of the Fourier transform and implies that a spatially broadly extended wave packet corresponds to a small spectrum of momentum values, whereas sharp wave packets can only be constructed from a broad band of Fourier components, i.e., they also contain components of short wavelength. We will give the general derivation of this later.

2.4.1 Illustration of the Uncertainty Principle

We would like to consider the following thought experiment to determine the position of an electron: The electron is illuminated with light of wavelength λ, and its image is projected onto a screen by means of an optical system. Figure 2.4 shows the simplified experimental apparatus in principle. The smallest distance which can be determined with a microscope is given by its resolving power $d = \lambda/\sin \varphi$. The inaccuracy of localization of the electron is thus $\Delta x \approx d = \lambda/\sin \varphi$. This uncertainty can thus be reduced with light of shorter wavelength. Now, the electron feels a back-reaction due to the collision with the photon. If we take the extreme values of the possible path of the photon, we see that the uncertainty of the x-component of the momentum of the electron and the photon is roughly

$$\Delta p_x \approx \left(\frac{2\pi}{\lambda}\hbar\right) \sin \varphi \quad .$$

We thus obtain

$$\Delta x\, \Delta p_x \approx 2\pi\hbar \quad .$$

In the experiment described above, position and momentum cannot in principle be determined simultaneously to greater accuracy than that permitted by this relation.

Two numerical examples will illustrate the uncertainty relation: The uncertainty relation holds even for macroscopic bodies. Consider for example a bullet with speed $v = 10^5$ cm/s (supersonic speed) and an uncertainty in the velocity of $\Delta v = 10^{-2}$ cm/s, corresponding to $\Delta p = m \times 10^{-2}$ cm/s. Now, the uncertainty relation says that the simultaneous determination of the position is only possible up to an uncertainty of

$$\Delta x = (1/m) \times 10^2 \hbar\,\mathrm{s\,cm}^{-1} \cong (1/m) \times 10^{-25}\,\mathrm{g\,cm} \quad ,$$

which becomes increasingly insignificant with growing mass. Even at a mass of only 10^{-6} kg $= 10^{-3}$ g, $\Delta x \cong 10^{-22}$ cm $\cong 10^{-14}$ atomic radii. On the

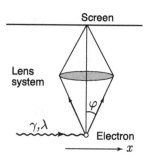

Fig. 2.4. Determination of position with a microscope

other hand, for electrons in an atom,

$$\Delta p \cong mv \cong 10^{-27} \times 10^{10}/137 \, \text{g cm/s} \quad \text{and} \quad \Delta x \cong a \cong 10^{-8} \, \text{cm}$$

(a: Bohr radius) which borders on what is permitted by the uncertainty relation. Because the given values are comparable to the dimensions of the effects being investigated, the uncertainties have considerable significance in the atomic domain.

2.4.2 Momentum in Coordinate Space

As we have seen, one can determine momentum expectation values, uncertainties, etc., in momentum space by means of the probability density $W(\boldsymbol{p}, t)$ defined in (2.20). Can these also be calculated in coordinate space? To this end we consider the familiar momentum expectation value

$$\langle \boldsymbol{p} \rangle = \int \frac{d^3p}{(2\pi\hbar)^3} \varphi(\boldsymbol{p}, t)^* \boldsymbol{p} \, \varphi(\boldsymbol{p}, t) \quad . \tag{2.25}$$

Substituting for $\varphi(\boldsymbol{p}, t)$ the Fourier transform, we obtain

$$\begin{aligned}
\langle \boldsymbol{p} \rangle &= \int \frac{d^3p}{(2\pi\hbar)^3} \int d^3x' \, e^{i\boldsymbol{p} \cdot \boldsymbol{x}'/\hbar} \, \psi^*(\boldsymbol{x}', t) \, \boldsymbol{p} \int d^3x \, e^{-i\boldsymbol{p} \cdot \boldsymbol{x}/\hbar} \, \psi(\boldsymbol{x}, t) \\
&= \int \frac{d^3p}{(2\pi\hbar)^3} \int d^3x' \, e^{i\boldsymbol{p} \cdot \boldsymbol{x}'/\hbar} \, \psi^*(\boldsymbol{x}', t) \int d^3x \left[-\frac{\hbar}{i} \boldsymbol{\nabla} \, e^{-i\boldsymbol{p} \cdot \boldsymbol{x}/\hbar} \right] \psi(\boldsymbol{x}, t) \\
&= \int d^3x \int d^3x' \frac{1}{(2\pi\hbar)^3} \psi^*(\boldsymbol{x}', t) \left(\frac{\hbar}{i} \boldsymbol{\nabla} \psi(\boldsymbol{x}, t) \right) \\
&\quad \times \int d^3p \, \exp\left\{ \frac{i}{\hbar}(\boldsymbol{x}' - \boldsymbol{x}) \cdot \boldsymbol{p} \right\} \quad .
\end{aligned}$$

In the preceding line, we have partially integrated under the assumption that $\psi(\boldsymbol{x})$ falls off sufficiently rapidly at infinity, that is, that the boundary terms are zero. If we also use the fact that the last integral is equal to $(2\pi\hbar)^3 \delta^3(\boldsymbol{x}' - \boldsymbol{x})$, we then finally obtain

$$\langle \boldsymbol{p} \rangle = \int d^3x \, \psi^*(\boldsymbol{x}, t) \frac{\hbar}{i} \boldsymbol{\nabla} \psi(\boldsymbol{x}, t) \quad . \tag{2.26}$$

Because of this connection, $(\hbar/i)\boldsymbol{\nabla}$ is called the momentum operator in coordinate space:

$$\boldsymbol{p} \longrightarrow \frac{\hbar}{i} \boldsymbol{\nabla} \quad \text{momentum operator in coordinate space} \quad . \tag{2.27}$$

2.4.3 Operators and the Scalar Product

In the previous section, we encountered the first example of the representation of physical quantities by operators in quantum mechanics. For this reason, we would like to summarize some of the properties of such objects here. We base our discussion on the space L^2 of square integrable functions (due to the normalization condition).

Definition. An *operator* A is defined by the prescription that for $\psi(x) \in L^2$ it follows that

$$A\psi(x) = \varphi(x) \in L^2 \quad .$$

Examples:

$$A\psi = \psi^2 + \frac{\partial}{\partial x_i}\psi \quad , \quad A\psi = e^\psi \psi \quad \text{(nonlinear)} \quad .$$

Definition. A is called a *linear operator* if $A\psi_1 = \varphi_1$ and $A\psi_2 = \varphi_2$ imply

$$A(c_1\psi_1 + c_2\psi_2) = c_1\varphi_1 + c_2\varphi_2 \quad , \tag{2.28}$$

where c_1, c_2 are complex numbers.

Examples:

$$x_i \quad , \quad \frac{\partial}{\partial x_i} \quad , \quad \nabla^2 \quad , \quad \frac{\partial}{\partial t} \quad , \quad f(x,t) \ \text{as multiplier} \quad .$$

Additional linear operators are obtained as the result of performing certain operations with linear operators:

- Multiplication by a number c gives the operator cA: $cA\psi := c(A\psi)$
- Sum of two operators $A + B$: $(A + B)\psi := A\psi + B\psi$ (2.29)
- Product of two operators AB: $AB\psi := A(B\psi)$.

Two special operators are the

- unit operator 1 $1\psi = \psi$
  and the   (2.30)
- zero operator 0 $0\psi = 0$.

One has

$$A1 = 1A = A \quad , \quad 0A = A0 = 0 \quad .$$

Generally, operators are not commutative, $AB \neq BA$, i.e., $AB\psi \neq BA\psi$.

Definition. *Commutator* $[A, B]$: Let A, B be operators; then the commutator is defined by

$$[A, B] = AB - BA \quad . \tag{2.31}$$

Examples:

$$\left[x_i, \frac{\partial}{\partial x_j}\right]\psi = \left(x_i\frac{\partial}{\partial x_j} - \frac{\partial}{\partial x_j}x_i\right)\psi = x_i\frac{\partial}{\partial x_j}\psi - \delta_{ij}\psi - x_i\frac{\partial}{\partial x_j}\psi = -\delta_{ij}\psi$$

$$\rightarrow \left[x_i, \frac{\partial}{\partial x_j}\right] = -\delta_{ij} \quad , \tag{2.32}$$

$$\left[f(\boldsymbol{x}), \frac{\partial}{\partial x_j}\right]\psi = f\frac{\partial}{\partial x_j}\psi - \left(\frac{\partial}{\partial x_j}f\right)\psi - f\frac{\partial}{\partial x_j}\psi = -\left(\frac{\partial}{\partial x_j}f\right)\psi$$

$$\rightarrow \left[f(\boldsymbol{x}), \frac{\partial}{\partial x_j}\right] = -\frac{\partial}{\partial x_j}f(\boldsymbol{x}) \quad , \tag{2.33}$$

$$[x_i, x_j] = 0 \quad \text{(real numbers commute)} \quad ,$$

$$\left[\frac{\partial}{\partial x_i}, \frac{\partial}{\partial x_j}\right] = 0 \quad \text{(order of differentiation for } \psi \in L^2 \text{ commutes)} \quad .$$

The basic commutators of the position and momentum operators are thus

$$[x_i, x_j] = 0 \quad ; \quad \left[\frac{\hbar}{i}\partial_i, \frac{\hbar}{i}\partial_j\right] = 0 \quad ; \quad \left[x_i, \frac{\hbar}{i}\partial_j\right] = i\hbar\delta_{ij} \quad , \tag{2.34}$$

with $\partial_i = \partial/\partial x_i$. If the commutator of two operators vanishes, one says "The two operators commute". Equation (2.34) shows that like spatial and momentum components do not commute, while different components do commute. One refers to x_j and $p_j = -i\hbar\partial_j$ as canonical variables and (2.34) as canonical commutation relations, which also can be written as

$$[x_i, x_j] = 0 \quad ; [p_i, p_j] = 0 \quad ; [x_i, p_j] = i\hbar\delta_{ij} \quad . \tag{2.34'}$$

Next, a scalar product is defined in L^2. Let $\varphi(\boldsymbol{x})$ and $\psi(\boldsymbol{x})$ be arbitrary functions in L^2:

Definition. *Scalar product* (φ, ψ): The scalar product of two wave functions ψ and φ is defined by

$$(\varphi, \psi) := \int d^3x\, \varphi^*(\boldsymbol{x})\psi(\boldsymbol{x}) \quad . \tag{2.35}$$

The scalar product has the following *properties*:

$$(\varphi, \psi)^* = (\psi, \varphi) \quad , \tag{2.36a}$$

$$(\varphi, c_1\psi_1 + c_2\psi_2) = c_1(\varphi, \psi_1) + c_2(\varphi, \psi_2) \quad , \tag{2.36b}$$

$$(c_1\varphi_1 + c_2\varphi_2, \psi) = c_1^*(\varphi_1, \psi) + c_2^*(\varphi_2, \psi) \quad . \tag{2.36c}$$

(The scalar product is linear in the second factor and antilinear in the first factor.)

Furthermore

$$(\varphi, \varphi) \geq 0 \quad \text{and thus} \quad (\varphi, \varphi) = 0 \Leftrightarrow \varphi \equiv 0 \quad . \tag{2.37}$$

Operators in the scalar product:

$$(\varphi, A\psi) = \int d^3x \, \varphi^*(\boldsymbol{x}) A\psi(\boldsymbol{x}) \quad . \tag{2.38}$$

Definition. $A^\dagger$ is called the *"adjoint operator to A"* if

$$(A^\dagger \varphi, \psi) = (\varphi, A\psi) \quad , \tag{2.39a}$$

i.e.,

$$\int d^3x (A^\dagger \varphi)^* \psi = \int d^3x \, \varphi^* A\psi \tag{2.39b}$$

holds for arbitrary φ and ψ.

Definition. The operator A is called *Hermitian* if

$$(A\varphi, \psi) = (\varphi, A\psi) \quad ; \tag{2.40}$$

we then write[2] $A^\dagger = A$.

The definition (2.39a) implies

$$(AB)^\dagger = B^\dagger A^\dagger \quad . \tag{2.41}$$

For later use, we state the following identities:

$$[AB, C] = A[B, C] + [A, C]B \quad , \tag{2.42}$$

$$[A, B]^\dagger = [B^\dagger, A^\dagger] \quad , \tag{2.43}$$

---

[2] In the mathematical literature the operator identity $A^\dagger = A$ is used only, if in addition to (2.40) the domains of A and $A^\dagger$ are the same and are dense. A is then termed self-adjoint.

and also mention the Baker–Hausdorff identity

$$e^A B e^{-A} = B + [A, B] + \tfrac{1}{2!}[A, [A, B]] + \cdots \quad , \tag{2.44}$$

where

$$e^A \equiv \sum_{\nu=0}^{\infty} \frac{1}{\nu!} A^\nu \tag{2.45}$$

is defined by the power series. See Problem 2.5(b).

If the commutator of two operators A and B commutes with them, i.e., $[[A, B], A] = [[A, B], B] = 0$, then

$$e^A e^B = e^B e^A e^{[A,B]} \tag{2.46}$$

and

$$e^{A+B} = e^A e^B e^{-[A,B]/2} \quad . \tag{2.47}$$

2.5 The Correspondence Principle and the Schrödinger Equation

2.5.1 The Correspondence Principle

We found in Sect. 2.4.2 that calculation of the momentum expectation value in coordinate space requires construction of the scalar product $(\psi, -i\hbar\boldsymbol{\nabla}\psi)$. We see further that the application of the operator $-i\hbar\boldsymbol{\nabla}$ on a plane wave $\psi(\boldsymbol{x}) = C \exp\{i(\boldsymbol{p}' \cdot \boldsymbol{x} - Et)/\hbar\}$ with wave number $\boldsymbol{p}'/\hbar$ just gives $\boldsymbol{p}'$ times the plane wave. Thus, the physical quantity *momentum* $\boldsymbol{p}$ is to be assigned to the operator $-i\hbar\boldsymbol{\nabla}$ in quantum mechanics. The procedure for the energy E is quite similar. From this, one obtains the correspondences

$$\begin{aligned} \text{momentum } \boldsymbol{p} &\longrightarrow \frac{\hbar}{i}\boldsymbol{\nabla} \quad , \\ \text{energy } E &\longrightarrow i\hbar\frac{\partial}{\partial t} \quad . \end{aligned} \tag{2.48}$$

To what extent can we assign quantum mechanical relations to classical ones on the basis of this correspondence? For example, does the classical energy–momentum relation for free particles $E = \boldsymbol{p}^2/2m$ imply the relation

$$i\hbar(\partial/\partial t) = -(\hbar^2/2m)\boldsymbol{\nabla}^2 \quad ?$$

This certainly cannot hold as an operator identity, but only when applied to a class of states, i.e.,

$$E = \boldsymbol{p}^2/2m \longrightarrow i\hbar\frac{\partial}{\partial t}\psi = -\frac{\hbar^2}{2m}\boldsymbol{\nabla}^2\psi \quad .$$

But this is precisely the Schrödinger equation (2.4) for free particles. So, from the classical energy–momentum relation, that is, from the classical Hamiltonian for a free particle, we have obtained the Schrödinger equation for a free particle.

This leads us to the quantum mechanical *correspondence principle:* In quantum mechanics, operators are assigned to physical quantities. Quantum mechanical relationships correspond to classical relationships.[3] This can be used to set up the Schrödinger equation on the basis of the classical Hamiltonian. We would now like to investigate this and use it to derive the equation of motion in a potential by assuming the classical Hamiltonian $\boldsymbol{p}^2/2m + V(\boldsymbol{x})$.

The Schrödinger equation for a particle in the potential $V(\boldsymbol{x})$ is obtained as follows. The assignment

$$E = \boldsymbol{p}^2/2m + V(\boldsymbol{x}) \longrightarrow i\hbar \frac{\partial}{\partial t}\psi(\boldsymbol{x},t) = \left(-\frac{\hbar^2}{2m}\boldsymbol{\nabla}^2 + V(\boldsymbol{x})\right)\psi(\boldsymbol{x},t)$$

implies the *Schrödinger equation of a particle in the potential* $V(\boldsymbol{x})$

$$i\hbar\frac{\partial}{\partial t}\psi(\boldsymbol{x},t) = H\psi(\boldsymbol{x},t) \tag{2.49}$$

with the *Hamiltonian operator* (or *Hamiltonian* for short)

$$H = -\frac{\hbar^2}{2m}\boldsymbol{\nabla}^2 + V(\boldsymbol{x}) \quad . \tag{2.50}$$

2.5.2 The Postulates of Quantum Theory

In a preliminary formulation of the theory developed up to now, which will be summarized more compactly in Sect. 2.9.4, the basic postulates of quantum mechanics are:

1. The state of a system is described by the wave function $\psi(\boldsymbol{x},t)$; $|\psi(\boldsymbol{x},t)|^2 d^3x$ expresses the probability of finding the particle at the time t at the position $\boldsymbol{x}$ in the volume element d^3x.
2. Quantum mechanically, operators $A, B, \ldots$ are assigned to the physically measurable quantities (observables) of classical mechanics.
3. The average values of operators are given by

$$\langle A \rangle = \int d^3x\, \psi^*(\boldsymbol{x},t) A\psi(\boldsymbol{x},t) \quad ,$$

if the system is in the state $\psi(\boldsymbol{x},t)$.

---

[3] In this use of the concept *correspondence principle* we deviate from the traditional use. Traditionally, one considers the *Bohr correspondence principle* as the statement that for large quantum numbers, quantum mechanical laws must reduce to their classical counterparts.

4. The time evolution of the states is described by the Schrödinger equation:

$$i\hbar\frac{\partial}{\partial t}\psi(\boldsymbol{x},t) = H\psi(\boldsymbol{x},t) \quad \text{with} \quad H = -\frac{\hbar^2}{2m}\boldsymbol{\nabla}^2 + V(\boldsymbol{x}) \quad .$$

Additionally:

(i) If one assigns to the classical quantity a the operator A, then the powers satisfy $a^2 \longrightarrow A^2$, $a^3 \longrightarrow A^3$, etc.

(ii) As we will see later, the operators corresponding to observables must be Hermitian.

(iii) From (2) and (3) it follows that $|\varphi(\boldsymbol{p},t)|^2/(2\pi\hbar)^3$ is the probability density in momentum space (i.e., for momentum measurements), as we will see in Sect. 2.9.3.

2.5.3 Many-Particle Systems

Finally, we seek the Schrödinger equation for a system of N particles. The state of this N-particle system is described by the wave function $\psi(\boldsymbol{x}_1, \boldsymbol{x}_2, \dots, \boldsymbol{x}_N, t)$, where $\boldsymbol{x}_i$ are the coordinates of the ith particle. Therefore, $|\psi(\boldsymbol{x}_1, \boldsymbol{x}_2, \dots, \boldsymbol{x}_N, t)|^2 \, d^3x_1 d^3x_2 \dots d^3x_N$ is the probability of finding the particles $1, \dots, N$ at time t in the volume elements $d^3x_1, \dots, d^3x_N$.

From the classical energy

$$E = \frac{\boldsymbol{p}_1^2}{2m_1} + \frac{\boldsymbol{p}_2^2}{2m_2} + \dots + \frac{\boldsymbol{p}_N^2}{2m_N} + V(\boldsymbol{x}_1, \boldsymbol{x}_2, \dots, \boldsymbol{x}_N)$$

we read off the Schrödinger equation for the N-particle system using the correspondence principle:

$$i\hbar\frac{\partial}{\partial t}\psi(\boldsymbol{x}_1, \boldsymbol{x}_2, \dots, \boldsymbol{x}_N, t)$$
$$= \left[-\frac{\hbar^2}{2m_1}\boldsymbol{\nabla}_1^2 - \dots - \frac{\hbar^2}{2m_N}\boldsymbol{\nabla}_N^2 + V(\boldsymbol{x}_1, \dots, \boldsymbol{x}_N)\right]\psi \quad . \tag{2.51}$$

Here, $\boldsymbol{\nabla}_i$, $i = 1, \dots N$, signifies the gradient operator with respect to $\boldsymbol{x}_i$.

2.6 The Ehrenfest Theorem

Classical Newtonian mechanics must be contained in quantum mechanics as a limiting case. In this section, we would like to investigate in what sense this is true.

We begin with the Schrödinger equation and its complex conjugate:

$$i\hbar\frac{\partial}{\partial t}\psi = H\psi \quad , \tag{2.52a}$$

$$-i\hbar\frac{\partial}{\partial t}\psi^* = H\psi^* \quad . \tag{2.52b}$$

For a linear operator A, the average value (= expectation value) in the state ψ is defined by

$$\langle A \rangle = \int d^3x\,\psi^*(\boldsymbol{x}, t)A\psi(\boldsymbol{x}, t) \quad . \tag{2.53}$$

This changes in time according to

$$\frac{d}{dt}\langle A \rangle = \int d^3x\left(\dot{\psi}^* A\psi + \psi^*\frac{\partial A}{\partial t}\psi + \psi^* A\dot{\psi}\right) \quad .$$

Using (2.52a,b), one then finds

$$\frac{d}{dt}\langle A \rangle = \frac{i}{\hbar}\langle [H, A] \rangle + \left\langle \frac{\partial A}{\partial t} \right\rangle \quad . \tag{2.54}$$

Remarks:

(i)   *Hermiticity of H:* Assuming that the wave functions ψ and φ vanish at infinity, one obtains after partially integrating twice

$$\int d^3x\left(-\frac{\hbar^2}{2m}\boldsymbol{\nabla}^2\psi\right)^*\varphi = \int d^3x\,\psi^*\left(-\frac{\hbar^2}{2m}\boldsymbol{\nabla}^2\varphi\right) \quad .$$

Since the operator $V(\boldsymbol{x})$ depends only on position, one has $\int d^3x(V\psi)^*\varphi = \int d^3x\,\psi^* V\varphi$.

(ii)   *Comparison with classical mechanics:* In classical mechanics, with respect to generalized momentum position coordinates p and q, the equations of motion

$$\frac{d}{dt}f(p, q, t) = \{H, f\} + \frac{\partial f}{\partial t}$$

hold, where the Poisson brackets are defined by

$$\{g, f\} = \frac{\partial g}{\partial p}\frac{\partial f}{\partial q} - \frac{\partial f}{\partial p}\frac{\partial g}{\partial q} \quad .$$

The Poisson brackets of classical mechanics evidently correspond in quantum mechanics to the commutator multiplied by $i/\hbar$.

(iii)   *Evaluation of the most important commutators:*

$$[H, x_i] = \left[\sum_j \frac{p_j^2}{2m}, x_i\right] = \frac{2}{2m}\sum_j p_j\frac{\hbar}{i}\delta_{ij} = \frac{-i\hbar p_i}{m} \quad . \tag{2.55a}$$

Here, we use (2.42); likewise, with (2.33),

$$[H, p_i] = \left[V(\boldsymbol{x}), \frac{\hbar}{i}\frac{\partial}{\partial x_i}\right] = i\hbar\frac{\partial V}{\partial x_i} \quad . \tag{2.55b}$$

(iv) *Application to x and p:* If the force $F(x) = -\nabla V(x)$ is introduced, then by means of the commutators (2.55a,b), one obtains from (2.54)

$$\frac{d}{dt}\langle x\rangle = \frac{1}{m}\langle p\rangle \quad , \tag{2.56a}$$

$$\frac{d}{dt}\langle p\rangle = -\langle \nabla V(x)\rangle = \langle F(x)\rangle \quad . \tag{2.56b}$$

If the two equations are combined, an equation which is analogous to the Newtonian equation of motion is obtained:

$$m\frac{d^2}{dt^2}\langle x\rangle = \langle F(x)\rangle \quad . \tag{2.56c}$$

This, and more generally, (2.54), constitute the *Ehrenfest theorem:* "Classical equations hold for the average values." However, this does not yet imply that the average values $\langle x\rangle$ and $\langle p\rangle$ obey the classical equations of motion. In order for this to hold, the average value of the force

$$\langle F(x)\rangle = \int d^3x\, \psi^*(x,t)F(x)\psi(x,t)$$

must be replaceable by its value $F(\langle x\rangle)$ at the position $\langle x\rangle$. To obtain a criterion for the validity of such an approximation, we expand the force F about the average value $\langle x\rangle$:

$$F_i(x) = F_i(\langle x\rangle) + (x_j - \langle x_j\rangle)F_{i,j}(\langle x\rangle)$$
$$+ \tfrac{1}{2}(x_j - \langle x_j\rangle)(x_l - \langle x_l\rangle)F_{i,jl}(\langle x\rangle) + \cdots \quad , \tag{2.57a}$$

where $f_{i,j} \equiv \partial f_i/\partial x_j$ and we sum over repeated indices. Since $\langle (x_j - \langle x_j\rangle)\rangle = 0$, one has

$$\langle F_i(x)\rangle = F_i(\langle x\rangle) + \tfrac{1}{2}\langle (x_j - \langle x_j\rangle)(x_l - \langle x_l\rangle)\rangle F_{i,jl}(\langle x\rangle) + \cdots \quad .$$

To replace $\langle F(x)\rangle$ by $F(\langle x\rangle)$ in (2.56c) is thus exact whenever the second and higher derivatives of the force vanish (for example, in the case of free particles or the harmonic oscillator). It is approximately valid whenever the wave packet is localized to such a degree that $F(x)$ changes slowly within the range of its extension, i.e.,

$$\frac{(\Delta x_j)^2 F_{i,jj}(\langle x\rangle)}{F_i(\langle x\rangle)} \ll 1 \quad , \tag{2.57b}$$

where for brevity we assume that the wave function is symmetric about the average value, $\langle (x_j - \langle x_j\rangle)(x_l - \langle x_l\rangle)\rangle = \delta_{jl}(\Delta x_j)^2$, and Δx_j is the spread of the packet in the j-direction.

Remarks:

(i)   The fact that $\langle x \rangle$ obeys the classical equation of motion does not imply that quantum effects are unimportant for harmonic oscillators.

(ii) By reversing the inequality (2.57b) one may conclude that quantum features will be significant if the characteristic length of the potential is smaller than that of the wave packet. Alternatively, for wavelengths larger than the length over which the potential varies, one finds quantum features.

2.7 The Continuity Equation for the Probability Density

Because of (2.52a,b), the time rate of change of the probability density (2.2) becomes

$$\frac{\partial}{\partial t} \varrho(\boldsymbol{x}, t) = \dot{\psi}^* \psi + \psi^* \dot{\psi} = \frac{1}{-i\hbar}(H\psi^*)\psi + \frac{1}{i\hbar}\psi^*(H\psi) \quad .$$

Since the terms involving $V(\boldsymbol{x})$ disappear, one obtains

$$\frac{\partial}{\partial t} \varrho(\boldsymbol{x}, t) = \frac{\hbar}{2mi}[(\boldsymbol{\nabla}^2 \psi^*)\psi - \psi^*(\boldsymbol{\nabla}^2 \psi)] \quad . \tag{2.58}$$

We define the *probability current density*

$$\boldsymbol{j}(\boldsymbol{x}, t) = \frac{\hbar}{2mi}[\psi^*(\boldsymbol{\nabla}\psi) - (\boldsymbol{\nabla}\psi^*)\psi] \quad . \tag{2.59}$$

From this, the *continuity equation* follows from (2.58)

$$\frac{\partial}{\partial t} \varrho(\boldsymbol{x}, t) + \boldsymbol{\nabla} \cdot \boldsymbol{j}(\boldsymbol{x}, t) = 0 \quad . \tag{2.60}$$

Its representation in integral form is obtained by means of Gauss's integral theorem for an arbitrary fixed volume V with surface S

$$\frac{d}{dt} \int_V d^3x \, \varrho(\boldsymbol{x}, t) = -\int_S d\boldsymbol{a} \cdot \boldsymbol{j}(\boldsymbol{x}, t) \quad . \tag{2.61}$$

Remark: We can now show that the norm of the wave function remains constant in time by allowing the volume in (2.61) to go to infinity. A normalized wave function must vanish faster than $1/|\boldsymbol{x}|^{3/2}$ at infinity in order for the integral over the probability density to be finite. With (2.59) and the assumption that possible periodic dependencies are only of the form $e^{i\boldsymbol{k} \cdot \boldsymbol{x}}$ for large $\boldsymbol{x}$ it follows that

$$\lim_{|\boldsymbol{x}| \to \infty} |\boldsymbol{j}| < \frac{1}{|\boldsymbol{x}|^3} \quad ,$$

and thus for a sphere whose radius R is allowed to go to infinity,

$$\lim_{V \to \infty} \left| \int_S da \cdot j \right| < \lim_{R \to \infty} \int d\Omega \, R^2 \frac{1}{R^3} = 0 \quad .$$

Thus

$$\frac{d}{dt} \int d^3x |\psi(\boldsymbol{x}, t)|^2 = 0 \quad ,$$

which shows that the normalization to unity does not change with time.

2.8 Stationary Solutions of the Schrödinger Equation, Eigenvalue Equations

2.8.1 Stationary States

Assuming that H is time independent, one can solve the Schrödinger equation by separation into a time dependent and a spatially dependent part:

$$\psi(\boldsymbol{x}, t) = f(t)\psi(\boldsymbol{x}) \quad . \tag{2.62}$$

The Schrödinger equation (2.49) then yields

$$\frac{1}{f(t)} i\hbar \frac{\partial}{\partial t} f(t) = \frac{1}{\psi(\boldsymbol{x})} H\psi(\boldsymbol{x}) \quad .$$

Since the left side only depends on t, and the right side only on $\boldsymbol{x}$, both sides must be equal to a constant, which we call E. Thus, $f(t)$ must satisfy the differential equation

$$i\hbar \frac{\partial}{\partial t} f(t) = E f(t) \tag{2.63}$$

with solution

$$f(t) = e^{-iEt/\hbar} \quad . \tag{2.64}$$

Similarly, we obtain for the spatially dependent part

$$H\psi(\boldsymbol{x}) = E\psi(\boldsymbol{x}) \quad . \tag{2.65}$$

This equation is known as the *time independent Schrödinger equation.*

Remarks:

(i)   The states $\psi(\boldsymbol{x}, t) = \exp\{-iEt/\hbar\}\psi(\boldsymbol{x})$ are called *stationary states*, since the corresponding probability densities $|\psi(\boldsymbol{x}, t)|^2 = |\psi(\boldsymbol{x})|^2$ are time independent.

(ii)   The condition of normalizability ($\int d^3x |\psi(\boldsymbol{x})|^2 < \infty$) will limit the allowed values of the energy E.

2.8.2 Eigenvalue Equations

Equation (2.65) is an eigenvalue equation. We would now like to give a general discussion of these.

The quantity ψ is an eigenfunction of an operator A with eigenvalue a when

$$A\psi = a\psi \qquad (2.66)$$

holds. This equation is called an eigenvalue equation. In the following it is assumed that the operator A is Hermitian.

Theorem 1. Eigenvalues of Hermitian operators are real.

Proof:

From (2.66), it follows that

$$(\psi, A\psi) = (\psi, a\psi) = a(\psi, \psi) \quad .$$

The complex conjugate equation is

$$(A\psi, \psi) = (a\psi, \psi) = a^*(\psi, \psi) \quad .$$

Since A is Hermitian, $(A\psi, \psi) = (\psi, A\psi)$ holds, and we find by taking the difference of the two equations

$$0 = (a - a^*)(\psi, \psi) \Rightarrow a = a^* \quad . \qquad (2.67)$$

Hermitian operators must be assigned to all measurable quantities (observables), in order that the expectation values and – as shown in Sect. 2.9 – the measured values be real. The operators H, p, and x which we have encountered up to now are indeed Hermitian, as follows immediately from the definition.

Theorem 2. Eigenfunctions of Hermitian operators belonging to different eigenvalues are orthogonal.

Proof:

Given the two eigenvalue equations

$$A\psi_m = a_m\psi_m \quad \text{and} \quad A\psi_n = a_n\psi_n \quad ,$$

we take the scalar product of the second equation with ψ_m, use the Hermiticity of A, and substitute the first eigenvalue equation:

$$a_n(\psi_m, \psi_n) = (\psi_m, A\psi_n) = (A\psi_m, \psi_n) = a_m(\psi_m, \psi_n) \quad .$$

From this it follows that

$$0 = (a_n - a_m)(\psi_m, \psi_n) \quad .$$

For $a_n \neq a_m$ it thus follows that

$$(\psi_m, \psi_n) = 0 \quad . \tag{2.68}$$

If several eigenfunctions belong to the same eigenvalue (degeneracy), one can orthogonalize as follows: We define

$$(\psi_m, \psi_n) = C_{mn} \quad \text{with} \quad C^*_{mn} = C_{nm} \quad . \tag{2.69}$$

As known from linear algebra, the Hermitian matrix C can be brought into the diagonal form $C^D = U^\dagger C U$ by means of a unitary transformation U. For this transformation, it follows from (2.69) that

$$\sum_{m,n} (U_{ma}\psi_m, \psi_n U_{n\beta}) = \sum_{m,n} U^*_{ma} C_{mn} U_{n\beta} = C^D_\alpha \delta_{\alpha\beta} \quad . \tag{2.70}$$

We now introduce the new functions $\varphi_\beta = \sum_n \psi_n U_{n\beta}$, which according to (2.70) are orthogonal. Using

$$\varphi_\alpha \longrightarrow \varphi_\alpha/(\varphi_\alpha, \varphi_\alpha)^{1/2}$$

one can normalize them to unity.

Thus, the eigenfunctions of a Hermitian operator can always be chosen such that the *orthogonality relation*

$$(\psi_m, \psi_n) = \delta_{mn} \tag{2.71}$$

is satisfied. Moreover, the eigenfunctions of the operators we are considering satisfy the *completeness relation*

$$\sum_n \psi^*_n(x')\psi_n(x) = \delta(x - x') \quad . \tag{2.72}$$

The ψ_n thus form a *complete set of orthonormal eigenfunctions*. Thus, a general state $\psi(x)$ can be expanded (represented) as

$$\psi(x) = \int dx' \delta(x - x')\psi(x') = \sum_n \int dx' \psi_n(x)\psi^*_n(x')\psi(x') \quad ,$$

that is

$$\psi(x) = \sum_n c_n \psi_n(x) \tag{2.73}$$

with

$$c_n = (\psi_n, \psi) \quad . \tag{2.74}$$

The normalization condition (2.3) and the orthogonality relation (2.71) imply that

$$\sum_n |c_n|^2 = 1 \quad .$$

(2.74')

Remark: The *Schmidt orthogonalization procedure* for degenerate eigenfunctions $\psi_1, \psi_2, \ldots$. Instead of diagonalizing the matrix C defined in (2.69) and thus determining the unitary matrix U of (2.70), it is sometimes more convenient to perform a stepwise orthogonalization of a system of degenerate but linearly independent eigenfunctions. Beginning with $\psi_1, \psi_2, \ldots$, the functions $\varphi_1, \varphi_2, \ldots$ are defined as follows: φ_j is constructed from ψ_j by projecting out all parts proportional to $\varphi_1, \ldots, \varphi_{j-1}$, i.e.,

$$C_1 \varphi_1 = \psi_1 \qquad\qquad C_1 = (\psi_1, \psi_1)^{1/2}$$

$$C_2 \varphi_2 = \psi_2 - \varphi_1 (\varphi_1, \psi_2) \qquad C_2 = ((\psi_2, \psi_2) - |(\psi_2, \varphi_1)|^2)^{1/2}$$

$$C_3 \varphi_3 = \psi_3 - \varphi_1 (\varphi_1, \psi_3) - \varphi_2 (\varphi_2, \psi_3)$$

$$\vdots \qquad\qquad\qquad \vdots$$

2.8.3 Expansion in Stationary States

Orthogonality and completeness hold in particular for the eigenfunctions of the Hamiltonian, i.e., for the stationary states ψ_n:

$$H\psi_n = E_n \psi_n \quad ,$$

(2.75)

$$\psi_n(\boldsymbol{x}, t) = e^{-iE_n t/\hbar} \psi_n(\boldsymbol{x}) \quad .$$

(2.76)

The E_n and ψ_n are called energy eigenvalues and energy eigenfunctions. For the solution of the Schrödinger equation, $\psi(\boldsymbol{x}, t)$ at the initial time $t = 0$, one has the expansion (2.73). Now, we know the time development of the individual $\psi_n(\boldsymbol{x}, t)$ and find for the sum

$$\psi(\boldsymbol{x}, t) = \sum_n c_n e^{-iE_n t/\hbar} \psi_n(\boldsymbol{x}) \quad \text{with} \quad c_n = (\psi_n, \psi(t = 0)) \quad .$$

(2.77)

It is easy to check that this $\psi(\boldsymbol{x}, t)$ is indeed a solution of the time dependent Schrödinger equation:

$$i\hbar \frac{\partial}{\partial t} \psi(\boldsymbol{x}, t) = \sum_n E_n c_n e^{-iE_n t/\hbar} \psi_n(\boldsymbol{x})$$

$$= H \sum_n c_n e^{-iE_n t/\hbar} \psi_n(\boldsymbol{x}) = H\psi(\boldsymbol{x}, t) \quad .$$

The expansion in terms of stationary states (2.77) solves the quantum mechanical initial value problem. The wave function $\psi(\boldsymbol{x}, t = 0)$ at the initial time $t = 0$ determines $\psi(\boldsymbol{x}, t)$.

2.9 The Physical Significance of the Eigenvalues of an Operator

Let an operator A with a complete, orthonormal system of eigenfunctions ψ_m be given with eigenvalues a_m and a wave function

$$\psi(\boldsymbol{x}) = \sum_m c_m \psi_m(\boldsymbol{x}) \quad . \tag{2.78}$$

What is the physical significance of the eigenvalues a_m and the expansion coefficients c_m? In order to clarify this question, we begin by introducing a few concepts from probability theory.

2.9.1 Some Concepts from Probability Theory

Let X be an arbitrary random variable[4] taking values x, and $w(x)dx$ the probability that the random variable takes a value in the interval $[x, x+dx]$.

Definition 1.

$$m_n = \int_{-\infty}^{+\infty} x^n w(x) dx = \langle X^n \rangle \tag{2.79}$$

is called the nth *moment* of the distribution $w(x)$.

Definition 2.

$$\chi(\tau) = \int_{-\infty}^{+\infty} e^{-i x \tau} w(x) dx \tag{2.80}$$

is called the *characteristic function*.

$\chi(\tau)$ is the Fourier transform of $w(x)$. Inverting the transformation, one obtains

$$w(x) = \int_{-\infty}^{+\infty} \frac{d\tau}{2\pi} e^{i x \tau} \chi(\tau) \quad . \tag{2.81}$$

---

[4] A quantity X is called a *random variable* if it assumes values x depending on the elements e of a "set of events" E. For each individual observation the event and thus the value of X is uncertain, and solely the probability for the appearance of a particular result (event) of E is known. For instance, in tossing a die the event is the appearance of a particular side of the die, and the random variable is the associated number of dots, which can assume values from 1 to 6, each with the probability $1/6$. If $e \in E$ is an event contained in E and P_e its probability, then for a large number of trials N, the number N_e of times the event e appears is related to P_e by $\lim_{N \to \infty} N_e/N = P_e$. See for instance W. Feller: *An Introduction to Probability Theory and Its Applications*, Vol. I (Wiley, New York, 1968).

Expansion of the exponential function in (2.80) together with the definition (2.79) gives

$$\chi(\tau) = \sum_n \frac{(-i)^n}{n!} \tau^n m_n \quad . \tag{2.82}$$

If all the moments are known, then by substitution of (2.82) into (2.81) one can determine $w(x)$.

Let $F(X)$ be a function of the random variable X. Then the average value of $F(X)$ is introduced by:

Definition 3.

$$\langle F(X) \rangle = \int_{-\infty}^{+\infty} F(x) w(x)\, dx \quad . \tag{2.79'}$$

Thus the nth moment defined in (2.79) equals the average value of X^n and the characteristic function can also be represented as

$$\chi(\tau) = \langle e^{-iX\tau} \rangle \quad . \tag{2.80'}$$

2.9.2 Application to Operators with Discrete Eigenvalues

(i)   Let the system be in an eigenstate ψ_m of A. Then, because of postulates 2 and 3,

$$\langle A^n \rangle = (\psi_m, A^n \psi_m) = (a_m)^n \tag{2.83}$$

and thus according to (2.82)

$$\chi(\tau) = \sum_n \frac{(-i)^n \tau^n (a_m)^n}{n!} = e^{-i\tau a_m} \quad . \tag{2.84}$$

Let $w(a)da$ be the probability that the observable represented by A takes values in the interval $[a, a + da]$. Then, by (2.81)

$$w(a) = \int \frac{d\tau}{2\pi} e^{ia\tau} e^{-i\tau a_m} = \delta(a - a_m) \quad , \tag{2.85}$$

i.e., one measures with certainty the value a_m.

(ii)   Let the system be in the state $\psi = \sum c_m \psi_m$. The quantities calculated in (i) then become

$$\langle A^n \rangle = (\psi, A^n \psi) = \left(\sum_m c_m \psi_m, A^n \sum_{m'} c_{m'} \psi_{m'} \right)$$

$$= \sum_m \sum_{m'} c_m^* c_{m'} (\psi_m, A^n \psi_{m'})$$

$$= \sum_m \sum_{m'} c_m^* c_{m'} (a_{m'})^n \delta_{mm'} \quad ,$$

$$\langle A^n \rangle = \sum_m |c_m|^2 (a_m)^n \quad , \tag{2.86}$$

$$\chi(\tau) = \sum_m |c_m|^2 e^{-i\tau a_m} \quad , \tag{2.87}$$

$$w(a) = \sum_m |c_m|^2 \delta(a - a_m) \quad . \tag{2.88}$$

Examination of (2.88) shows that the probability density vanishes when a does not coincide with one of the eigenvalues. Thus, the result of a measurement can only be one of the eigenvalues a_m. The probability of measuring a_m is $|c_m|^2$. (Thus, for example, the possible values of the energy are the energy eigenvalues E_n, and the expansion coefficients in (2.77) determine the probability $|c_n|^2$.)

What is the state after a particular result has been measured? After a measurement with the result a_m, the system must be in the eigenstate ψ_m, because repetition of the measurement should give the same result. If the system was originally not in an eigenstate of the observable, then the measurement changes the state! The fact that a measurement can change the state of a system was seen in Sect. 2.4.1.

If after the measurement the wave function is precisely known and is not further altered by the measurement in some uncontrolled way, one refers to this as an ideal measurement. After an ideal measurement, the system finds itself in an eigenstate of the operator corresponding to the observable.

2.9.3 Application to Operators with a Continuous Spectrum

As an example of an operator with a continuous spectrum, we consider the momentum p (one-dimensional). Integration of the eigenvalue equation

$$\frac{\hbar}{i} \frac{\partial}{\partial x} \psi_p(x) = p \psi_p(x) \tag{2.89}$$

yields for the *momentum eigenfunctions* the ubiquitous plane waves

$$\psi_p(x) = (2\pi\hbar)^{-1/2} e^{ipx/\hbar} \quad . \tag{2.90}$$

The $\psi_p(x)$ form a complete, orthonormal system, where the sum over the discrete index n in (2.72) is replaced by an integral over p, and the Kronecker δ in (2.71) is replaced by a Dirac δ-function.

Orthogonality relation

$$\int dx \, \psi_p^*(x) \psi_{p'}(x) = \delta(p - p') \quad . \tag{2.91}$$

Completeness relation

$$\int dp\, \psi_p^*(x')\psi_p(x) = \delta(x - x') \quad . \tag{2.92}$$

The expansion of an arbitrary wave function in momentum eigenfunctions (Fourier transform) takes the form

$$\psi(x) = \int dp'\, \frac{\varphi(p')}{\sqrt{2\pi\hbar}}\, \frac{\exp\{ip'x/\hbar\}}{\sqrt{2\pi\hbar}} \quad . \tag{2.93}$$

If we compare with the formulas for a discrete spectrum, the replacements

$$c_m \;\rightarrow\; (2\pi\hbar)^{-1/2}\varphi(p') \quad , \quad \sum_m \;\rightarrow\; \int dp' \tag{2.94}$$

result. If these are inserted into the results of the previous section, (2.88) yields for the *probability density* of the *momentum*

$$w(p) = \int dp'\, \left| \frac{\varphi(p')}{\sqrt{2\pi\hbar}} \right|^2 \delta(p - p') \tag{2.95a}$$

for

$$w(p) = \frac{|\varphi(p)|^2}{2\pi\hbar} \quad . \tag{2.95b}$$

From this we see that the expression $W(p)$, introduced in the development of our conceptual system as a temporary hypothesis (2.20), is not a separate axiom, as mentioned above, but rather a consequence of the form of the momentum operator and axiom III.

Now let us consider the *position eigenfunctions*

$$\psi_\xi(x) = \delta(x - \xi) \quad , \tag{2.96}$$

which evidently satisfy

$$x\psi_\xi(x) = \xi\psi_\xi(x) \quad . \tag{2.97}$$

The quantity $\psi_\xi(x)$ is sharply localized in space. These are also eigenfunctions with a continuous spectrum; they satisfy the following orthogonality and completeness relations:

$$(\psi_\xi, \psi_{\xi'}) = \delta(\xi - \xi') \quad , \tag{2.98}$$

$$\int d\xi\, \psi_\xi(x)\psi_\xi(x') = \delta(x - x') \quad . \tag{2.99}$$

Now, evidently,

$$\psi(x) = \int d\xi\, \psi(\xi)\psi_\xi(x) \quad , \tag{2.100}$$

that is, the expansion coefficients of the position eigenfunctions are just given by the wave function. It *follows* that $|\psi(\xi)|^2$ is the *probability density* for the *position ξ*.

Remark: One also refers to $\psi(x)$ as the wave function in the x-representation or coordinate representation, while $\varphi(p)$, which contains the same information, is referred to as the p-representation or momentum representation.

In many cases, the spectrum of an operator consists of a discrete part (eigenvalues a_n, eigenfunctions ψ_n) and a continuous part (eigenvalues a, eigenfunctions ψ_a). The expansion of a wave function is then

$$\psi(x) = \sum_n c_n \psi_n(x) + \int da\, c(a)\psi_a(x) \equiv \mathop{S}_n c_n \varphi_n \quad . \tag{2.101}$$

The symbol $\mathop{S}_n$ is used to cover both summation and integration. (We will use the symbol $\mathop{S}_n$ only when making special reference to the simultaneous existence of a continuous and a discrete part of the spectrum; otherwise, we will simply use the discrete representation $\sum_n$.) The probability density is then

$$w(a) = \sum_n |c_n|^2 \delta(a - a_n) + |c(a)|^2 \quad . \tag{2.102}$$

We can now formulate the axioms of quantum theory in final form.

2.9.4 Axioms of Quantum Theory

I.   The state is described by the wave function $\psi(x)$.
II. The observables are represented by Hermitian operators $A \ldots$, with functions of observables being represented by the corresponding functions of the operators.
III. The expectation value of the observable represented by the operator A is given in the state ψ by

$$\langle A \rangle = (\psi, A\psi) \quad .$$

IV. The time evolution of the states is given by the Schrödinger equation

$$i\hbar \frac{\partial}{\partial t}\psi = H\psi \quad , \quad H = -\frac{\hbar^2}{2m}\nabla^2 + V(x) \quad .$$

V.   If in a measurement of A the value a_n is found, the wave function changes to the corresponding eigenfunction ψ_n.

From axioms II and III it follows that the only possible results of a measurement of an observable are the eigenvalues of the corresponding operator A, and the probabilities are given by $|c_n|^2$, where c_n are the expansion coefficients of $\psi(\boldsymbol{x})$ in the eigenfunctions of A. In particular, it follows that $|\psi(\boldsymbol{x})|^2$ is the probability density for the position.

Remarks:

(i)   We add a comment on *terminology*. Measurable physical quantities (also referred to as dynamical variables) are called *observables*. We distinguish these from their mathematical counterparts, the operators they are represented by. In a number of expositions of quantum mechanics the term "observable" is used instead to denote any Hermitian operator which possesses a complete set of eigenfunctions.

For convenience we use the same symbol for the observable as for the operator. The one to one correspondence between observables and operators allows terminology such as "the observable A" or even "the average value of the operator A".

(ii)   Experimentally, in principle, the average value of an observable (e.g. the momentum) is determined as follows. One prepares a large number, say N, of identical systems, all in one and the same state $\psi(x)$. One then measures the observable in question for each system. In general, a range of measured values is obtained (the distribution depending on the state). The observable is truly a random variable. The experimental average value of the observable is the sum of the measured values divided by N. For this series of measured values, any function (e.g., a power) of the measured values can be computed. Again taking the average value, one obtains the experimental average value of a function of the observable. Axioms III and II state how these average values are computed quantum mechanically by means of the operator corresponding to the observable and the wave function. The process of measurement is studied in detail in Chap. 20, see Sect. 20.3, 20.3.4.

2.10 Additional Points

2.10.1 The General Wave Packet

We saw examples of wave packets in Sects. 2.3 and 2.4. The general wave packet has the form

$$\psi(\boldsymbol{x}, t) = \int \frac{d^3 p}{(2\pi\hbar)^3} \, g(\boldsymbol{p}) \, \exp\left\{ i(\boldsymbol{p} \cdot \boldsymbol{x} - E(\boldsymbol{p})t + \hbar\alpha(\boldsymbol{p}))/\hbar \right\} \quad . \tag{2.103}$$

Comparison with (2.5) shows that $\varphi(\boldsymbol{p}) = g(\boldsymbol{p}) \, \exp\{i\alpha(\boldsymbol{p})\}$. Suppose that the real weight function $g(\boldsymbol{p})$ has a maximum at $\boldsymbol{p}_0$ and differs significantly from zero only in a region with $|\boldsymbol{p} - \boldsymbol{p}_0| \lesssim \Delta p$. For most of the values of the position variable $\boldsymbol{x}$, the phase factor will vary rapidly over this momentum region and therefore give $\psi(\boldsymbol{x}, t) = 0$. Now, $\psi(\boldsymbol{x}, t)$ is maximal at that position $\boldsymbol{x}(t)$ for which the phase is stationary, i.e., where the condition

$$\boldsymbol{\nabla}_{\boldsymbol{p}} \left(\boldsymbol{p} \cdot \boldsymbol{x} - E(\boldsymbol{p})t + \hbar\alpha(\boldsymbol{p}) \right)|_{\boldsymbol{p}_0} = 0 \tag{2.104}$$

holds. This yields the "stationarity condition"

$$x(t) = x_0 + v_0 t \tag{2.105a}$$

with

$$x_0 = -\hbar \nabla_p \alpha(p)|_{p_0} \quad \text{and} \quad v_0 = \nabla_p E(p)|_{p_0} \quad . \tag{2.105b}$$

Since for this value $x(t)$ the phase does not vary rapidly as a function of momentum p in the vicinity of p_0, i.e., is stationary, $\psi(x, t)$ is large at the position $x(t)$. Thus $x(t)$ describes approximately the center of the wave packet and therefore can be compared to the classical position of the particle, which moves with the group velocity v_0. In order to calculate the integral (2.103), we expand the phase about p_0:

$$\begin{aligned}
p \cdot x &- E(p)t + \hbar\alpha = p_0 \cdot x - E(p_0)t + \hbar\alpha(p_0) \\
&+ (x - \nabla_p E(p)t + \hbar\nabla_p \alpha(p))|_{p_0} \cdot (p - p_0) \\
&+ \frac{1}{2}\sum_{i,j}\left[-\frac{t}{m}\delta_{ij} + \hbar\frac{\partial^2\alpha}{\partial p_i \partial p_j}\bigg|_{p_0}\right](p_i - p_{0i})(p_j - p_{0j}) + \cdots \quad ,
\end{aligned}$$

where we substitute $E(p) = p^2/2m$. Using (2.105) we then obtain

$$\begin{aligned}
p \cdot x &- E(p)t + \hbar\alpha(p) \\
&= p_0 \cdot x - E(p_0)t + \hbar\alpha(p_0) + (x - x(t)) \cdot (p - p_0) \\
&+ \frac{1}{2}\sum_{i,j}\left[-\frac{t}{m}\delta_{ij} + \hbar\partial_i\partial_j\alpha(p)\bigg|_{p_0}\right](p_i - p_{0i})(p_j - p_{0j}) + \cdots \quad .
\end{aligned}$$

Then, *in one dimension,*

$$\begin{aligned}
\psi(x, t) \cong \; &\exp\{i(p_0 x - E(p_0)t + \hbar\alpha(p_0))/\hbar\} \\
&\times \int \frac{dp}{2\pi\hbar} g(p) \exp\left\{i(x - x(t))\frac{p - p_0}{\hbar}\right. \\
&\left. + \frac{i}{2}\left[-\frac{t}{m} + \hbar\alpha''(p_0)\right]\frac{(p - p_0)^2}{\hbar}\right\} \quad . \tag{2.106}
\end{aligned}$$

The spreading of the wave packet in position space is found by considering the x-dependence in the integrand of (2.106) (or its three-dimensional counterpart) in the following way: if Δp is the width of $g(p)$, then there is constructive interference everywhere in the domain of integration if the real part (or imaginary part) of the exponential function $\exp\{i(x - x(t))(p - p_0)/\hbar\}$ does not change sign, i.e., $(x - x(t))$ roughly satisfies the condition

$$|x - x(t)|\Delta p/\hbar \lesssim \frac{\pi}{2} \quad .$$

On the other hand, there is destructive interference for

$$|x - x(t)|\Delta p/\hbar \gtrsim \pi \quad .$$

Thus, in order of magnitude, the spreading of the wave packet, i.e., the positional uncertainty Δx is related to the uncertainty of the momentum Δp by

$$\Delta x \Delta p \cong \pi \hbar \quad ; \tag{2.107}$$

we obtained this result earlier in the discussion of the Gaussian wave packet. The second phase factor $\exp\{-it(p-p_0)^2/2m\hbar\}$ increasingly counteracts the first with growing t. For this reason, one still gets a finite result for the wave function $\psi(x,t)$ even for larger values of $|x - x(t)|$, i.e., the wave packet spreads out, as we showed earlier in the case of the Gaussian wave packet.[5] Subject to the condition

$$\frac{t(\Delta p)^2}{m\hbar} \ll 1 \quad , \tag{2.108}$$

it is possible to neglect the spreading in time.

2.10.2 Remark on the Normalizability of the Continuum States

The eigenfunctions of the momentum operator fulfill the orthogonality condition (2.91). The $\psi_p(x)$ are thus evidently not normalized to 1. By superposition of such states, however, one can construct wave packets which are quadratically integrable and thus normalizable to 1. An analogous situation arises in the case of the continuum states $(E > 0)$ of the Hamiltonian. For the bound states $(E < 0)$, only normalized eigenfunctions with discrete eigenvalues are to be considered. The remaining solutions of the time independent Schrödinger equation with $E < 0$ are divergent at infinity, and thus cannot be normalized even using superposition.

One can avoid the problem of normalizability of continuum states by introducing a finite volume $V = L^3$. The states

$$\psi_p^V(x) = \frac{1}{\sqrt{V}} e^{i p \cdot x/\hbar}$$

---

[5] The positional uncertainty of (2.106) is most easily calculated in the momentum representation (see Chap. 8). The calculation is given for example in W. Pauli: *General principles of quantum mechanics*, Springer, Berlin, New York 1980. Translation of: *Die allgemeinen Prinzipien der Wellenmechanik I*, in Encyclopedia of Physics, Vol. V/1, p. 19, ed. by S. Flügge (Springer, Berlin, Heidelberg 1958).

with

$$p = \frac{2\pi\hbar}{L}(n_1, n_2, n_3) \quad , \quad (n_i \text{ are integers})$$

then form a complete, orthonormal system:

$$(\psi_{\boldsymbol{p}}^V, \psi_{\boldsymbol{p}'}^V) = \frac{1}{V}\int d^3x\, e^{-i(\boldsymbol{p}-\boldsymbol{p}')\cdot\boldsymbol{x}/\hbar} = \delta_{\boldsymbol{p},\boldsymbol{p}'} \quad .$$

In such a finite normalization volume, the continuum states become discrete and are normalized to unity as well.

The non-normalizable solutions of the time independent Schrödinger equation for $E < 0$ remain unacceptable even if one encloses the system in a finite box, because they do not satisfy the boundary conditions.

Problems

2.1 (a) Show that, for complex α with $\mathrm{Re}\,\alpha > 0$,

$$\int_{-\infty}^{+\infty} dx\, e^{-\alpha x^2} = \sqrt{\pi}/\alpha$$

holds.

(b) Compute

$$\int_{-\infty}^{+\infty} d^3k\, e^{i\boldsymbol{k}\cdot\boldsymbol{x}}\, e^{-k^2\alpha^2} \quad .$$

2.2 Investigate the one-dimensional wave packet

$$\varphi(p) = A\Theta\left[(\hbar/d) - |p - p_0|\right] \quad .$$

(a) Determine the constant of normalization and $\psi(x)$.

(b) Determine in coordinate space the wave function $\psi(x, t)$ in the limit $\hbar/dp_0 \ll 1$.

(c) Compute in this limit $\langle p \rangle$, $\langle x \rangle$, Δp, and Δx.

2.3 Using the Bohr–Sommerfeld quantization rules, determine the energy eigenstates of a particle of mass m moving in an infinitely high potential well:

$$V(x) = \begin{cases} 0 & \text{for } 0 \le x \le a \\ \infty & \text{otherwise} \end{cases} \quad .$$

2.4 Show that

(a) $p^\dagger = p$, $(p^2)^\dagger = p^2$, $V(\boldsymbol{x})^\dagger = V(\boldsymbol{x})$

(b) $(AB)^\dagger = B^\dagger A^\dagger$

(c) $[AB, C] = A[B, C] + [A, C]B$.

2.5 (a) Show that

$$[A, B^n] = nB^{n-1}[A, B]$$

under the assumption $[[A, B], B] = 0$.

(b) Demonstrate the Baker–Hausdorff relation (2.44). Hint: Consider the Taylor expansion of the function $f(\lambda) = e^{\lambda A} B e^{-\lambda A}$.

(c) Using (a), show that (2.47) holds under the conditions stated in the main text. Hint: Discuss the derivative of the function $e^{\lambda A} e^{\lambda B}$.

(d) Assuming $\lambda \ll 1$, expand the operator expression $(A - \lambda B)^{-1}$ in a power series in λ.

2.6 Compute the probability current density for the one-dimensional Gaussian wave packet (2.8) and verify that the continuity equation is satisfied.

2.7 Let $\psi_a(x)$ and $\psi_b(x)$ be two orthonormal solutions of the time independent Schrödinger equation for a given potential with energy eigenvalues E_a and E_b. At the time $t = 0$, suppose that the system is in the state

$$\psi(x, t = 0) = \frac{1}{\sqrt{2}} \left(\psi_a(x) + \psi_b(x) \right) \quad .$$

Discuss the probability density at a later time t.

2.8 For the operators

$$p_i = \frac{\hbar}{i} \partial_i \text{ and } L_i = \varepsilon_{ijk} x_j p_k \quad ,$$

compute the commutators

$$[p_i^2, f(x)] \text{ and } [L_i, L_j] \quad .$$

3. One-Dimensional Problems

In this chapter, we will solve the Schrödinger equation for characteristic one-dimensional potentials and discuss some typical quantum mechanical effects.

3.1 The Harmonic Oscillator

The Hamiltonian of the classical harmonic oscillator (Fig. 3.1) with mass m and frequency ω is

$$H = \frac{p^2}{2m} + \frac{m\omega^2}{2}x^2 \quad . \tag{3.1}$$

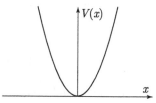

Fig. 3.1. The potential of the harmonic oscillator

With the momentum operator p, (3.1) represents the Hamiltonian operator. The time independent Schrödinger equation is thus

$$\left[-\frac{\hbar^2}{2m}\frac{d^2}{dx^2} + \frac{m\omega^2}{2}x^2 \right]\psi(x) = E\psi(x) \quad , \tag{3.2}$$

and it evidently contains

$$x_0 = \sqrt{\frac{\hbar}{\omega m}} \tag{3.3}$$

as a characteristic length. The standard method of analysis for the solution of the differential equation (3.2) subject to the auxiliary condition that $\psi(x)$ be square integrable leads to the Hermite polynomials. However, here we would like to use an algebraic method in which we try to represent H as the (absolute) square of an operator.

3.1.1 The Algebraic Method

For this, we define the non-Hermitian operator a by

$$a = \frac{\omega m x + i p}{\sqrt{2\omega m \hbar}} \quad , \tag{3.4a}$$

$$a^\dagger = \frac{\omega m x - i p}{\sqrt{2\omega m \hbar}} \quad , \tag{3.4b}$$

and, inverting these relations, we obtain

$$x = \sqrt{\frac{\hbar}{2\omega m}} (a + a^\dagger) \quad , \tag{3.5a}$$

$$p = -i\sqrt{\frac{\hbar \omega m}{2}} (a - a^\dagger) \quad . \tag{3.5b}$$

As one can easily derive from the commutators for x and p,

$$[a, a^\dagger] = 1 \quad , \tag{3.6}$$

while a and $a^\dagger$ commute with themselves. With the characteristic length x_0, one obtains

$$a = \frac{1}{\sqrt{2}} \left(\frac{x}{x_0} + x_0 \frac{d}{dx} \right) \quad , \tag{3.7a}$$

$$a^\dagger = \frac{1}{\sqrt{2}} \left(\frac{x}{x_0} - x_0 \frac{d}{dx} \right) \quad . \tag{3.7b}$$

Substituting the relations (3.5a,b) into (3.1), one gets for the Hamiltonian

$$H = \tfrac{1}{2} \hbar \omega (a^\dagger a + a a^\dagger) \quad ,$$

and, using the commutator (3.6),

$$H = \hbar \omega (a^\dagger a + \tfrac{1}{2}) \quad . \tag{3.8}$$

This reduces the problem to that of finding the eigenvalues of the *occupation number operator*

$$\hat{n} = a^\dagger a \quad . \tag{3.9}$$

Let ψ_ν be eigenfunctions with eigenvalue ν:

$$\hat{n} \psi_\nu = \nu \psi_\nu \quad . \tag{3.10}$$

Calculation of the eigenfunction ψ_0

From

$$\nu(\psi_\nu, \psi_\nu) = (\psi_\nu, a^\dagger a \psi_\nu) = (a\psi_\nu, a\psi_\nu) \geq 0 \tag{3.11}$$

it follows that

$$\nu \geq 0 \quad .$$

The lowest possible eigenvalue is thus $\nu = 0$. To calculate the corresponding eigenfunction ψ_0, we remark that by (3.11) the norm of $a\psi_0$ vanishes, and therefore

$$a\psi_0 = 0 \quad , \tag{3.12}$$

i.e.

$$\left(\frac{d}{dx} + \frac{x}{x_0^2} \right) \psi_0 = 0 \quad . \tag{3.13}$$

The solution of this differential equation, normalized to unity, is

$$\psi_0(x) = (\sqrt{\pi} x_0)^{-1/2} \exp\left\{ -\frac{1}{2} \left(\frac{x}{x_0} \right)^2 \right\} \quad . \tag{3.14}$$

Calculation of the remaining eigenfunctions

From (3.6) and (3.9) one obtains, using $[a^\dagger a, a^\dagger] = a^\dagger[a, a^\dagger] + [a^\dagger, a^\dagger]a = a^\dagger$, the important commutators

$$[\hat{n}, a^\dagger] = a^\dagger \quad , \tag{3.15}$$

$$[\hat{n}, a] = -a \quad . \tag{3.16}$$

Lemma. $a^\dagger \psi_\nu$ is an eigenfunction with the eigenvalue $\nu + 1$.

Proof:

$$\hat{n} a^\dagger \psi_\nu = (a^\dagger \hat{n} + a^\dagger) \psi_\nu = (\nu + 1) a^\dagger \psi_\nu \quad . \tag{3.17}$$

Norm:

$$\begin{aligned}
(a^\dagger \psi_\nu, a^\dagger \psi_\nu) &= (\psi_\nu, a a^\dagger \psi_\nu) = (\psi_\nu, (a^\dagger a + 1)\psi_\nu) \\
&= (\nu + 1)(\psi_\nu, \psi_\nu) > 0 \quad .
\end{aligned} \tag{3.18}$$

Thus, for ψ_ν and $\psi_{\nu+1}$ normalized to unity,

$$a^\dagger \psi_\nu = \sqrt{(\nu + 1)} \psi_{\nu+1} \quad . \tag{3.19}$$

Table 3.1. Eigenstates of the harmonic oscillator, eigenvalues of $\hat{n}$ and H

	$\hat{n}$	H
Ground state ψ_0	0	$\hbar\omega/2$
1st excited state $\psi_1 = a^\dagger\psi_0$	1	$3\hbar\omega/2$
2nd excited state $\psi_2 = \frac{1}{\sqrt{2}}(a^\dagger)^2\psi_0$	2	$5\hbar\omega/2$
$\vdots$	$\vdots$	$\vdots$
nth excited state $\psi_n = \frac{1}{\sqrt{n!}}(a^\dagger)^n\psi_0$	n	$(n+1/2)\hbar\omega$
$\vdots$	$\vdots$	$\vdots$

Beginning with ψ_0, we obtain from (3.19)

$$\psi_n = \frac{1}{\sqrt{n}}a^\dagger\psi_{n-1} = \frac{1}{\sqrt{n!}}(a^\dagger)^n\psi_0 \quad , \tag{3.19'}$$

the infinite sequence of eigenstates given in Table 3.1 together with the eigenvalues of $\hat{n}$ and H. One calls ψ_0 the ground state wave function and ψ_n for $n = 1, 2, \ldots$ the wave function of the nth excited state.

Lemma. $a\psi_\nu$ is an eigenfunction with eigenvalue $\nu - 1$.

Proof:

$$\hat{n}a\psi_\nu = (a\hat{n} - a)\psi_\nu = (\nu - 1)a\psi_\nu \quad . \tag{3.20}$$

Norm:

$$(a\psi_\nu, a\psi_\nu) = (\psi_\nu, a^\dagger a\psi_\nu) = \nu(\psi_\nu, \psi_\nu) = \nu \underset{(=)}{>} 0 \quad \text{for} \quad \nu \underset{(=)}{>} 0 \quad . \tag{3.21}$$

This implies for $\nu = 0$ (3.12) and for $\nu \geq 1$

$$a\psi_\nu = \sqrt{\nu}\psi_{\nu-1} \quad . \tag{3.22}$$

We can now show in addition that in Table 3.1 we have already found all the states.

Lemma. With ψ_n, $n = 0, 1, 2, \ldots$, all the eigenfunctions have been found.

Proof by contradiction:

Suppose there were an eigenvalue $\nu = n + \alpha$ with $0 < \alpha < 1$ and n a natural number:

$$\hat{n}\psi_\nu = (n + \alpha)\psi_\nu \quad .$$

Then, from (3.22),

$$\hat{n}(a^n \psi_\nu) = \alpha(a^n \psi_\nu) \ ,$$

$$\hat{n}(a^{n+1} \psi_\nu) = (\alpha - 1)(a^{n+1} \psi_\nu) \ .$$

Since by (3.21) the norm of the wave function $a^{n+1} \psi_\nu$ exists and $\alpha - 1 < 0$, there would be an eigenfunction of $\hat{n}$ with finite norm and negative eigenvalue, in contradiction to the positivity of the eigenvalues.

Thus, all the stationary states of the harmonic oscillator are given by Table 3.1 or (3.19'). The operators $a^\dagger (a)$ raise (lower) the energy eigenvalue by $\hbar\omega$ and are therefore called *creation (annihilation) operators* and occasionally ladder operators.

Remarks:

(i)   There can however exist non-square-integrable eigenfunctions with negative eigenvalues. For example

$$\left(-\frac{d^2}{dx^2} \right) \cosh x = -\cosh x \ ,$$

although

$$-\frac{d^2}{dx^2} = \left(i\frac{d}{dx} \right)^2$$

is a positive definite operator. Similarly, $\tilde{\psi} = \exp((x/x_0)^2/2)$ is a non-normalizable eigenfunction of $a^\dagger$ ($a^\dagger \tilde{\psi} = 0$) and also of $\hat{n}$ with the eigenvalue -1.

(ii)   The ground state, Eqn. (3.14), is not degenerate. One can deduce from this that none of the remaining eigenvalues of $\hat{n}$ are degenerate, because if corresponding to n there were another orthonormal eigenfunction φ_n besides ψ_n, then $a^n \varphi_n$ would be a ground state eigenfunction orthogonal to ψ_0, which is a contradiction.

The energy eigenstates of the harmonic oscillator (Fig. 3.2) are thus

$$\psi_n = (n!\sqrt{\pi}x_0)^{-1/2}(a^\dagger)^n \ \exp\left\{ -\frac{1}{2}\left(\frac{x}{x_0} \right)^2 \right\} \tag{3.23}$$

or

$$\psi_n = (2^n n!\sqrt{\pi}x_0)^{-1/2} \ \exp\left\{ -\frac{1}{2}\left(\frac{x}{x_0} \right)^2 \right\} H_n\left(\frac{x}{x_0} \right) \ , \tag{3.23'}$$

and the corresponding eigenvalues are

$$E_n = \hbar\omega(n + \tfrac{1}{2}) \tag{3.24}$$

with $n = 0, 1, 2, \ldots$. The spectrum of the energy eigenvalues is discrete. In a moment, we will further investigate the polynomials H_n introduced here and defined by (3.23) and (3.23'), and we will show that they are identical with the Hermite polynomials.

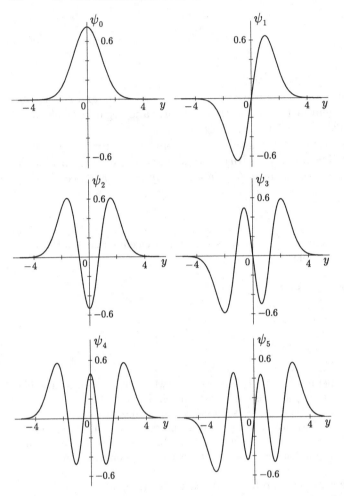

Fig. 3.2. The eigenfunctions of the harmonic oscillator for the quantum numbers $n = 0$ to 5: $y = x/x_0 = \sqrt{m\omega/\hbar}\,x$

3.1.2 The Hermite Polynomials

The polynomials H_n are defined according to (3.23) and (3.23') by

$$H_n(x) = \mathrm{e}^{x^2/2}(\sqrt{2}a^\dagger)^n\bigg|_{x_0=1}\ \mathrm{e}^{-x^2/2}$$

$$= \mathrm{e}^{x^2}\mathrm{e}^{-x^2/2}\left(x - \frac{d}{dx}\right)^n \mathrm{e}^{x^2/2}\mathrm{e}^{-x^2}\ .$$

If one further uses the operator identity

$$\mathrm{e}^{-x^2/2}\left(x - \frac{d}{dx}\right)^n \mathrm{e}^{x^2/2} = (-1)^n \frac{d^n}{dx^n}\ ,$$

which follows from

$$e^{-x^2/2}\left(x - \frac{d}{dx}\right)e^{x^2/2} = -\frac{d}{dx} \quad ,$$

one gets the usual representation for the Hermite polynomials

$$H_n(x) = (-1)^n e^{x^2} \frac{d^n}{dx^n} e^{-x^2} \quad . \tag{3.25}$$

Table of the first six Hermite polynomials:

$$
\begin{aligned}
H_0(x) &= 1 \\
H_1(x) &= 2x \\
H_2(x) &= 4x^2 - 2 \\
H_3(x) &= 8x^3 - 12x \\
H_4(x) &= 16x^4 - 48x^2 + 12 \\
H_5(x) &= 32x^5 - 160x^3 + 120x
\end{aligned}
\tag{3.26}
$$

Orthogonality relation:

$$\int_{-\infty}^{+\infty} dx\, e^{-x^2} H_n(x) H_m(x) = \sqrt{\pi} 2^n n! \delta_{mn} \tag{3.27}$$

Generating function:

$$e^{-t^2 + 2tx} = \sum_{n=0}^{\infty} \frac{1}{n!} t^n H_n(x) \tag{3.28}$$

Differential equation:

$$\left[\frac{d^2}{dx^2} - 2x\frac{d}{dx} + 2n\right] H_n(x) = 0 \tag{3.29}$$

Completeness:

$$\sum_{n=0}^{\infty} \psi_n(x)\psi_n(x') = \delta(x - x') \tag{3.30}$$

Note that H_n, and thus ψ_n, has n nodes (simple real zeros).

3.1.3 The Zero-Point Energy

Classically, the lowest energy of the harmonic oscillator is $E = 0$; quantum mechanically, it is $\hbar\omega/2$. Since $\psi_0(x)$ does not minimize $V(x)$ alone, but rather the sum of the kinetic and potential energy, a finite ground state energy or "zero-point energy" results. To illustrate this further, we calculate the uncertainty product $\Delta x \Delta p$. For the average value and the square fluctuation of position, we find

$$\langle x \rangle = (\psi_n, x\psi_n) \propto (\psi_n, (a + a^\dagger)\psi_n) = 0 \quad ,$$

$$(\Delta x)^2 = \langle x^2 \rangle = \frac{\hbar}{2\omega m}(\psi_n, (a^2 + aa^\dagger + a^\dagger a + a^{\dagger 2})\psi_n) = x_0^2(n + 1/2) \quad ,$$

and similarly for the momentum

$$\langle p \rangle = 0 \quad , \quad (\Delta p)^2 = \langle p^2 \rangle = \frac{\hbar^2}{x_0^2}(n + 1/2) \quad .$$

For the uncertainty product, one thus gets

$$\Delta x \Delta p = (n + 1/2)\hbar \quad . \tag{3.31}$$

This is a minimum for the ground state. The ground state wave function is not concentrated at $x = 0$ – the minimum of the potential – but rather has the spatial extension x_0 and a corresponding finite positional uncertainty (= fluctuation). This behavior, which is characteristic for quantum theory, is known as the "zero-point fluctuation". We would also like to derive an inequality for the zero-point energy without explicit calculation of the wave function, but only using the uncertainty relation

$$\Delta x \Delta p \geq \frac{\hbar}{2} \quad . \tag{3.32}$$

Since by symmetry $\langle p \rangle = \langle x \rangle = 0$ must hold for the ground state (see Sect. 3.6),

$$\langle p^2 \rangle \langle x^2 \rangle \geq \frac{\hbar^2}{4} \quad ,$$

and thus the following inequality for the energy is obtained:

$$E = \langle H \rangle = \frac{\langle p^2 \rangle}{2m} + \frac{1}{2}m\omega^2\langle x^2 \rangle \geq \frac{\langle p^2 \rangle}{2m} + \frac{m\omega^2}{2}\frac{\hbar^2}{4}\frac{1}{\langle p^2 \rangle} \quad .$$

The derivative with respect to $\langle p^2 \rangle$ gives as a condition for the minimum

$$\frac{1}{2m} - \frac{m\omega^2\hbar^2}{8}\frac{1}{(\langle p^2 \rangle_{min})^2} = 0$$

and thus

$$\langle p^2 \rangle_{\min} = \frac{m\hbar\omega}{2} \quad .$$

The energy then satisfies

$$E \geq \frac{m\hbar\omega}{2m \times 2} + \frac{m\omega^2\hbar^2}{8}\frac{2}{m\hbar\omega} = \frac{\hbar\omega}{2} \quad .$$

The zero-point energy is the smallest energy eigenvalue which is consistent with the uncertainty relation.

Remark: Comparison with the *classical oscillator*. The classical motion satisfies

$$x = q_0 \sin \omega t \quad , \tag{3.33}$$

$$E = \tfrac{1}{2}m\omega^2 q_0^2 \quad . \tag{3.34}$$

We define a classical "position probability"

$$W_{\text{class}}(x)dx = 2\frac{dt}{T} \quad ,$$

where dt is the amount of time spent within dx and $T = 2\pi/\omega$. From (3.33) it follows that

$$dx = q_0\omega \cos \omega t\, dt = q_0\omega\sqrt{1 - (x/q_0)^2}dt \quad ,$$

and thus

$$W_{\text{class}} = \frac{1}{\pi q_0 \sqrt{1 - (x/q_0)^2}} \quad . \tag{3.35}$$

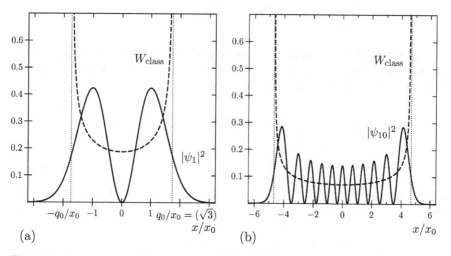

(a)

(b)

Fig. 3.3. Comparison of the quantum mechanical (——) with the "classical" (- - -) position probability for **(a)** energy E_1 and **(b)** E_{10}; turning points ($\cdots\cdots$)

Example:

The first excited state has the energy $E_1 = (3/2)\hbar\omega$. With (3.34), it follows (Fig. 3.3) that

$$q_0 = \sqrt{\frac{3\hbar\omega}{m\omega^2}} = \sqrt{3}x_0 \quad .$$

3.1.4 Coherent States

The position expectation value vanishes for the stationary states, i.e., $\langle x \rangle = 0$; these states therefore individually have nothing in common with the classical oscillatory motion. We now determine solutions of the time dependent Schrödinger equation representing periodic motion. We first determine wave functions for which at the initial time $t = 0$ the average value of x is different from zero $\langle x \rangle \neq 0$. This is certainly the case for states φ_α which satisfy

$$a\varphi_\alpha = \alpha\varphi_\alpha \tag{3.36}$$

with some complex number α, i.e., which are eigenfunctions of the annihilation operator. Since

$$(\psi_n, \varphi_\alpha) = \frac{1}{\sqrt{n!}}(a^{\dagger n}\psi_0, \varphi_\alpha) = \frac{1}{\sqrt{n!}}(\psi_0, a^n\varphi_\alpha) = \frac{\alpha^n}{\sqrt{n!}}(\psi_0, \varphi_\alpha) \quad ,$$

the expansion of φ_α in terms of the ψ_n results in

$$\varphi_\alpha(x) = C\sum_{n=0}^{\infty}\frac{\alpha^n}{\sqrt{n!}}\psi_n = C\sum_{n=0}^{\infty}\frac{(\alpha a^{\dagger})^n}{n!}\psi_0 \quad . \tag{3.37}$$

For $\alpha \neq 0$, the φ_α are not eigenfunctions of the Hamiltonian H, i.e., they are not stationary solutions. If one uses the orthogonality of the ψ_n, (3.37) yields for the normalization constant C

$$1 = (\varphi_\alpha, \varphi_\alpha) = C^2\sum_{n=0}^{\infty}\frac{|\alpha|^{2n}}{n!} = C^2 e^{|\alpha|^2} \quad ,$$

$$C = e^{-|\alpha|^2/2} \quad . \tag{3.38}$$

By the known time evolution of the stationary states, the time evolution of the $\varphi_\alpha(x, t)$ becomes

$$\varphi_\alpha(x, t) = e^{-|\alpha|^2/2}\sum_{n=0}^{\infty}\frac{(\alpha e^{-i\omega t})^n}{\sqrt{n!}}\psi_n\, e^{-i\omega t/2} \tag{3.37'}$$

or

$$\varphi_\alpha(x,t) = \varphi_{\alpha(t)}(x)e^{-i\omega t/2} \quad , \tag{3.39}$$

with

$$\alpha(t) = \alpha e^{-i\omega t} \quad . \tag{3.40}$$

$\varphi_\alpha(x,t)$ is a solution of the time dependent Schrödinger equation. These states are called *coherent states,* since they are of significance in optics for the representation of coherent light waves. Incidentally, they were derived by Schrödinger in one of his first papers[1] and have become very popular since the invention of the laser. The position expectation value is time dependent in the state $\varphi_\alpha(x,t)$:

$$\langle x \rangle = (\varphi_{\alpha(t)}, x\varphi_{\alpha(t)}) = \frac{x_0}{\sqrt{2}}(\varphi_{\alpha(t)}, (a + a^\dagger)\varphi_{\alpha(t)})$$

$$= \frac{x_0}{\sqrt{2}}(\alpha(t) + \alpha^*(t)) \quad .$$

If we write α in the form $\alpha = |\alpha|e^{i\delta}$, then we finally get

$$\langle x \rangle = \sqrt{2}x_0|\alpha| \cos(\omega t - \delta) \quad . \tag{3.41}$$

The position expectation value thus has the same time dependence as the classical oscillation.

From (3.37) and (3.39), one can easily calculate with the help of (2.47) the representation

$$\begin{aligned}
\varphi_\alpha(x,t) &= e^{-i\omega t/2}e^{\alpha(t)a^\dagger - \alpha^*(t)a}\psi_0(x)\\
&= \frac{1}{\sqrt[4]{\pi}\sqrt{x_0}} \exp\left\{-i\left[\frac{\omega t}{2} - \frac{|\alpha|^2}{2}\sin 2(\omega t - \delta)\right.\right.\\
&\quad \left.\left. + \frac{\sqrt{2}|\alpha|x}{x_0}\sin(\omega t - \delta)\right] - \frac{1}{2x_0^2}[x - x_0\sqrt{2}|\alpha|\cos(\omega t - \delta)]^2\right\}
\end{aligned} \tag{3.37''}$$

of $\varphi_\alpha(x,t)$. The probability density then becomes

$$|\varphi_\alpha(x,t)|^2 = \frac{1}{\sqrt{\pi}x_0}\exp\left\{-\frac{(x - x_0\sqrt{2}|\alpha|\cos(\omega t - \delta))^2}{x_0^2}\right\} \quad . \tag{3.42}$$

A coherent state is a Gaussian wave packet which does not spread out because all of the terms in (3.37') are in phase.

Classical limit: For large α (without loss of generality: $\delta = 0$ and $\alpha > 0$), i.e., large oscillation amplitudes, the coefficients $\alpha^n/n!$ have a sharp maximum for $n_0 = \alpha^2$; the relative width of the n-values contributing to (3.37) decreases

---

[1] E. Schrödinger: Die Naturwissenschaften **28**, 664 (1926)

like $(n - n_0)/n_0 \sim 1/\sqrt{n_0}$. Since $n_0 \sim \alpha^2$, the energy of the oscillator is then $(\varphi_\alpha, H\varphi_\alpha) = \hbar\omega\alpha^2 = m\omega^2 A^2/2$, where $A = \sqrt{2}x_0\alpha$ is the amplitude of the oscillation, consistent with classical mechanics.

3.2 Potential Steps

In both nuclear and solid-state physics, one often deals with potentials which can be divided up into regions inside which they are more or less constant, but where the transition from one region to another occurs within a very short distance. Since we are familiar with the solution of the free Schrödinger equation, we would now like to study, as an idealization of such physical situations, motion in potential steps and related problems.

3.2.1 Continuity of $\psi(x)$ and $\psi'(x)$ for a Piecewise Continuous Potential

Let us consider a one-dimensional potential with a discontinuity at the position a (Fig. 3.4). The time independent Schrödinger equation for this problem takes the form

$$\frac{d^2\psi(x)}{dx^2} = -\frac{2m}{\hbar^2}(E - V(x))\psi(x) \quad . \tag{3.43}$$

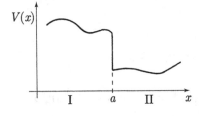

Fig. 3.4. A potential with a discontinuity

Now, suppose that $\psi(x)$ or $\psi'(x)$ were discontinuous at a. Then, the behavior $\psi(x) \sim \Theta(x-a)$ would have the consequence $\psi''(x) \sim \delta'(x-a)$, and similarly the behavior $\psi'(x) \sim \Theta(x-a)$ would imply $\psi''(x) \sim \delta(x-a)$. However, since by the right-hand side of (3.43) $\psi''(x)$ may make at most a finite jump at the position a, our supposition leads to a contradiction. Consequently, $\psi(x)$ and $\psi'(x)$ must be continuous even if the potential $V(x)$ is merely piecewise continuous. At the discontinuity a of $V(x)$, one thus obtains the *continuity conditions*

$$\psi_I(a) = \psi_{II}(a) \quad , \tag{3.44a}$$

$$\psi_I'(a) = \psi_{II}'(a) \quad . \tag{3.44b}$$

It is often convenient to use, in place of the second equation,

$$\frac{\psi_I'(a)}{\psi_I(a)} = \frac{\psi_{II}'(a)}{\psi_{II}(a)} \quad , \tag{3.44c}$$

which results upon division of (3.44b) by (3.44a) and which implies the continuity of the logarithmic derivative. ($\psi_{I,II}$ represent the solutions for the regions I and II, respectively.)

The arguments given above and (3.44b) lose their validity if the potential behaves like a Dirac δ-function.

3.2.2 The Potential Step

As an example of a problem of the type just discussed, we study the potential step (Fig. 3.5a), that is, the potential

$$V(x) = V_0\Theta(x) \quad , \quad \Theta(x) = \begin{cases} 1 & , \quad x > 0 \\ 0 & , \quad x < 0 \end{cases} \quad , \tag{3.45}$$

where the constant $V_0 \geq 0$. Let us consider the Schrödinger equation separately in the regions I ($x < 0$) and II ($x > 0$):

$$\frac{d^2\psi}{dx^2} = -\frac{2mE}{\hbar^2}\psi \quad ; \quad \frac{d^2\psi}{dx^2} = -\frac{2m(E - V_0)}{\hbar^2}\psi \quad . \tag{3.46a,b}$$

The continuity requirements for ψ and ψ' will give us relations between the free constants of the solutions in the regions I and II. We distinguish the cases $E > V_0$ and $E < V_0$, since they correspond to different physical situations.

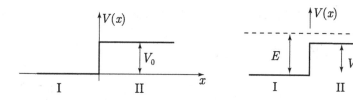

Fig. 3.5a. The potential step   **Fig. 3.5b.** The potential step, $E > V_0$

3.2.2.1 Particle Energy Above the Potential Step ($E > V_0$, Fig. 3.5b)

Defining the two wave numbers k and q, one sees that (3.46a) and (3.46b) become

$$\text{I:} \quad \frac{d^2\psi}{dx^2} = -k^2\psi \quad , \quad k = \sqrt{2mE}/\hbar \quad ; \tag{3.47a}$$

$$\text{II:} \quad \frac{d^2\psi}{dx^2} = -q^2\psi \quad , \quad q = \sqrt{2m(E - V_0)}/\hbar \quad . \tag{3.47b}$$

These are classical oscillator equations with the fundamental solutions

$$e^{iKx} \quad , \quad e^{-iKx} \quad , \quad K = \begin{cases} k & , \quad x < 0 \\ q & , \quad x > 0 \end{cases} .$$

Suppose that the particle is incident from the left[2]; the wave function in region I is then the superposition of a wave incident from the left, whose amplitude can be set to unity without loss of generality, and a reflected wave, whereas in region II it is a transmitted wave:

$$\psi_I(x) = e^{ikx} + R e^{-ikx} \quad , \tag{3.48a}$$

$$\psi_{II}(x) = T e^{iqx} \quad , \tag{3.48b}$$

$$\psi(x) = \Theta(-x)\psi_I(x) + \Theta(x)\psi_{II}(x) \quad . \tag{3.48c}$$

The coefficients R and T are determined by the continuity conditions for ψ and ψ' at $x = 0$:

$$1 + R = T \quad , \quad ik(1 - R) = iqT \quad ,$$

so that

$$R = \frac{k - q}{k + q} \quad , \quad T = \frac{2k}{k + q} \quad . \tag{3.49}$$

In order to give these coefficients a physical interpretation, we compute the probability current density in the regions I and II by (2.59):

$$j_I(x) = \frac{\hbar}{2mi}[(e^{-ikx} + R^* e^{ikx})(ik)(e^{ikx} - R e^{-ikx}) - \text{c.c.}]$$

$$= \frac{\hbar}{2mi}[ik(1 - |R|^2 - R e^{-2ikx} + R^* e^{2ikx}) - \text{c.c.}] \quad ,$$

$$j_I(x) = \frac{\hbar k}{m}(1 - |R|^2) \equiv j_{in} - j_{refl} \quad , \tag{3.50a}$$

$$j_{II}(x) = \frac{\hbar q}{m}|T|^2 \equiv j_{trans} , \tag{3.50b}$$

where "c.c." stands for "complex conjugate". Here, we have decomposed the current densities in regions I and II into incident, reflected, and transmitted current densities. Hence, one obtains for the reflection and transmission coefficients r and t:

---

[2] Although we solve just the time independent Schrödinger equation, we use the terminology incoming, reflected and transmitted waves. This is justified, if one recalls the time dependence $e^{-\hbar k^2 t/2m}$ of the stationary states. Furthermore, one can build wave packets (see Sect. 3.7.2), which behave in the following manner. Initially there is a wave packet incident from the left, which after having reached the barrier splits into a transmitted and reflected packet. It should also be mentioned that in addition to (3.48), there are solutions representing waves incident from the right.

$$r = \frac{j_{\text{refl}}}{j_{\text{in}}} = |R|^2 \quad ; \quad t = \frac{j_{\text{trans}}}{j_{\text{in}}} = \frac{q}{k}|T|^2 \ . \tag{3.51}$$

Remarks:

(i)   The particle is reflected with probability r. Classically, on the other hand, there would be no reflection, but rather to the right of the step the particle would just continue to move, albeit with smaller velocity. This reflection is a wave phenomenon analogous to the reflection of light at the boundary surface of media with different indices of refraction.

(ii) In the limiting case $E \to \infty$ ($E \gg V_0$), one obtains

$$q \to k: \quad R \to 0 \quad , \quad T \to 1 \quad ,$$

i.e., the reflected wave vanishes. Since here we are confronted with an infinitely sharp potential step, the Ehrenfest theorem, and hence the transition to the classical case, is not valid until $E \to \infty$, i.e., $\lambda \to 0$, is satisfied. If the transition from zero to V_0 were to occur continuously over a length d, particles with $k \gg 1/d$ would then be completely transmitted. This can be verified explicitly in the case of the potential $V(x) = V_0(1 + \tanh(x/2d))/2$, which leads to hypergeometric functions.

(iii) Conservation of particle number: From (3.49) follows

$$\frac{\hbar k}{m}(1 - |R|^2) = \frac{\hbar q}{m}|T|^2 \quad ,$$

and hence

$$j_{\text{I}} = j_{\text{II}} \quad ,$$

so that

$$j_{\text{in}} = j_{\text{refl}} + j_{\text{trans}} \quad ;$$

i.e., the incident particle flux is equal to the sum of the transmitted and the reflected particle flux. This result also follows from the continuity equation (2.60) (particle number conservation), which gives $\partial j(x)/\partial x = 0$, since $\varrho(x)$ is time independent.

(iv) Eq. (3.49) implies $R > 0$, i.e., the reflected and the incoming wave are in phase. However, if the potential step descends towards the right, i.e., $V_0 < 0$, then $q = \sqrt{2m(E + |V_0|)}/\hbar$ in Eqs. (3.48) and (3.49) and hence $R < 0$. In this case the reflected wave experiences a phase jump π.

We show the real and imaginary part of $\psi(x)$ and the probability density $|\psi(x)|^2$ in Fig. 3.6. For the incident energy, we choose $E = 4V_0/3$; the ratio of the wave numbers is then $q/k = 1/2$.

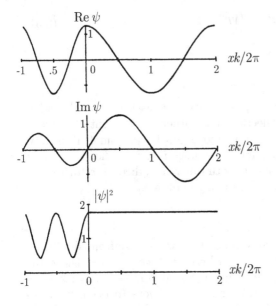

Fig. 3.6. The real and imaginary parts of $\psi(x)$ and the probability density $|\psi(x)|^2$ vs. $xk/2\pi$, for incident energy $E = 4V_0/3$, i.e. $q/k = 1/2$

3.2.2.2 Particle Energy Below the Potential Step $E < V_0$, Fig. 3.5c)

Now let the energy E of the particle incident from the left be smaller than the height V_0 of the potential step. The Schrödinger equation (3.46a) or (3.47a) in region I then remains unchanged and is solved as in the case $E > V_0$ by (3.48a). In region II, on the other hand, (3.47b) is replaced by

$$\psi'' = \kappa^2\psi , \quad \text{with} \quad \kappa = \sqrt{2m(V_0 - E)}/\hbar \quad , \tag{3.52}$$

whose solutions either increase or fall exponentially. However, we do not need to find the solutions of (3.47a) and (3.52) together with continuity conditions all over again; we need only observe that q in (3.47b–3.49) becomes purely imaginary:

$$q = i\kappa \quad . \tag{3.53}$$

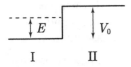

I   II   **Fig. 3.5c.** The potential step, $E < V_0$

The solution leading to a finite probability density in region II is then

$$\psi_{\mathrm{II}}(x) = T\,e^{-\kappa x} \quad , \tag{3.54}$$

and the reflection and transmission amplitudes from (3.49) become

$$R = \frac{k - i\kappa}{k + i\kappa} \quad , \quad T = \frac{2k}{k + i\kappa} \quad . \tag{3.55}$$

Remarks:

(i)   According to (3.55)

$$|R|^2 = 1 \quad , \tag{3.56}$$

that is, complete reflection occurs.

(ii)   Since $T \neq 0$, the particles penetrate into the step up to a depth of about κ^{-1}. However, no particle flux to the right takes place, as one sees either from (3.56) or from $j_{\mathrm{II}} = 0$.

The wave function

$$\psi(x) = \left\{ \left(\cos kx - \frac{\kappa}{k} \sin kx \right)\Theta(-x) + e^{-\kappa x}\Theta(x) \right\} \frac{2}{1 + i\kappa/k} \tag{3.57}$$

without the factor $2/(1 + i\kappa/k)$ is shown in Fig. 3.7a.

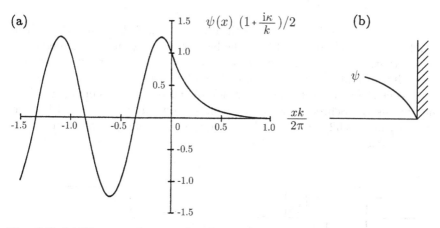

Fig. 3.7. (a) The wave function (3.57) for $\kappa/k = 3/4$. (b) The wave function at an infinity

3.2.2.3 The Limiting Case of an Infinitely High Potential Step ($V_0 \to \infty$)

We now consider the important limiting case of an infinitely high potential step $V_0 \to \infty$. In this case, one has

$$\kappa \to \infty \quad , \quad T = 0 \quad , \quad R = -1 \quad .$$

Hence,

$$\psi_{\mathrm{I}}(x) = e^{ikx} - e^{-ikx} \quad ,$$

and thus

$$\psi_{\mathrm{I}}(0) = 0 \quad .$$

This yields the *general boundary condition* at an *infinitely high step* (Fig. 3.7b):

$$\psi|_{\mathrm{step}} = 0 \quad . \tag{3.58}$$

3.3 The Tunneling Effect, the Potential Barrier

3.3.1 The Potential Barrier

We now investigate motion in the presence of the square potential barrier

$$V(x) = V_0 \Theta(a - |x|) \tag{3.59}$$

shown in Fig. 3.8a. We consider only the case $E < V_0$. A classical particle then would be completely reflected from the barrier. After seeing in the preceding section that, in quantum mechanics, particles can slip part way into the classically forbidden region, we should not be surprised that particles can even penetrate through to the other side of the barrier. Assuming $E < V_0$, one obtains the Schrödinger equation (3.47a) outside the potential barrier ($|x| > a$) and (3.52) inside the barrier ($|x| < a$). The general form of the solution is thus

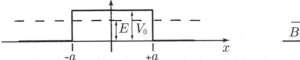

Fig. 3.8a. The potential barrier

Fig. 3.8b. The incident, reflected, and transmitted waves

$$\psi(x) = \begin{cases} A\,\mathrm{e}^{\mathrm{i}kx} + B\,\mathrm{e}^{-\mathrm{i}kx} & x < -a \\ C\,\mathrm{e}^{-\kappa x} + D\,\mathrm{e}^{+\kappa x} & -a < x < a \\ F\,\mathrm{e}^{\mathrm{i}kx} + G\,\mathrm{e}^{-\mathrm{i}kx} & x > a \end{cases} \tag{3.60}$$

with wave numbers $k = \sqrt{2mE}/\hbar$, $\kappa = \sqrt{2m(V_0 - E)}/\hbar$.

In order to obtain the relation between the constants $A, B, \ldots, G$ (Fig. 3.8b), one again uses the continuity requirements for the wave function ψ and its derivative.

Matching condition at $x = -a$:

$$A\,\mathrm{e}^{-\mathrm{i}ka} + B\,\mathrm{e}^{\mathrm{i}ka} = C\,\mathrm{e}^{\kappa a} + D\,\mathrm{e}^{-\kappa a} \quad,$$

$$\mathrm{i}k(A\,\mathrm{e}^{-\mathrm{i}ka} - B\,\mathrm{e}^{\mathrm{i}ka}) = -\kappa(C\,\mathrm{e}^{\kappa a} - D\,\mathrm{e}^{-\kappa a}) \quad.$$

In matrix notation

$$\begin{pmatrix} \mathrm{e}^{-\mathrm{i}ka} & \mathrm{e}^{\mathrm{i}ka} \\ \mathrm{e}^{-\mathrm{i}ka} & -\mathrm{e}^{\mathrm{i}ka} \end{pmatrix} \begin{pmatrix} A \\ B \end{pmatrix} = \begin{pmatrix} \mathrm{e}^{\kappa a} & \mathrm{e}^{-\kappa a} \\ \dfrac{\mathrm{i}\kappa}{k}\mathrm{e}^{\kappa a} & -\dfrac{\mathrm{i}\kappa}{k}\mathrm{e}^{-\kappa a} \end{pmatrix} \begin{pmatrix} C \\ D \end{pmatrix} \quad,$$

$$\begin{pmatrix} A \\ B \end{pmatrix} = \frac{1}{2} \begin{pmatrix} \mathrm{e}^{\mathrm{i}ka} & \mathrm{e}^{\mathrm{i}ka} \\ \mathrm{e}^{-\mathrm{i}ka} & -\mathrm{e}^{-\mathrm{i}ka} \end{pmatrix} \begin{pmatrix} \mathrm{e}^{\kappa a} & \mathrm{e}^{-\kappa a} \\ \dfrac{\mathrm{i}\kappa}{k}\mathrm{e}^{\kappa a} & -\dfrac{\mathrm{i}\kappa}{k}\mathrm{e}^{-\kappa a} \end{pmatrix} \begin{pmatrix} C \\ D \end{pmatrix} \quad,$$

$$\begin{pmatrix} A \\ B \end{pmatrix} = M(a) \begin{pmatrix} C \\ D \end{pmatrix} \quad, \tag{3.61}$$

where

$$M(a) \equiv \frac{1}{2} \begin{pmatrix} \left(1 + \dfrac{\mathrm{i}\kappa}{k}\right)\mathrm{e}^{\kappa a + \mathrm{i}ka} & \left(1 - \dfrac{\mathrm{i}\kappa}{k}\right)\mathrm{e}^{-\kappa a + \mathrm{i}ka} \\ \left(1 - \dfrac{\mathrm{i}\kappa}{k}\right)\mathrm{e}^{\kappa a - \mathrm{i}ka} & \left(1 + \dfrac{\mathrm{i}\kappa}{k}\right)\mathrm{e}^{-\kappa a - \mathrm{i}ka} \end{pmatrix} \quad. \tag{3.62}$$

Matching condition at $x = +a$:

$$\begin{pmatrix} F \\ G \end{pmatrix} = M(-a) \begin{pmatrix} C \\ D \end{pmatrix} \quad. \tag{3.63}$$

Thus, the connection between $\begin{pmatrix} A \\ B \end{pmatrix}$ and $\begin{pmatrix} F \\ G \end{pmatrix}$ then becomes

$$\begin{pmatrix} A \\ B \end{pmatrix} = M(a)M(-a)^{-1} \begin{pmatrix} F \\ G \end{pmatrix} \quad. \tag{3.64}$$

With

$$M(-a)^{-1} = \frac{1}{2} \begin{pmatrix} \left(1 - \dfrac{\mathrm{i}k}{\kappa}\right)\mathrm{e}^{\kappa a + \mathrm{i}ka} & \left(1 + \dfrac{\mathrm{i}k}{\kappa}\right)\mathrm{e}^{\kappa a - \mathrm{i}ka} \\ \left(1 + \dfrac{\mathrm{i}k}{\kappa}\right)\mathrm{e}^{-\kappa a + \mathrm{i}ka} & \left(1 - \dfrac{\mathrm{i}k}{\kappa}\right)\mathrm{e}^{-\kappa a - \mathrm{i}ka} \end{pmatrix} \quad, \tag{3.65}$$

this leads to the equation

$$
\begin{pmatrix} A \\ B \end{pmatrix}
= \begin{pmatrix} (\cosh 2\kappa a + \dfrac{i\varepsilon}{2} \sinh 2\kappa a)e^{2ika} & \dfrac{i\eta}{2} \sinh 2\kappa a \\[2ex] -\dfrac{i\eta}{2} \sinh 2\kappa a & (\cosh 2\kappa a - \dfrac{i\varepsilon}{2} \sinh 2\kappa a)e^{-2ika} \end{pmatrix} \begin{pmatrix} F \\ G \end{pmatrix} ,
$$
(3.66)

where ε and η are defined by

$$
\varepsilon = \frac{\kappa}{k} - \frac{k}{\kappa} \quad ,
$$
(3.67a)

$$
\eta = \frac{\kappa}{k} + \frac{k}{\kappa} \quad .
$$
(3.67b)

We now specialize to the case of a *particle incident from the left*, i.e., let $G = 0$. Equation (3.66) then simplifies to

$$
A = F\left(\cosh 2\kappa a + \frac{i\varepsilon}{2} \sinh 2\kappa a \right) e^{2ika} \quad ,
$$

$$
B = F\left(-\frac{i\eta}{2} \right) \sinh 2\kappa a \quad .
$$
(3.68)

In order to characterize the transmission, one defines the *transmission amplitude*

$$
S(E) \equiv \frac{F}{A} = \frac{e^{-2ika}}{\cosh 2\kappa a + (i\varepsilon/2) \sinh 2\kappa a} \quad ,
$$
(3.69)

and hence the *transmission coefficient*

$$
|S(E)|^2 = \frac{1}{1 + (1 + (\varepsilon^2/4)) \sinh^2 2\kappa a} \quad .
$$
(3.70)

$|S(E)|^2$ expresses the probability that a particle incident on the potential barrier passes through it.

In the *limiting case of a very high and wide barrier*, $\kappa a \gg 1$, and since $\sinh 2\kappa a \cong (1/2)e^{2\kappa a} \gg 1$,

$$
|S(E)|^2 \cong \left(1 + \frac{\varepsilon^2}{4} \right)^{-1} 4e^{-4\kappa a} = \frac{16(\kappa k)^2}{(\kappa^2 + k^2)^2} e^{-4\kappa a} \quad ,
$$

$$
|S(E)|^2 = \frac{16E(V_0 - E)}{V_0^2} \exp\left\{ -4\sqrt{2m(V_0 - E)}\frac{a}{\hbar} \right\} \quad .
$$
(3.71)

Absorbing the prefactor into the exponential,

$$
|S(E)|^2 = \exp\left\{ -4\sqrt{2m(V_0 - E)}\frac{a}{\hbar} + \log\left(\frac{16E(V_0 - E)}{V_0^2} \right) \right\} \quad ,
$$

one can carry out a further approximation by neglecting the logarithmic term, and one finally obtains for $|S(E)|^2$

$$|S(E)|^2 \cong \exp\left\{-4\sqrt{2m(V_0 - E)}\frac{a}{\hbar}\right\} . \tag{3.72}$$

A classical particle would be reflected from the barrier for $E < V_0$. In contrast, according to quantum theory, we even find a finite transmission probability in this case, described by (3.70) and (3.72). This purely quantum mechanical phenomenon is known as the *tunneling effect*. Important examples of the tunneling effect are alpha decay of nuclei (see Sect. 3.3.3) and the cold emission of electrons from metals.

Finally, let us discuss the wave function in the interior of the barrier. According to (3.63) and (3.65),

$$C = \frac{1}{2}\left(1 - \frac{ik}{\kappa}\right)e^{(\kappa+ik)a}F \quad , \quad D = \frac{1}{2}\left(1 + \frac{ik}{\kappa}\right)e^{(-\kappa+ik)a}F \quad ,$$

and for $\kappa a \gg 1$ therefore $C \sim \exp\{-\kappa a - ika\}$ and $D \sim \exp\{-3\kappa a - ika\}$. Here,

$$Ce^{-\kappa x}\Big|_{x=a} \sim e^{-2\kappa a} \sim De^{\kappa x}\Big|_{x=a} \quad ,$$

using (3.68) and $A = 1$. In Fig. 3.8c, the exponentially decreasing and growing parts of the wave function are shown separately (without phase factors).

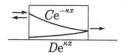

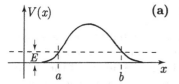

Fig. 3.8c. Absolute values of the exponentially decreasing and increasing parts of the wave function within the barrier

3.3.2 The Continuous Potential Barrier

In realistic tunneling processes, the potential has a continuous shape, as illustrated in Fig. 3.9a. The probability of tunneling through this potential barrier can be calculated approximately with the help of (3.72) by approximating $V(x)$ between a and b by N individual square barriers of width dx. The step width $2a$ is now to be replaced by dx (Fig. 3.9b). The total transmission probability is then the product

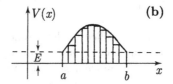

Fig. 3.9. (a) A continuous potential barrier, and (b) its decomposition into rectangular barriers

$$|S(E)|^2 = \prod_{i=1}^{N} \exp\left\{-\frac{\sqrt{2m(V(x_i) - E)}}{\hbar} 2dx\right\}$$

$$= \exp\left\{-2\sum_{i=1}^{N} \frac{\sqrt{2m(V(x_i) - E)}}{\hbar} dx\right\} ,$$

which approaches

$$|S(E)|^2 = \exp\left\{-2\int_a^b \frac{\sqrt{2m(V(x) - E)}}{\hbar} dx\right\} \tag{3.73}$$

in the limit $N \to \infty$. This result can be justified more precisely by the WKB method (see Sect. 11.3).

3.3.3 Example of Application: α-decay

The potential of an α-particle in a nucleus has approximately the shape shown in Fig. 3.10. Here, the range of the nuclear force is $R \approx 10^{-12}$ cm, the charge number of an α-particle is $Z_2 = 2$, and the nuclear charge number of the daughter nucleus is Z_1. In the classical case, energy would have to be supplied to the particle in order for it to climb over the Coulomb barrier and leave the nucleus. Quantum mechanically, one obtains a finite tunneling probability, which can be calculated using (3.73) by setting $V(x) \cong Z_1 Z_2 e^2/x$ between the classical turning points. The limits of integration are $a = R$, $b = Z_1 Z_2 e^2/E$, and it follows that

$$2\int_a^b dx \frac{\sqrt{2m[V(x) - E]}}{\hbar} = 2\frac{\sqrt{2mE}}{\hbar} \int_R^b dx \sqrt{\frac{b}{x} - 1}$$

$$= 2\frac{\sqrt{2mE}}{\hbar} b\left[\arccos\sqrt{\frac{R}{b}} - \left(\frac{R}{b} - \left(\frac{R}{b}\right)^2\right)^{1/2}\right] .$$

For $b \gg R$, i.e., $E \ll$ Coulomb barrier, this becomes

$$2\left(\frac{\sqrt{2m}Z_1 Z_2 e^2}{\sqrt{E}\hbar}\right)\left(\frac{\pi}{2} - 2\sqrt{\frac{R}{b}}\right) ,$$

so that the transmission probability finally becomes

$$|S|^2 = \exp\left\{-\pi\frac{\sqrt{2m}Z_2 e^2}{\hbar}\left(\frac{Z_1}{\sqrt{E}} - \frac{4}{\pi}\frac{Z_1^{1/2}R^{1/2}}{Z_2^{1/2}e}\right)\right\} . \tag{3.74}$$

This does not yet suffice to determine the decay probability: one must also take into account how often the particle is able to tunnel, i.e., how often it "collides" with the nuclear surface. As a semiclassical estimate of the decay probability, we suppose: *decay probability per second* = frequency of wall collisions × transmission coefficient = $(v_i/2R)|S(E)|^2$. Here, v_i is the speed of the α-particle in the interior of the nucleus, an only vaguely defined quantity.

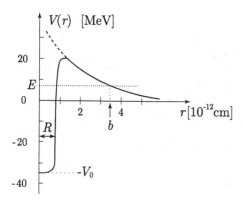

Fig. 3.10. The potential of an α-particle in a nucleus which is unstable to α-decay

Letting the number of radioactive nuclei be N and the change in the number of radioactive nuclei be dN, one then finds the decay law

$$dN = -N \times \text{decay probability per second} \times dt = -\frac{N}{\tau}dt \quad . \qquad (3.75a)$$

The mean *lifetime* τ is hence given by

$$\tau = \frac{2R}{v_i}|S(E)|^{-2} \quad , \qquad (3.75b)$$

and the decay law becomes

$$N(t) = N(0)\,e^{-t/\tau} \quad .$$

The *half-life* T of a substance is the time it takes for the number of nuclei to be reduced to half of the initial value. It is related to the lifetime by

$$e^{-T/\tau} = \tfrac{1}{2} \quad \text{or} \quad T = (\ln 2)\tau = 0.693\tau \quad .$$

Numerical formula for T:

With $Z_2 = 2, R = 1.5 \times 10^{-13}A^{1/3}\,\text{cm} \cong 2 \times 10^{-13}Z_1^{1/3}\,\text{cm}$ for T in years, E in MeV, the numerical formula of Taagepera and Nurmia follows from (3.74, 3.75b):

$$\log_{10} T = 1.61\left(\frac{Z_1}{E^{1/2}} - Z_1^{2/3}\right) - 28.9 \quad . \qquad (3.76)$$

Just this dependence of $\log \tau$ on the energy of the emitted α-particle was found experimentally by Geiger and Nutall. In three dimensions, the factor 2 in (3.75b) is replaced by 1.[3] In heavy nuclei, v_i is of the order of $10^9\,\text{cm s}^{-1}$, and R is about $10^{-12}\,\text{cm}$, and thus the quantity $\log_{10}(0.69R/v_i)$ varies only slightly from nucleus to nucleus and may be replaced by the constant -28.9.

---

[3] E. Segré: *Nuclei and Particles,* 2nd ed. (Benjamin, New York, Amsterdam 1977)

Figure 3.11[4] shows experimental curves for various nuclei and demonstrates the confirmation of the $\sqrt{E}$-dependence. Different isotopes of a radioactive element have different mass number A and hence in the ground state the energy E of the α-particle differs. In Fig. 3.11 the points corresponding to the various isotopes of an element lie on a straight line.

The factor $\exp\{-\sqrt{2m}\pi Z_1 Z_2 e^2/\hbar\sqrt{E}\}$ also determines the probability for the inverse process, i.e., the fusion of two nuclei of charge Z_1 and Z_2. This implies that nuclear fusion occurs preferably for low Z, that is, for hydrogen (or rather heavy hydrogen, i.e., deuterium and tritium)

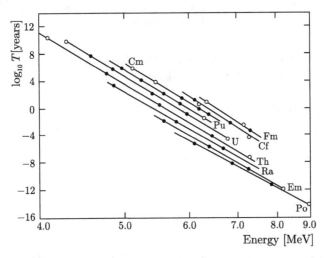

Fig. 3.11. The logarithm of the half-life T as a function of the inverse square root of the energy of the emitted α-particle plus the recoil energy (after Hyde et al.)

$$_1H^2 + {}_1H^2 \rightarrow {}_2He^3 + n \quad (3.27\,\text{MeV}) \quad,$$

$$\rightarrow {}_1He^3 + p \quad (4\,\text{MeV}) \quad,$$

$$_1H^2 + {}_1H^3 \rightarrow {}_2He^4 + n \quad (17.6\,\text{MeV}) \quad,$$

$$_1H^3 + {}_1H^3 \rightarrow {}_2He^4 + 2n \quad (11\,\text{MeV}) \quad.$$

The numbers in parentheses give the reaction energy (Q-value). For higher atomic number Z the Coulomb barrier rises and thus the temperature required to overcome it. In stellar evolution, the lightest elements are used up one after the other, and controlled fusion experiments also concentrate on $Z = 1$. Further important examples of the tunneling effect are the cold

---

[4] E.K. Hyde, I. Perlman, G.T. Seaborg: *The Nuclear Properties of the Heavy Elements I* (Prentice Hall, Englewood Cliffs 1964)

emission of electrons from a metal in an electric field and the tunneling between two metals separated by an insulating layer, an example of which is the Josephson effect, that is, the tunneling of Cooper pairs in superconductors.

3.4 The Potential Well

Let us now determine the *bound states* of the potential well

$$V(x) = -V_0 \Theta(a - |x|) \tag{3.77}$$

(Fig. 3.12). The potential well serves as a model for short-range forces such as those found in nuclear physics or at screened defects in solids.

The tendency to bind a particle will certainly grow with the depth and width of the potential well and the particle mass. A dimensionless parameter constructed from V_0, a, and m which characterizes the strength is

$$\zeta = \sqrt{2mV_0}\,a/\hbar \quad . \tag{3.78}$$

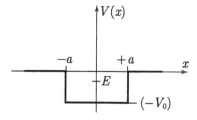

Fig. 3.12. The potential well

As in the preceding sections, one considers the Schrödinger equation for regions of different potential strength separately. Since the energies of the bound states lie in the interval

$$-V_0 \le E \le 0 \quad , \tag{3.79}$$

we have

$$\psi'' = \kappa^2 \psi \quad \text{with} \quad \kappa = \sqrt{2m(-E)}/\hbar \quad \text{for} \quad |x| > a \tag{3.80a}$$

and

$$\psi'' = -q^2 \psi \quad \text{with} \quad q = \sqrt{2m(E + V_0)}/\hbar \quad \text{for} \quad |x| < a \quad . \tag{3.80b}$$

In order to guarantee the normalizability of the wave function in the region $|x| > a$, one chooses from the two fundamental solutions $e^{\pm \kappa x}$ the one that falls off exponentially. Inside the well, the solutions are oscillatory: $\cos qx, \sin qx$, and possibly linear combinations of these two fundamental solutions. However, the reflection symmetry of the potential (3.77) strongly suggests seeking solutions that are purely even,

$$\psi(x) = \begin{cases} A \cos qx & |x| < a \\ e^{\mp \kappa x} & x \gtrless \pm a \end{cases} \quad , \tag{3.81}$$

and odd,

$$\psi(x) = \begin{cases} A \sin qx & |x| < a \\ \pm e^{\mp \kappa x} & x \gtrless \pm a \end{cases} \quad . \tag{3.82}$$

Later we will show generally, in Sects. 3.5 and 3.6, that the bound states for a reflection symmetric potential are either even or odd. From the continuity conditions (3.44a–c) for the wave function and its derivative, we obtain transcendental equations for the energy eigenvalues E and the amplitudes A in (3.81) and (3.82). We consider even and odd solutions in turn.

3.4.1 Even Symmetry

Here, the continuity conditions imply

$$A \cos qa = e^{-\kappa a} \quad , \quad Aq \sin qa = \kappa e^{-\kappa a} \quad , \tag{3.83}$$

and, after division, they yield

$$\tan qa = \frac{\kappa}{q} \quad , \tag{3.84}$$

or, explicitly,

$$\tan qa = \frac{[\zeta^2 - (qa)^2]^{1/2}}{qa} \quad , \tag{3.84'}$$

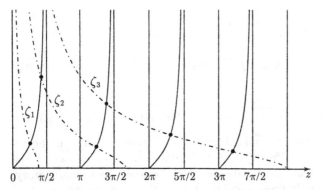

Fig. 3.13. Graphical solution of the transcendental equation (3.84); (—): $\tan z$, $(-\cdot-)$: $(\zeta^2 - z^2)^{1/2}/z$ for various values of ζ ($\zeta_1 < \zeta_2 < \zeta_3$)

where the dimensionless parameter ζ introduced in (3.78) enters. By (3.79), the wave numbers q lie in the interval

$$0 \leq qa \leq \zeta \quad . \tag{3.85}$$

Equation (3.84′) is the transcendental equation mentioned above for the wave number q, or, by (3.80b), for the binding energy E. It can easily be solved graphically (Fig. 3.13). Here, $\tan z$ and $(\zeta^2 - z^2)^{1/2}/z$ are plotted for three values of ζ against $z \equiv qa$. The permitted values of z are found from the intersections of these two curves. For these values qa, the energy eigenvalues according to (3.80b) are then

$$E = -V_0 + \frac{(q\hbar)^2}{2m} = -V_0 \left(1 - \frac{(qa)^2}{\zeta^2} \right) \quad . \tag{3.86}$$

One reads off the following characteristic properties of the eigenvalues from Fig. 3.13.

(i) Since $(\zeta^2 - z^2)^{1/2}/z$ vanishes at $z = \zeta$, the number of intersections n_e can be read off from the value of ζ and is

$$n_e = [\zeta/\pi] \quad , \tag{3.87}$$

where $[\alpha]$ is the nearest natural number greater than α.

(ii) Thus, in any case, there is always an even bound state for $\zeta > 0$; the number of even bound states grows with increasing ζ according to (3.87).

3.4.2 Odd Symmetry

The continuity conditions now take the form

$$A \sin qa = e^{-\kappa a} \quad , \quad Aq \cos qa = -\kappa e^{-\kappa a} \quad , \tag{3.88}$$

and, after division,

$$- \cot qa = \frac{\kappa}{q} \equiv \frac{(\zeta^2 - (qa)^2)^{1/2}}{qa} \quad . \tag{3.89}$$

The graphical solution of (3.89) is illustrated in Fig. 3.14.

If ζ lies in the interval

$$\frac{\pi}{2}(2n_o - 1) < \zeta < \frac{\pi}{2}(2n_o + 1) \quad , \tag{3.90}$$

then (3.89) has exactly n_o solutions. In particular, we see from this that odd solutions only exist if

$$2mV_0 a^2/\hbar^2 > \pi^2/4 \quad , \tag{3.91}$$

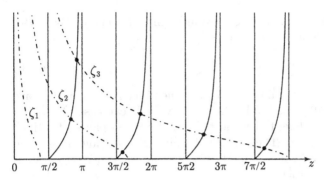

Fig. 3.14. Graphical solution of the transcendental equation (3.89); (—): $-\cot z$, $(-\cdot-)$: $(\zeta^2 - z^2)^{1/2}/z$

that is, the potential must exceed a minimum strength. Of course, the connection between the wave number determined from (3.89) and the energy is also given for odd states by (3.86). We can now summarize our results for even and odd states in Table 3.2. We give the wave number interval, the symmetry, and the number of nodes, i.e., the zeros of the corresponding wave function. The sequence of states terminates at the $(n_e + n_o - 1)$th excited state. Even and odd states alternate with increasing energy and number of nodes. We can evidently number the states with the number of nodes. In the next section we will show in general that the bound states for a symmetric potential can be represented by real even and odd functions. Once we have found all of these, we then know all the bound states for the potential well.

Table 3.2. State, wave number, symmetry, and number of nodes

State	qa	Symmetry	Number of nodes
Ground state	$[0, \frac{\pi}{2}]$	even	0
1st excited state	$[\frac{\pi}{2}, \pi]$	odd	1
2nd excited state	$[\pi, \frac{3}{2}\pi]$	even	2
$\vdots$			

We illustrate our results in Fig. 3.15 for a potential of strength $\zeta = 5$. Here, according to (3.87) and (3.90), there are $[\zeta/\pi] = 2$ even and 2 odd solutions.

Finally, let us consider the limit of an *infinitely deep potential well*, $V_0 \to \infty$. Then $\zeta \to \infty$, and the solutions of (3.84') and (3.89) (the intersections in Figures 3.13 and 3.14) move to the asymptotes of $\tan qa$ and $\cot qa$. Thus the even states are

$$\varphi_q(x) = \Theta(a - |x|) \cos qx \quad \text{with} \quad qa = (s + \tfrac{1}{2})\pi \quad , \tag{3.92}$$

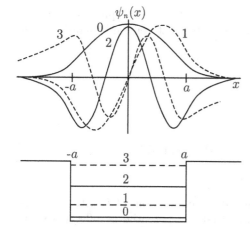

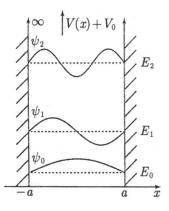

Fig. 3.15. Bound states and energy eigenvalues for a potential well with $\zeta = 5$. The states are designated by the number of nodes. *Solid curves* indicate even solutions, *dashed curves* odd solutions

and the odd ones

$$\varphi_q(x) = \Theta(a - |x|) \sin qx \quad \text{with} \quad qa = s\pi \quad . \tag{3.93}$$

Both solutions satisfy the general boundary conditions $\psi(\text{step}) = 0$ found in (3.58). Normalized, they can also be expressed in the common form

$$\psi_n(x) = \frac{1}{\sqrt{a}} \sin(x+a)k_n \quad \text{with} \quad k_n = \frac{(n+1)\pi}{2a} , n = 0, 1, 2, \ldots, \tag{3.94a}$$

having energy

$$E_n = \frac{\hbar^2 k_n^2}{2m} = \frac{\hbar^2}{2m} \left(\frac{\pi}{2a}\right)^2 (n+1)^2 \quad . \tag{3.94b}$$

The lowest eigenstates of the infinitely deep potential well are shown in Fig. 3.16.

Fig. 3.16. The lowest states in an infinitely deep potential well, shifted along the ordinate

Now that we have solved a few special, characteristic problems, in the next two sections we would like to investigate what general conclusions can be drawn about stationary states.

3.5 Symmetry Properties

3.5.1 Parity

The parity operator (reflection operator) P is defined by

$$Pf(x) = f(-x) \quad . \tag{3.95}$$

Evidently, the even and odd functions are eigenfunctions of P with eigenvalues ± 1 (see (3.100)). We now assume that the potential is reflection symmetric,

$$PV(x) = V(x) \quad . \tag{3.96}$$

Since the kinetic energy contains only a second derivative, for any complex function $f(x)$ one has

$$PHf(x) = Hf(-x) = HPf(x) \quad ,$$

i.e., P and H commute, so that

$$[P, H] = 0 \tag{3.97}$$

for symmetric potentials. If we now consider the time independent Schrödinger equation

$$H\psi(x) = E\psi(x) \quad , \tag{3.98}$$

then upon applying P it follows that

$$H\psi(-x) = E\psi(-x) \quad ,$$

so that both $\psi(x)$ and $\psi(-x)$ are solutions with the same eigenvalue E. By forming the sum and the difference of these two equations, we see that

$$\psi_{e/o}(x) = \psi(x) \pm \psi(-x) \tag{3.99}$$

are eigenfunctions of H as well, with eigenvalue E. Since

$$P\psi_{e/o} = \pm \psi_{e/o} \quad , \tag{3.100}$$

they are also eigenfunctions of P.

It thus follows that for symmetric potentials we can choose a basis of stationary states consisting only of even and odd states. If an energy eigenvalue is nondegenerate, then the corresponding eigenfunction is automatically even or odd.

3.5.2 Conjugation

If we take the complex conjugate of the time independent Schrödinger equation, we get

$$H\psi^*(x) = E\psi^*(x) \quad .$$
(3.101)

Combining (3.98) and (3.101), one sees that $\psi + \psi^*$ and $(\psi - \psi^*)/i$ are real eigenfunctions with eigenvalue E. If, as always assumed, the potential is real, then one can always choose real eigenfunctions. Nondegenerate energy eigenfunctions are, aside from trivial factors, automatically real.

3.6 General Discussion of the One-Dimensional Schrödinger Equation

In this section, we would like to discuss the one-dimensional Schrödinger equation for a general potential in order to see to what extent the properties found for our model potentials (oscillator, potential well, etc.) are generally valid. We will base our considerations on a qualitative, mostly graphical discussion of the solutions of the Schrödinger equation, in order to underscore the key elements. It should be clear from the discussion that this argumentation could also be presented in the form of an analytic proof.

For clarity, we rewrite the one-dimensional Schrödinger equation:

$$\frac{d^2\psi}{dx^2} = \frac{2m}{\hbar^2}(V(x) - E)\psi(x) \quad .$$
(3.102)

If for example we are presented with the continuous, short-range potential of Fig. 3.17, there can be regions with E larger or smaller than $V(x)$. These are associated with different solution elements.

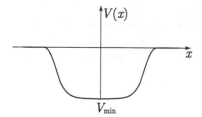

Fig. 3.17. A symmetric short range potential

$V(x) - E > 0$. Then ψ'' has the same sign as ψ, i.e., ψ is convex to the x-axis, that is, curved away from the x-axis (Fig. 3.18a).

Fig. 3.18. Solution elements of the Schrödinger equation in regions with $V(x) - E > 0$

$V(x) - E < 0$. Here, the sign of ψ'' is opposite to that of ψ, and the wave function ψ is concave to the x-axis (Fig. 3.18b).

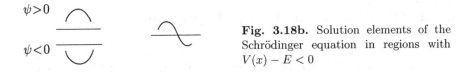

Fig. 3.18b. Solution elements of the Schrödinger equation in regions with $V(x) - E < 0$

The solutions of the Schrödinger equation are constructed from the elements of Fig. 3.18a and 3.18b in a continuous and continuously differentiable manner.

Let us now consider the potential of Fig. 3.17 and the possible solutions of the Schrödinger equation (3.102) for different values of E.

I. $E < V_{\min}$. Then the case $V(x) - E > 0$ would be realized everywhere, and the solution would diverge at infinity. For $E < V_{\min}$, there is thus no acceptable solution.

II. $V_{\min} < E < 0$. Here, we are in the domain of the bound states. In Fig. 3.19, we show a typical bound state of this sort. The horizontal line E intersects the potential at the turning points TP (shown dotted). Outside of the turning points, the solution is curved away from the x-axis, while inside of the turning points the solution is curved toward the x-axis. Roughly speaking, the solution outside the turning point has the form

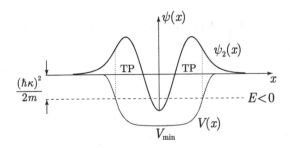

Fig. 3.19. A typical bound state for the potential of Fig. 3.17

$e^{\pm \kappa x}$ with $\kappa = [2m(V(x) - E)]^{1/2}/\hbar$, and inside the form $-\cos qx$ with $q = [2m(E - V(x))]^{1/2}/\hbar$. If we increase the energy, the turning points shift outwards. The function which then results, $\psi(x)(x > 0)$, is curved away less in the outer region from, and in the inner region more strongly toward, the x-axis. The problem of finding bound states now consists in combining symmetrically or antisymmetrically solutions that fall off exponentially at infinity (Sect. 3.5) in such a way that ψ and ψ' are continuous at the origin. We draw the resulting solution sequences for four values of E in the interval $[V_{min}, 0]$ in Fig. 3.20. The value of E increases from (a) to (d), whilst the turning points shift outward. First of all, it is clear that for $E = V_{min}$ the two turning points coincide at the origin, and the continuity conditions cannot be fulfilled. Figure 3.20a corresponds to the situation where E is somewhat larger than V_{min}. The negative curvature in the region inside of the turning points does not yet suffice to bring the derivative of the solution to the same value, namely 0, at the origin, and as for the value of ψ to the right of the origin, it is nowhere near being coincident with the dashed curve to the left of the origin. If E is raised, we finally get to Fig. 3.20b. Here, the solid curves to the left and right have the same slope and value at the origin. It is clear that one reaches this situation for any potential, no matter how weak, as long as E gets sufficiently close to $E = 0$. Thus, there always exists a bound state. This state is symmetric. If one further increases the energy above E of Fig. 3.20b, one arrives at Fig. 3.20c. The wave function in the interior has been curved so strongly downward that the derivatives of the solid branches on the left

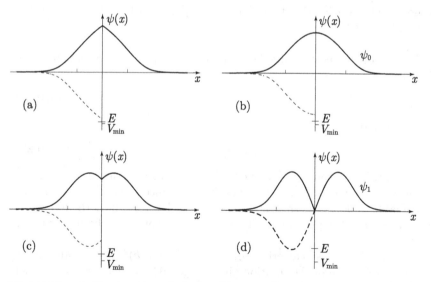

Fig. 3.20a–d. Construction of the bound states for a continuous, symmetric potential for increasing energy E. The marks on the x-axis indicate the position of the turning points

and right are different. This is not an allowable wave function. If the potential is sufficiently strong, then upon additional raising of E the situation of Fig. 3.20d is finally reached, in which ψ vanishes at the origin and the dashed curve, together with the curve on the right, gives a continuous, differentiable solution. This is the first excited state with one node. Clearly such a state does not exist if the potential is too weak. In the case of a strong potential there can be further states in which alternately solid and dashed curves fit together.[5]

We can summarize these considerations as follows:

(i)   There exists at least one bound state ψ_0; it has no zeros.
(ii) The spectrum of the bound states is discrete.
(iii) The bound states are nondegenerate.
(iv) To the extent that the potential possesses multiple bound states, they are alternately symmetric and antisymmetric with increasing energy. The state ψ_n, $n = 0, 1, \ldots$ has n nodes.

Remark: We state without proof that for potentials $V(x)$ behaving at large distances like $\lim_{x \to \infty} V(x) \sim -1/x^s$, one has: For $s < 2$, the energy eigenvalues have a point of accumulation for $E = 0$. For $s > 2$, $E = 0$ is not a point of accumulation; the next bound state is located at a finite distance from 0. If the potential is singular for small distances like $\lim_{x \to 0} V(x) = -1/x^r$, then for $r < 2$, the energy spectrum is bounded from below, and for $r > 2$, the energy spectrum goes on to $-\infty$.

We now continue with the discussion of the potential from Fig. 3.17 in the energy region $E > 0$.

III. _$E > 0$_ . In the regions outside the potential $(V(x) = 0)$ we can write the solution as

$$\psi_{\mathrm{I}} = \mathrm{e}^{\mathrm{i}kx} + R\,\mathrm{e}^{-\mathrm{i}kx} \tag{3.103a}$$

on the left and

$$\psi_{\mathrm{II}} = T\,\mathrm{e}^{\mathrm{i}kx} \tag{3.103b}$$

on the right. In the domain of the potential, the solution is more complicated; the effective wave number is larger than k there. Since two constants R and T are available to us, however, it is possible for each k to achieve the following: the solutions resulting from ψ_{I} and ψ_{II} match in a continuously differentiable way at $x = 0$.

The energy of this scattering solution is $E = (\hbar k)^2/2m$. Finally, if we recall that reflection of the solution given by (3.103a,b) gives rise to another solution, we can summarize: For each $E > 0$, there are two stationary states.

---

[5] For the potential in Fig. 3.17 this is the solution shown in Fig. 3.19 and also a solution having three nodes.

In general, the reflection amplitude differs from zero, $R \neq 0$. Only for particular potentials is $R = 0$ for all k. These reflectionless potentials have a very intimate connection to the solitons of classical mechanics (see Problem 3.6 and Sect. 19.2.1). In the case of resonances, $R \to 0$ can occur for particular k-values. A wave packet, which represents the superposition of many k-values, will be partially reflected even in this domain.

3.7 The Potential Well, Resonances

We now return to the potential well described by (3.77) and Fig. 3.12. We treated *bound states* $(-V_0 < E < 0)$ in Sect. 3.4. We found that a finite potential well has a finite number of bound states.

It remains for us to derive and discuss the stationary states for $E > 0$, the *scattering states*. We find these immediately from Sect. 3.3 by replacing V_0 by $-V_0$ in the stationary solutions for the potential barrier. The wave number in the exterior remains

$$k = \sqrt{2mE}/\hbar \quad . \tag{3.104}$$

Now, in the interior of the potential, the solution is oscillatory too, and κ should be replaced by $\kappa = iq$, where

$$q = \sqrt{2m(E + V_0)}/\hbar \quad , \tag{3.105}$$

which is the wave number in the interior. Then, by (3.60), the solution is composed of the following elements: In the interior, we have

$$\psi(x) = C\,e^{-iqx} + D\,e^{iqx} \quad \text{for} \quad |x| \leq a \quad . \tag{3.106a}$$

To the left of the potential well, $\psi(x) = \psi_{in}(x) + \psi_r(x)$, the sum of an incident wave

$$\psi_{in} = A e^{ikx} \quad \text{for} \quad x < -a \quad , \tag{3.106b}$$

and a reflected wave

$$\psi_r = AS(E)\frac{i}{2}\left(\frac{q}{k} - \frac{k}{q}\right) \sin(2qa) e^{-ikx} \quad \text{for} \quad x < -a \quad . \tag{3.106c}$$

To the right of the potential well we have the transmitted wave

$$\psi_t = AS(E)e^{ikx} \quad \text{for} \quad x > a \quad . \tag{3.106d}$$

We assume from the outset that $G = 0$, i.e., no waves are incident from the right. From (3.69), $S(E)$ becomes with $\cosh(iq) = \cos q$ and $\sinh(iq) = i \sin q$

$$S(E) = \frac{e^{-2ika}}{\cos 2qa - (i/2)((q/k) + (k/q)) \sin 2qa} \quad . \tag{3.107}$$

Therefore, the *transmission coefficient* is

$$|S(E)|^2 = \frac{1}{1 + (1/4)((q/k) - (k/q))^2 \sin^2(2qa)}$$

with

$$\left(\frac{q}{k} - \frac{k}{q}\right)^2 = \left(\frac{\sqrt{E + V_0}}{\sqrt{E}} - \frac{\sqrt{E}}{\sqrt{E + V_0}}\right)^2 = \frac{V_0^2}{E(E + V_0)} \quad,$$

i.e.

$$|S(E)|^2 = \left[1 + \frac{\sin^2(2qa)}{4(E/V_0)(1 + (E/V_0))}\right]^{-1} \quad . \tag{3.108}$$

One easily recognizes from (3.106c) that

$$\left|S(E)\frac{i}{2}\left(\frac{q}{k} - \frac{k}{q}\right)\sin 2qa\right|^2 = 1 - |S(E)|^2 \tag{3.109}$$

holds for the *reflection coefficient*.

Next, we would like to investigate the properties of the transmission coefficient more thoroughly. First of all, one sees that $|S(E)|^2$ is restricted to the interval $[0,1]$. The transmission is equal to 1 and is thus maximal for

$$2qa = n\pi \quad . \tag{3.110}$$

This is equivalent to energy values

$$E_R = \frac{\hbar^2 q^2}{2m} - V_0 = n^2 \frac{\hbar^2 \pi^2}{8ma^2} - V_0 \quad, \tag{3.111}$$

where n is an integer and large enough so that $E_R > 0$. For particles with this incident energy, the potential is completely penetrable. Minima of the transmission coefficient are found approximately at $2qa = (2n + 1)\pi/2$, because here $\sin^2(2qa) = 1$. They thus lie approximately on the curve

$$\left[1 + \frac{1}{4(E/V_0)(1 + E/V_0)}\right]^{-1} = \frac{4(E/V_0)(1 + E/V_0)}{1 + 4(E/V_0)(1 + E/V_0)} \quad .$$

We recall the infinitely deep potential well of Sect. 3.4. Comparison of (3.110) and (3.111) with (3.94b) and (3.86) shows that the energies of the maxima of the transmission coefficients coincide with the energy eigenvalues of the infinitely deep potential well. For a deep well, i.e., $\zeta \gg 1$ and $E \ll V_0$, the maxima and minima are strongly pronounced. The maxima of the transmission coefficients are known as *resonances*. We can understand the resonance phenomenon qualitatively using our knowledge of reflection at a step: The wave ψ_{in}, incident from the left, penetrates into the potential, changing its wave vector and its amplitude, and is reflected at $x = +a$. Because $4aq = 2\pi n$, the incident and reflected wave are in phase, whereas the wave reflected at $-a$ makes a phase jump π ($R < 0$, see remark iv on page 61). Thus, the wave reflected at $-a$ and the waves reflected again and again at a are out of phase and interfere destructively.

In Fig. 3.21 we have displayed $|S(E)|^2$ as a function of E for two values of the strength ζ. One sees from the figure and from (3.108) that for large ζ the resonances become narrower or "sharper". The n-values of the resonances, according to (3.110) and (3.111), for $\zeta = 6$, are $n = 4, 5, \ldots$, and for the stronger potential $\zeta = 370$, they are $n = 236, 237, \ldots$. We note here that for potentials of the kind shown in Fig. 3.26, the resonances are even more pronounced. We will encounter these again in Chap. 18 when scattering is discussed.

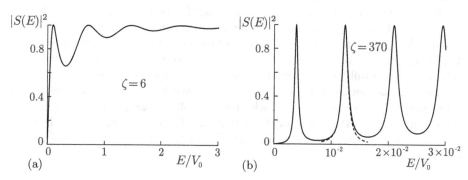

(a)

(b)

Fig. 3.21. The transmission coefficient $|S(E)|^2$ for a potential well for (a) potential strength $\zeta = 6$ and (b) $\zeta = 370$. *Solid* curve: $|S(E)|^2$ from (3.108); *dotted* Lorentz curve from (3.120)

3.7.1 Analytic Properties of the Transmission Coefficient

We now turn to the properties of $S(E)$ in the complex E-plane. In order to find the poles of $S(E)$, we must set the denominator to zero, which leads to the condition $\cos 2qa = (\mathrm{i}/2)(q/k + k/q) \sin 2qa$. With the identity $\cot(2qa) = (\cot qa - \tan qa)/2$ the condition for a pole becomes

$$\cot qa - \tan qa = \frac{\mathrm{i}k}{q} + \frac{\mathrm{i}q}{k} \quad . \tag{3.112}$$

This equation is of the form $f - f^{-1} = g - g^{-1}$, and it is satisfied if and only if either

$$\tan qa = -\frac{\mathrm{i}k}{q} \tag{3.113a}$$

or

$$\cot qa = \frac{\mathrm{i}k}{q} \tag{3.113b}$$

holds.

For $E > 0$, i.e., for real k and q, there are no solutions of (3.113). If $E < -V_0$, and thus k and q are both pure imaginary, there are no solutions either. The region $-V_0 < E < 0$ remains, in which k is imaginary and q is real. With $E = |E|e^{i\varphi}$, $\sqrt{E} = |E|^{1/2}e^{i\varphi/2}$, one obtains for $E < 0$, i.e., $\varphi = \pi$,

$$k = i\frac{(2m|E|)^{1/2}}{\hbar} \ .$$

Therefore (3.113a,b) take the form

$$\tan qa = \frac{(2m|E|)^{1/2}}{\hbar q} \quad \text{and} \quad \cot qa = -\frac{(2m|E|)^{1/2}}{\hbar q} \quad ,$$

which we are already familiar with from the potential well (3.84) and (3.89) as conditions for the energy eigenvalues of even and odd bound states.

$S(E)$ thus has a pole at the positions of the bound states E_b of the potential well, i.e., it goes to infinity there. For this property of $S(E)$ there is a plausible interpretation as well. Because of the pure imaginary wave number k, the amplitude A of the incident wave must go to zero for negative energy; otherwise, ψ_{in} would be divergent for $x \to -\infty$. However, since $S(E_b) = \infty$ one still has for energy E_b a "reflected" wave for $x < -a$ and and a "transmitted" wave for $x > a$, which now falls off exponentially because of the imaginary k. In the region of the potential, a finite wave function is then present even without an incident wave ψ_{in}, representing the bound state (see Fig. 3.22).

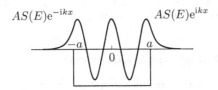

$AS(E)e^{-ikx}$ $\qquad\qquad$ $AS(E)e^{ikx}$

Fig. 3.22. The wave function of a bound state: For $E \to E_b$ such that, for $A \to 0$, $AS(E)$ remains finite, one obtains the bound states at the poles of $S(E)$ along the negative real axis

Let us now consider $S(E)$ near the resonances E_R:

$$S(E)e^{2ika} = \frac{1}{\cos 2qa}\frac{1}{1 - (i/2)((q/k) + (k/q))\tan 2qa} \ . \tag{3.114}$$

At the resonances E_R one has $2qa = n\pi$ and therefore

$$\cos 2qa|_{E_R} = (-1)^n \quad , \quad \tan 2qa|_{E_R} = 0 \ .$$

Taylor expansion of the denominator in $E - E_R$ leads to

$$\frac{1}{2}\left(\frac{q}{k} + \frac{k}{q}\right)\tan 2qa = \frac{2}{\Gamma}(E - E_R) + \cdots \quad , \tag{3.115}$$

where Γ is determined by

$$\frac{2}{\Gamma} = \left[\frac{1}{2}\left(\frac{q}{k} + \frac{k}{q}\right)\frac{d(2qa)}{dE}\right]_{E_R} \ . \tag{3.116}$$

Substitution of q and k gives

$$\frac{2}{\Gamma} = \frac{1}{2} \frac{\sqrt{2m}\,a}{\hbar} \frac{2E_R + V_0}{\sqrt{E_R}\,(E_R + V_0)} \quad . \tag{3.117}$$

Finally, for a very deep well $(V_0 \gg E_R)$, one gets

$$\frac{2}{\Gamma} \approx \frac{1}{2} \frac{\sqrt{2mV_0}\,a}{\hbar} \frac{1}{\sqrt{V_0 E_R}} = \frac{a}{v_R \hbar} \,, \tag{3.118}$$

where $v_R = \sqrt{2E_R/m}$ is the incident velocity at the resonant energy. One recognizes that Γ can become very small if $|E_R| \ll V_0$ and $\zeta \gg 1$ (Eq. (3.78)). We thus arrive at the following representation of $S(E)$ in the neighborhood of the resonances:

$$S(E)e^{2ika} = (-1)^n \frac{i\Gamma/2}{E - E_R + i\Gamma/2} \quad . \tag{3.119}$$

The transmission amplitude $S(E)$ has poles at the complex energy values

$$E = E_R - i\frac{\Gamma}{2} \quad .$$

Since because of our definition of $\sqrt{E}$ the complex plane has a branch cut along the positive real axis, i.e., $\sqrt{E}$ is discontinuous as one goes from above to below the positive real axis ($\sqrt{E} = |E|^{1/2}$ for $E = |E|$, but $\sqrt{E} = -|E|^{1/2}$ for $E = |E|\,e^{2\pi i}$), $S(E)$, as a function of $\sqrt{E}$, also possesses this branch cut along the positive real axis. The poles which are related to the resonances belong to the analytic continuation of $S(E)$ into the second Riemannian sheet.

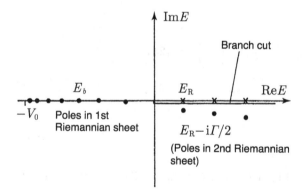

Fig. 3.23. Poles of $S(E)$ in the complex energy plane

The analytic properties of $S(E)$ are summarized in Fig. 3.23. In the first Riemannian sheet, $S(E)$ has poles at the binding energies of the potential well, and in the second Riemannian sheet it has poles at the energy values $E_R - i\Gamma/2$, which lead to the resonances on the positive real axis. The precise determination of the poles of $S(E)$ – which can be achieved by continuing

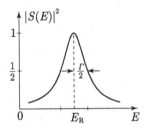

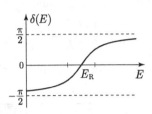

Fig. 3.24. Equation (3.120)   **Fig. 3.25.** The phase shift (3.122)

the Taylor expansion (3.115) to higher order in the complex plane – shows that the real part of the poles is displaced somewhat to the left with respect to E_R.

Now we take again E as real. Near resonance, the transmission coefficient becomes, with (3.119),

$$|S(E)|^2 = \frac{(\Gamma/2)^2}{(E - E_R)^2 + (\Gamma/2)^2} \quad . \tag{3.120}$$

This function is called the *Lorentz* curve or *Breit–Wigner* function. It is shown in Fig. 3.24. Comparison with the exact form in Fig. 3.21 shows that the scattering amplitude in the resonance region is represented very precisely by (3.120). The width of the resonance is determined by Γ. The quantity $S(E)$ can also be described by a real amplitude and a phase:

$$S(E) = |S(E)| \exp\{i\delta(E) - 2ika\} \quad , \tag{3.120'}$$

where

$$\tan \delta(E) = \frac{\text{Im}\,(S(E)e^{2ika})}{\text{Re}\,(S(E)e^{2ika})} = \frac{1}{2}\left(\frac{q}{k} + \frac{k}{q}\right)\tan(2qa) \quad , \tag{3.121}$$

so that according to (3.119), near resonance, $\tan \delta(E) = \frac{2}{\Gamma}(E - E_R)$, and thus

$$\delta(E) = \arctan\left[\frac{2}{\Gamma}(E - E_R)\right] \quad . \tag{3.122}$$

The quantity $\delta(E)$, shown in Fig. 3.25, gives the phase shift of the transmitted wave compared to the incident wave

$$\psi_t(x = a) = |S(E)|\,e^{i\delta(E)}\,\psi_{in}(x = -a) \quad .$$

Here as well, the amplitude and phase show the typical behavior near resonance which is familiar from phenomena of classical mechanics and electrodynamics.

On the basis of our earlier physical considerations, it is clear that resonances also occur for more general potentials, for which one can derive (3.119) and (3.122) in an analogous fashion. We would now like to investigate how a particle whose energy is near a resonance behaves under the influence of such a potential.

3.7.2 The Motion of a Wave Packet Near a Resonance

Let the wave packet incident on the potential be given as the superposition of plane waves by

$$\psi_{in}(x,t) = \int_0^\infty \frac{dp}{2\pi\hbar} g(p) \exp\left\{\frac{i}{\hbar}(px - E(p)t)\right\}, \quad E(p) = p^2/2m \quad (3.123)$$

(the integration is performed only from 0 to ∞, since for $x < -a$ only plane waves propagating to the right $(p > 0)$ should be superposed). Assuming that $g(p)$ has a maximum at the value p_0 with $E(p_0) \approx E_R$, it follows for the velocity and the position of the maximum of $\psi_{in}(x,t)$ that

$$v_0 = p_0/m \quad, \quad x(t) = v_0 t \quad. \quad (3.124)$$

According to (3.106d) and (3.120′), the transmitted wave packet then has the form $(p = \hbar k)$

$$\psi_t(x,t) = \int_0^\infty \frac{dp}{2\pi\hbar} g(p)$$

$$\times \exp\left\{\frac{i}{\hbar}(px - E(p)t - 2pa + \delta(E)\hbar)\right\}|S(E)| \quad, \quad x > a. \quad (3.125)$$

We obtain the center of mass of the wave packet from the stationarity of the phase according to Sect. 2.10.1:

$$x(t) = v_0 t + 2a - \frac{d}{dp}\delta(E)\hbar\big|_{p_0}$$

$$= v_0 t + 2a - \hbar\frac{dE}{dp}\frac{d}{dE}\delta(E)\big|_{p_0} \quad. \quad (3.126)$$

If we substitute $(dE/dp|_{p_0} = v_0)$ and (3.122), it follows that

$$x(t) = v_0 t + 2a - \frac{\hbar\,2/\Gamma}{1 + [(2/\Gamma)(E(p_0) - E_R)]^2} v_0 \quad. \quad (3.127)$$

We can read off the time the particle spends in the potential well from the contributions to (3.127). The first term corresponds to free motion without a potential. The second term corresponds to the infinitely fast transmission through the potential which would be applicable to a classical particle in a very deep potential. Finally the last term divided by v_0 gives the time spent

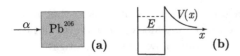

(a)   (b)

Fig. 3.26. (a) Capture of an α-particle in Pb^{206}. (b) Potential well + Coulomb barrier

in the well. Just at the resonance $E(p_0) = E(p_R)$, it is $2\hbar/\Gamma$. For sharp resonances, as in the potential of Fig. 3.26b, the time spent is very large, and the resonance becomes similar to a bound state. Hence one calls $2\hbar/\Gamma$ the life time of the resonance. In the potential well considered previously for which the resonances are not particularly sharp, it follows from (3.118) with the "internal velocity" $v_i = \sqrt{2/m}(E_R + V_0)^{1/2}$ that the particle oscillates back and forth in the potential region about $(v_i 2\hbar/\Gamma)/4a \approx (1/4)\sqrt{1 + V_0/E_R}$ times, before it finally leaves this region due to its positive energy.

Equation (3.127) describes the maximum of the wave packet if the phase $\delta(E)$ does not vary too strongly over the extension of the packet, i.e., when the spread of the wave packet is small compared to the resonance. In the more interesting case in which the resonance is much sharper than the incident wave packet $g(p)$, (3.126) gives only the average position to the right of the potential. Because of the sharp resonance, the packet is very strongly deformed. We will discuss this at the end of this section. First, we would like to discuss several physical applications based on the results thus far. There are a large number of examples of such quantum mechanical resonance phenomena.

In *nuclear physics*, for example, when Pb^{206} nuclei are bombarded with α-particles of energy $E_\alpha = 5.4$ MeV, the element Po^{210} is formed, which decays again by α-emission with a half-life of 138 days, corresponding to a width Γ of 3.8×10^{-23} eV (Fig. 3.26).

Resonances also play an important role in *elementary particle physics*. In the scattering of pions from nucleons, $\pi + N \rightarrow N^* \rightarrow \pi + N$, corresponding to the maximum of the scattering cross section, a temporary bound state of width $\Gamma \approx 120$ MeV or lifetime $\tau \approx 0.5 \times 10^{-23}$ s (Fig. 3.27) is formed. The resonance can be identified with an elementary particle of finite lifetime. The J/ψ-particle was discovered in 1974 as a very sharp resonance with the mass

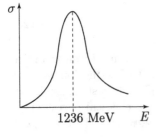

Fig. 3.27. Resonance in π-meson–nucleon scattering cross section (qualitative)

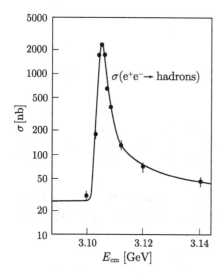

Fig. 3.28. The scattering cross section for hadron production in electron–positron scattering as a function of the center of mass energy E_{cm} of e^- and e^+. The resonance J/ψ represents a bound state of quarks with the quantum number "charm"

3.1 GeV by Ting in the elementary particle reaction $p + p \rightarrow e^+ + e^- + X$ and by Richter in the elementary particle reaction $e^+ + e^- \rightarrow$ hadrons (Fig. 3.28). Here, one is dealing with orthocharmonium $c\bar{c}$, the bound state of one quark and one antiquark with the quantum number "charm".

We now return to (3.125). We would like to determine the precise form of the transmitted wave packet. Here, we assume that the wave packet overlaps with only one of the resonances and that the width Γ of this resonance is much narrower than the extension of the wave packet. Then $g(p)$ is practically constant in the region over which the resonance extends, and we can replace $g(p)$ by $g(p_R)$ (Fig. 3.29), where $p_R = \sqrt{2mE_R}$. We further substitute for $S(E)$ the Breit–Wigner formula (3.119):

$$\psi_t(x,t) \approx \pm i g(p_R)$$
$$\times \int_0^\infty \frac{dp}{2\pi\hbar} \exp\left\{\frac{i}{\hbar}(px - E(p)t - 2pa)\right\} \frac{\Gamma/2}{E - E_R + i\Gamma/2} \ .$$

$$(3.128)$$

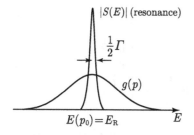

$|S(E)|$ (resonance)

$\frac{1}{2}\Gamma$

$g(p)$

$E(p_0) = E_R$ E

Fig. 3.29. The wave packet $g(p)$ and a sharp resonance

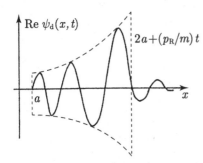

Fig. 3.30. The wave packet at a resonance after transmission through the potential; (- - -): exponential envelope

Because of the resonance structure of the second factor in the integrand, we introduce for the remainder of this discussion the following approximations:

$$dp = \frac{dp}{dE}dE = \frac{m}{p}dE \approx \frac{m}{p_R}dE$$

and

$$p \cong p_R + \left.\frac{dp}{dE}\right|_{E_R}(E - E_R) = p_R + \frac{m}{p_R}(E - E_R) \quad ,$$

where $v_R = p_R/m$. The integration over the new variable E is performed over an interval around the resonance energy which is much larger than Γ. However, since the denominator in the integrand grows with $|E - E_R|$ and the exponential function oscillates strongly with E at great distances from E_R, we extend the integration from $-\infty$ to $+\infty$ for mathematical convenience, without making an unacceptably large error. Thus, we finally obtain

$$\psi_t(x,t) \approx \pm\, ig(p_R)\frac{\Gamma m}{4\pi\hbar p_R}\exp\left\{\frac{i}{\hbar}[p_R(x - 2a) - E_R t]\right\}$$

$$\times \int_{-\infty}^{+\infty} dE\, \frac{\exp\left\{\frac{i}{\hbar}[(x-2a)/v_R - t](E - E_R)\right\}}{E - E_R + i\Gamma/2} \quad .$$

This integral can now be evaluated with the help of the residue theorem with the result

$$\psi_t(x,t) \cong \begin{cases} 0 & \text{for } x > 2a + \dfrac{p_R}{m}t \\[2mm] \pm\, g(p_R)\ \dfrac{\Gamma m}{2\hbar p_R}\exp\left\{\dfrac{i}{\hbar}[p_R(x - 2a) - E_R t]\right\} \\[2mm] \qquad \times \exp\left\{-\dfrac{\Gamma}{2\hbar}\left[t - \dfrac{m}{p_R}(x - 2a)\right]\right\} \\[2mm] \qquad \text{for } x < 2a + \dfrac{p_R}{m}t \end{cases} \quad . \tag{3.129}$$

The fact that, according to (3.129), ψ_t falls abruptly to 0 at the position $x = 2a + (p_R/m)\,t$ comes from our extending the integration from $-\infty$

to $+\infty$. The real part of the wave function is displayed in Fig. 3.30; it is deformed characteristically compared to the incident wave packet. We recall that $\psi_t(x,t)$ – the transmitted wave function – is confined to $x > a$. By (3.124), the maximum of the incident wave packets arrives at the left side of the potential region at the time $t_{-a} = -a/v_R$. The nonoscillatory part of ψ_t is

$$\psi_t(x,t) \propto g(p_R)\frac{\Gamma}{2\hbar v_R}\exp\left\{\frac{-\Gamma}{2\hbar}\left(t - t_{-a} - \frac{x-a}{v_R}\right)\right\}$$
$$\times\;\Theta((t - t_{-a})v_R - (x - a))\;.\tag{3.130}$$

The total position probability to the right of the potential well is given by

$$P_t(t) = \int_a^\infty dx|\psi_t(x,t)|^2$$
$$= |g(p_R)|^2\frac{\Gamma/\hbar}{4v_R}\left(1 - \exp\left\{\frac{-\Gamma}{\hbar}(t - t_{-a})\right\}\right)\;.\tag{3.131}$$

The height of the wave front is $g(p_R)\Gamma/2\hbar v_R$. Although it propagates with velocity v_R, the growth of the position probability to the right is only proportional to $(1 - \exp\{-\Gamma(t - t_{-a})/\hbar\})$, i.e., the decay rate of the resonantly bound particle is $\Gamma/\hbar$. A resonance is thus an "almost bound state" of the potential with a finite lifetime $\hbar/\Gamma$; its energy uncertainty is $\Delta E \approx \Gamma/2$. We will encounter the relation between energy and time uncertainty of a state in the following chapter in a more general context.

The special functions of mathematical physics needed in this and the subsequent chapters can be found in:

M. Abramowitz and I. Stegun: Handbook of Mathematical Functions (National Bureau of Standards, Dover Publications, New York 1964)

P. Dennery and A. Krzywicki: Mathematics for Physicists (A. Harper International Edition, New York 1969)

P. M. Morse and H. Feshbach: Methods of Theoretical Physics, 2 vols (McGraw-Hill, New York 1953)

V. I. Smirnov: A Course of Higher Mathematics (Pergamon Press, Oxford 1964, 16th revised edition)

A. Sommerfeld: Lectures on Theoretical Physics Vol. 6: Partial Differential Equations in Physics (Academic Press, New York 1967)

E. T. Whittaker and G. N. Watson: A Course of Modern Analysis (Cambridge University Press, 1963, 4th ed.)

Problems

3.1 The Hermite polynomials $H_n(x)$ are defined in (3.25).

(a) Show that e^{-t^2+2tx} is a generating function of the Hermite polynomials, i.e.,

$$e^{-t^2+2tx} = \sum_{n=0}^{\infty} \frac{t^n}{n!} H_n(x) \quad .$$

(b) By taking the derivative of this expression, show the validity of the recursion relations

$$H'_n(x) = 2nH_{n-1}(x)$$

and

$$H_{n+1}(x) = 2xH_n(x) - 2nH_{n-1}(x) \quad .$$

(c) Prove the completeness relation (3.30). Hint: Express $H_n(x)$ and $H_n(x')$ in terms of (3.25) and represent the Gaussian by its Fourier transform.

(d) Show that the Hermite polynomials and hence the wave functions ψ_n have n nodes (n simple real zeros) for finite x. Hint: proof by induction, using Rolle's theorem.

3.2 (a) Show that coherent states can be written in the form (3.37''). Hint: Use (2.47).

(b) Demonstrate that

$$(\varphi_\alpha, \varphi_\beta) = e^{-\frac{1}{2}(|\alpha|^2 + |\beta|^2) + \alpha^* \beta}$$

holds.

(c) Prove the relation

$$\int \frac{d^2\alpha}{\pi} \varphi_\alpha(x)\varphi_\alpha^*(x') = \delta(x - x') \quad ,$$

where

$$d^2\alpha = (\text{Re } \alpha)(\text{Im } \alpha) \quad .$$

3.3 The classical limit of the harmonic oscillator (see end of Sect. 3.1).

(a) Show using the Stirling formula

$$n! \approx \sqrt{2\pi n}\, n^n e^{-n} \quad \text{for } n \gg 1$$

that the absolute value of the amplitude $\alpha^n/\sqrt{n!}$ is maximal for

$$n_0 = |\alpha|^2 \quad .$$

(b) Show that the expectation value $\langle \hat{n} \rangle$ of the operator $\hat{n}$ in the state φ_α is equal to n_0.

(c) Calculate $\Delta n/\langle \hat{n} \rangle$.

(d) Show that in this limit the energy expectation value is

$$(\varphi_\alpha, H\varphi_\alpha) = \hbar\omega|\alpha|^2 \quad .$$

Compare this answer with the classical result. See E. Schrödinger, quoted in footnote 4.

3.4 In this problem examples of non normalizable and hence unphysical solutions of the Schrödinger equation of the harmonic oscillator are discussed.

(a) Is $e^{x^2/(2x_0^2)}$ an eigenfunction of the operator $\hat{n} = a^\dagger a$?

(b) Can you give a divergent eigenfunction of $\hat{n}$ with eigenvalue 0?

(c) Why don't these unphysical solutions occur in the algebraic method?

3.5 The state of the art of semiconductor technology allows one to manufacture hardware in which the electrons move freely in two dimensions (e.g., the yz-plane). In the third dimension (x-axis), by means of appropriate coatings of material – even single-atom monolayers are "under control" – very narrow, high potential wells can be achieved. These lead to a sharp quantization of the motion in this direction.
In this problem, a single, infinitely high potential well (this almost exists) is to be treated. Imagine that the free motion in the yz-plane has been separated off and solve the following subproblems.

(a) Determine the eigenfunctions and energy eigenvalues for motion in the infinitely high potential box

$$V(x) = \begin{cases} 0 & \text{for } x \in (-a, a) \\ \infty & \text{otherwise} \end{cases} \quad .$$

(b) Show that the system of eigenfunctions is complete.

(c) Determine the momentum distribution for a particle in the n-th energy eigenstate.

3.6 Solve the time independent Schrödinger equation for the potential

$$V(x) = -\hbar^2 / \left(ma^2 \cosh^2 \frac{x}{a} \right) \quad .$$

Hint: Introduce $y = x/a$, make the ansatz

$$\psi(y) = e^{iky}\, \varphi(y) \quad ,$$

and substitute $z = \tanh y$. The resulting equation can be solved in elementary fashion by a power series together with a condition for the series to terminate. Discuss the solutions that occur.

3.7 (a) Calculate the bound state and the scattering states for the one-dimensional potential

$$V(x) = -\lambda \, \delta(x)$$

with $\lambda \geq 0$.

(b) Prove the completeness relation for these states.

3.8 In the manner described in the introduction to Problem 3.5, it is possible to manufacture "double heterostructures". Consider as the simplest realization of this an infinitely high potential well containing a δ-barrier (see schematic figure).

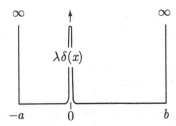

(a) Determine the normalized eigenfunctions and give a formula for the energy eigenvalues.

(b) Discuss the limiting cases $\lambda \to 0$ and $\lambda \to \infty$.

(c) Discuss the special case $a = b$.

3.9 A one-dimensional, narrow Gaussian wave packet is scattered from a high potential barrier V_0 of width $2a$.

$$\frac{p_0^2}{2m} \ll V_0 \quad , \qquad \frac{\Delta p}{p_0} \ll 1 \quad .$$

Compute the time t_p required for the packet to penetrate the potential barrier. Hint: Expand the transmission coefficient in the limit

$$a\sqrt{2m(V_0 - E)}/\hbar \gg 1 \quad .$$

Result:

$$t_p = \frac{\hbar}{\sqrt{V_0 E}}$$

3.10 For a general, one-dimensional wave packet

$$\psi(x,t) = \int \frac{dp}{2\pi\hbar} \, g(p) \, \exp\{i[px - E(p)t + \hbar\alpha(p)]/\hbar\} \quad ,$$

whose real weighting function $g(p)$ is symmetric about its maximum at p_0 and which differs substantially from zero only in a region $|p - p_0| \leq \Delta p$, calculate the expectation values $\langle p \rangle$, $\langle p^2 \rangle$, $\langle x \rangle$, and $\langle x^2 \rangle$. It is advisable to carry out the computation of the positional uncertainty in the momentum representation.

3.11 (a) Calculate $(\psi_m, x\psi_n)$ and $(\psi_m, p\psi_n)$ for the linear harmonic oscillator.

(b) Calculate $(\psi_m, p^2\psi_n)$ and Δp for the linear harmonic oscillator.

3.12 What are the energy levels and the eigenfunctions for a potential of the form

$$V(x) = \begin{cases} \infty & x \leq 0 \\ \frac{m\omega^2 x^2}{2} & x > 0 \end{cases} \quad ?$$

3.13 Compute the transmission coefficient for the potential barrier

$$V(x) = \begin{cases} V_0 & \text{for } |x| \leq a \\ 0 & \text{for } |x| > a \end{cases}$$

if the energy E exceeds the height V_0 of the barrier. Plot the transmission coefficient as a function of E/V_0 (up to $E/V_0 = 3$) for $\sqrt{2mV_0a^2/\hbar^2} = 0.75$.

3.14 Carry out the integration and expansion leading to (3.74).

3.15 Using the uncertainty relation, find an estimate of the ground state energy of a particle (mass m) located in a potential $V(x) = cx^4$.

3.16 Using the approximate formula given in (3.73), calculate the transmission coefficient for electrons from a metal under the action of a large electric field $\mathcal{E}_x$ ("cold emission" or "field emission").

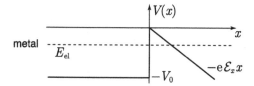

Use the potential shown in the figure: $V(x) = -e\mathcal{E}_x x = e_0\mathcal{E}_x x$. Compute the transmission coefficient for $E_{el} = -4.5\,\text{eV}$ and for fields $\mathcal{E}_x = -5 \times 10^6\,\text{V/cm}$, $-10^7\,\text{V/cm}$, and $-5 \times 10^7\,\text{V/cm}$. (Remark: This calculation also plays a role in the derivation of the current–voltage characteristics of a Schottky diode in semiconductor physics.)

3.17 Given a potential $V(x)$ of the form shown in the figure, explain qualitatively why every energy eigenvalue E with $0 < E < V_1$ is allowed.

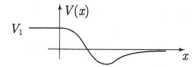

3.18 A particle of mass m moves in the potential

$$V(x) = -\lambda\,\delta(x-a) - \lambda\,\delta(x+a) \quad , \quad \lambda > 0 \quad .$$

(a) Give the (transcendental) equations for the two bound states of the system and estimate the difference of their energy levels for large a.

(b) Calculate the transmission coefficient for the above potential for $\lambda < 0$.

3.19 Determine the transmission amplitude $S(E)$ for the one-dimensional attractive δ-potential $V(x) = -\lambda\,\delta(x)$. Determine the poles of $S(E)$ and discuss their physical significance.

3.20 Calculate the transmission coefficient $S(E)$ for scattering from the potential

$$V(x) = -V_0\Theta(a - |x|) + \lambda\,(\delta(x-a) + \delta(x+a)) \quad ,$$

where $\lambda \geq 0$ and $\Theta(x)$ is the step function. (The δ-peaks simulate Coulomb barriers in realistic problems.) Investigate the change in the form of the resonances in comparison with the case $\lambda = 0$ and discuss $|\lambda/V_0| \to \infty$.

4. The Uncertainty Relation

4.1 The Heisenberg Uncertainty Relation

4.1.1 The Schwarz Inequality

Theorem 1. For the scalar product of two wave functions, the *Schwarz inequality*

$$|(\varphi, \psi)|^2 \leq (\varphi, \varphi)(\psi, \psi) \tag{4.1}$$

holds.

Proof:

(i)   For $\varphi = 0$, the inequality is obviously satisfied.

(ii) For $\varphi \neq 0$, we decompose ψ into a part parallel to φ and a part perpendicular to φ : $\psi = z\varphi + \chi$ with $(\varphi, \chi) = 0$. It then follows that $(\varphi, \psi) = z(\varphi, \varphi)$, and thus for the proportionality factor $z = (\varphi, \psi)/(\varphi, \varphi)$. Furthermore,

$$(\psi, \psi) = (z\varphi + \chi, z\varphi + \chi) = z^* z(\varphi, \varphi) + (\chi, \chi) \geq z^* z(\varphi, \varphi) \quad .$$

Substitution of z gives

$$(\psi, \psi) \geq \frac{|(\varphi, \psi)|^2}{(\varphi, \varphi)} \quad \text{q.e.d.}$$

The equal sign holds only for $\chi = 0$, i.e., for $\psi = z\varphi$.

4.1.2 The General Uncertainty Relation

Let two Hermitian operators H_1 and H_2 and an arbitrary state ψ be given. We define the operators $\hat{H}_i$ by subtracting from H_i the expectation value in the state ψ:

$$\hat{H}_i = H_i - \langle H_i \rangle = H_i - (\psi, H_i \psi) \quad , \tag{4.2}$$

and we substitute $\hat{H}_1 \psi$ and $\hat{H}_2 \psi$ into the Schwarz inequality (4.1):

$$(\hat{H}_1\psi, \hat{H}_1\psi)(\hat{H}_2\psi, \hat{H}_2\psi) \geq |(\hat{H}_1\psi, \hat{H}_2\psi)|^2 \quad . \tag{4.1'}$$

Using Hermiticity, we obtain

$$(\psi, \hat{H}_1^2\psi)(\psi, \hat{H}_2^2\psi) \geq |(\psi, \hat{H}_1\hat{H}_2\psi)|^2 \quad . \tag{4.3}$$

By means of the anticommutator

$$\{A, B\} = AB + BA \quad , \tag{4.4}$$

we decompose the product $\hat{H}_1\hat{H}_2$ into a Hermitian and an anti-Hermitian part

$$\hat{H}_1\hat{H}_2 = \tfrac{1}{2}\{\hat{H}_1, \hat{H}_2\} + \tfrac{1}{2}[\hat{H}_1, \hat{H}_2] \quad ,$$

where

$$\{\hat{H}_1, \hat{H}_2\}^\dagger = \{\hat{H}_1, \hat{H}_2\}$$

is Hermitian and

$$[\hat{H}_1, \hat{H}_2]^\dagger = -[\hat{H}_1, \hat{H}_2]$$

is anti-Hermitian. The expectation value of a Hermitian operator is real, that of an anti-Hermitian operator is purely imaginary. The decomposition of an operator into Hermitian and anti-Hermitian parts implies for the expectation value a decomposition into real and imaginary parts:

$$|(\psi, \hat{H}_1\hat{H}_2\psi)|^2 = \tfrac{1}{4}(\psi, \{\hat{H}_1, \hat{H}_2\}\psi)^2 + \tfrac{1}{4}|(\psi, [\hat{H}_1, \hat{H}_2]\psi)|^2 \quad . \tag{4.5}$$

Since expectation values $\langle H_i \rangle$ are ordinary numbers, one has

$$[\hat{H}_1, \hat{H}_2] = [H_1, H_2] \quad ,$$

and it thus follows from (4.5) that

$$|(\psi, \hat{H}_1\hat{H}_2\psi)|^2 \geq \tfrac{1}{4}|(\psi, [H_1, H_2]\psi)|^2 \quad . \tag{4.6}$$

The uncertainty ΔA is defined as the root-mean-square deviation, i.e., the positive square root of the mean-square deviation (also called variance), which in turn is defined by

$$(\Delta A)^2 = (\psi, (A - \langle A \rangle)^2\psi) \quad . \tag{4.7}$$

Thus we obtain from (4.3) and (4.6) for the product of the uncertainties

$$\Delta H_1 \Delta H_2 \geq \tfrac{1}{2}|\langle [H_1, H_2] \rangle| \quad . \tag{4.8}$$

This inequality represents the general formulation of the *Heisenberg uncertainty relation* for noncommuting operators. An important special case of (4.8) is the position–momentum uncertainty relation

$$H_1 = x_i, \quad H_2 = p_j \quad,$$

$$\Delta x_i \Delta p_j \geq \frac{\hbar}{2} \delta_{ij} \quad . \tag{4.9}$$

Remark: Under what conditions does the product of the uncertainties attain a minimum?

In the Schwarz inequality (4.1′), the equal sign holds when

$$\hat{H}_2 \psi = z \hat{H}_1 \psi \quad . \tag{4.10}$$

The equal sign in the estimate of $\langle \hat{H}_1 \hat{H}_2 \rangle$ in (4.6) holds when the expectation value of the anticommutator vanishes:

$$(\psi, \hat{H}_1 \hat{H}_2 \psi) + (\psi, \hat{H}_2 \hat{H}_1 \psi) = 0 \quad .$$

Substitution of (4.10) leads to

$$0 = (\psi, \hat{H}_1 z \hat{H}_1 \psi) + (z \hat{H}_1 \psi, \hat{H}_1 \psi) = (z + z^*)(\hat{H}_1 \psi, \hat{H}_1 \psi) \quad,$$

that is, z must be pure imaginary. Substituting into (4.10), one finds that the condition for the state ψ to minimize the uncertainty product $\Delta H_1 \, \Delta H_2$ is

$$\hat{H}_2 \psi = i \lambda \hat{H}_1 \psi \quad, \quad \lambda \text{ real} \quad . \tag{4.11}$$

For the operators $H_1 = x$ and $H_2 = p$, the differential equation

$$\left(\frac{\hbar}{i} \frac{\partial}{\partial x} - \langle p \rangle \right) \psi = i \lambda (x - \langle x \rangle) \psi \tag{4.12}$$

results. The solution of this equation is a Gaussian wave packet. At the initial time, this has a minimal uncertainty product. Depending on the potential, it disperses more or less rapidly, deforms, and $\Delta x \Delta p$ will no longer be minimal. Only for the harmonic oscillator do the minimal wave packets coincide with the coherent states and remain minimal in the course of the time development of the system.

4.2 Energy–Time Uncertainty

The uncertainty relations considered up to now concern fluctuations of observables at a particular time and result from their commutators. In addition to these, there are also uncertainty relations concerning energy and time, whose derivation cannot be given in this simple, formal manner. Here, Δt usually refers to a time interval, and ΔE to the difference between energies at times separated by Δt. In any case, one should specify precisely in what sense these quantities are defined. Since t is only a parameter in quantum theory, and since the energy at a particular time can be measured with arbitrary precision, it is clear that energy–time uncertainty relations can only hold in this modified sense. We would like to present a few variants of the energy–time uncertainty relation.

4.2.1 Passage Time and Energy Uncertainty

The energy uncertainty of a free wave packet is $\Delta E = p_0 \Delta p / m$. We define the time uncertainty Δt as the time during which the particle can be found at the position x, i.e., the time that a wave packet with the linear extension Δx needs to pass through the point x:

$$\Delta t = \frac{\Delta x}{v_0} = \frac{m \Delta x}{p_0} \quad .$$

With this, the product becomes

$$\Delta E \Delta t = \Delta x \Delta p \gtrsim \hbar \quad . \tag{4.13}$$

The physical meaning of energy–time uncertainty is different from that of, e.g., position–momentum uncertainty. Position and momentum are observables to which Hermitian operators are assigned and which are measured at a particular time t. The time merely plays the role of a parameter. The uncertainty ΔE of the dynamic variable E in the inequality (4.13) is connected with a characteristic time interval Δt for changes in the system.

The energy–time uncertainty relation (4.13) can also be found by analyzing a wave packet at a fixed position:

$$\begin{aligned}
\psi(x,t) &= \int \frac{dp}{2\pi\hbar} \varphi(p) \exp\left\{\frac{i}{\hbar}(px - Et)\right\} \\
&= \int \frac{dE}{2\pi\hbar} \tilde{\varphi}(E,x) \exp\left\{-\frac{i}{\hbar}Et\right\} \quad .
\end{aligned}$$

Here, $\psi(x,t)$ is the Fourier transform of

$$\tilde{\varphi}(E,x) = \Theta(E) \sum_{\pm} \exp\{\pm i\sqrt{2mE}x/\hbar\} \times \varphi(\pm\sqrt{2mE})\sqrt{m/2E}$$

and vice versa, i.e., the energy uncertainty ΔE and the time spent Δt at the position x are related by (4.13). If $\varphi(p)$ is restricted to positive momenta, then only the "+" part contributes to $\tilde{\varphi}(E,x)$.

4.2.2 Duration of an Energy Measurement and Energy Uncertainty

An energy measurement with accuracy ΔE requires at least a time $\Delta t = \hbar/\Delta E$, for the following reason: The measuring time required to determine the properties as for instance the energy distribution of a wave packet is at least as large as the time the packet needs to pass the set up

$$\Delta t \gtrsim \frac{\Delta x}{v_0} \geq \frac{\hbar}{\Delta p \, v_0} = \frac{\hbar}{\Delta E} \quad .$$

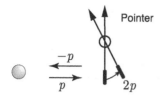

Fig. 4.1. Momentum transfer to a pointer

Figure 4.1 shows an idealized momentum or energy measurement. The measuring apparatus has momentum zero before the measurement; the momentum transferred to the apparatus (mass M) in the course of the measurement is $2p$. The apparatus traverses during the time T a distance $2pT/M$. In order to measure the "pointer displacement", it must be larger than the position uncertainty:

$$\frac{2p}{M}T > \Delta x > \frac{\hbar}{\Delta p} = \frac{\hbar(p/M)}{\Delta E_{\text{meas}}} \quad .$$

Since the energy uncertainty ΔE_{meas} of the measuring apparatus is proportional to the energy uncertainty ΔE of the particles ($\Delta E_{\text{meas}} \approx \Delta E$), one obtains from this inequality

$$\Delta E\, T = \Delta E \Delta t \gtrsim \hbar \quad .$$

4.2.3 Lifetime and Energy Uncertainty

A connection also exists between the average lifetime τ of an excited state (e.g., an excited atom, a radioactive nucleus, or an unstable elementary particle) and the energy width ΔE of the particle emitted in the transition:

$$\Delta E \sim \frac{\hbar}{\tau} \quad . \tag{4.14}$$

We can see this intuitively from the previous discussion if we regard the emitted particle as a measuring apparatus which interacts with the unstable object only during the period τ and therefore acquires the energy difference of the "bare" (noninteracting) levels only up to the precision $\hbar/\tau$. Quantitatively, this relation follows from time dependent perturbation theory. The probability of a transition after the time t from energy level E to E' with the emission of a quantum of energy ε (Fig. 4.2) is proportional to (see Sect. 16.3)

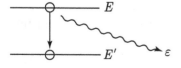

Fig. 4.2. Transition of an excited state E into the state E' and emission of a quantum with energy ε

$$\sin^2\left[(E-(E'+\varepsilon))\frac{t}{\hbar}\right]/(E-(E'+\varepsilon))^2 \quad ;$$

after a time t, the typical energy difference is

$$|E - E' - \varepsilon| \sim \frac{\hbar}{t} \quad .$$

If we replace the time by the characteristic lifetime, we then recover (4.14).

4.3 Common Eigenfunctions of Commuting Operators

Let two Hermitian operators A and B be given. Let the $\psi_n(x)$ be eigenfunctions of A with eigenvalues a_n. Let the state of the physical system be described by the wave function $\psi(x)$, which can be represented in terms of eigenfunctions of A:

$$\psi(x) = \sum_n c_n \psi_n(x) \quad .$$

According to Sect. 2.9, in a measurement of the observable represented by A, the probability of measuring the eigenvalue a_n is just $|c_n|^2$. For $c_n = \delta_{nn_0}$, one measures with certainty a_{n_0}; i.e., if the system is in an eigenstate of A, the corresponding observable has a precise value, namely the eigenvalue.

Under what conditions are the $\psi_n(x)$ eigenfunctions of B as well, and under what conditions do A and B both have precise values?

Theorem 2. If $[A, B] = 0$, then A and B have a common set of eigenfunctions.

Proof:

(i) Let ψ be a nondegenerate eigenfunction of A:

$$A\psi = a\psi \quad . \tag{4.15}$$

Since A and B commute,

$$AB\psi = BA\psi$$

and with (4.15)

$$A(B\psi) = a(B\psi) \quad .$$

Since ψ is the only eigenfunction of A with eigenvalue a, $B\psi$ must be proportional to ψ. Let us denote the proportionality factor by b. The result is then

$$B\psi = b\psi \quad , \tag{4.16}$$

just the eigenvalue equation of the operator B.

(ii) Let the eigenvalue a be m-fold degenerate:

$$A\psi_j = a\psi_j \quad , \quad j = 1, \ldots, m \quad , \tag{4.17a}$$

whereby

$$(\psi_j, \psi_k) = \delta_{jk} \quad . \tag{4.17b}$$

It then follows from the vanishing of the commutator

$$AB\psi_j = BA\psi_j$$

and the eigenvalue equation (4.17a) that

$$A(B\psi_j) = a(B\psi_j) \quad .$$

Thus, $B\psi_j$ is also an eigenfunction of A with the eigenvalue a and thus a linear combination of the functions ψ_j; i.e.,

$$B\psi_j = \sum_k C_{jk}\psi_k \quad \text{with coefficients} \quad C_{jk} = (\psi_k, B\psi_j) = C_{kj}^* \quad . \tag{4.18}$$

The matrix (C_{jk}) is Hermitian and can be diagonalized by a unitary transformation U:

$$U^\dagger C U = C_{\mathrm{D}} \quad \text{with} \quad U^\dagger U = UU^\dagger = \mathbb{1} \quad . \tag{4.19}$$

From this it follows that $CU = UC_{\mathrm{D}}$, or, in components,

$$\sum_j C_{ij}U_{jk} = U_{ik}C_{\mathrm{D}k} \tag{4.20a}$$

and

$$\sum_i U_{ir}^* C_{ik} = C_{\mathrm{D}r}U_{kr}^* \quad . \tag{4.20b}$$

This means that the kth column vector of the matrix U, i.e.,

$$\begin{pmatrix} U_{1k} \\ \vdots \\ U_{mk} \end{pmatrix} \quad ,$$

is an eigenvector of the matrix C with eigenvalue $C_{\mathrm{D}k}$.

Multiplying (4.18) by U_{jr}^*, one finds using (4.20b)

$$\sum_j BU_{jr}^*\psi_j = \sum_{j,k} U_{jr}^* C_{jk}\psi_k = \sum_k C_{\mathrm{D}r}U_{kr}^*\psi_k \quad . \tag{4.21}$$

The linear combinations

$$\varphi_r = \sum_k U_{kr}^*\psi_k \tag{4.22}$$

of the degenerate eigenfunctions ψ_k therefore represent eigenfunctions of both A and B. The eigenvalues of B are given by the diagonal elements C_{Dr} of the diagonal matrix C_D.

Theorem 3. Let a complete set of eigenfunctions $\psi_n(\boldsymbol{x})$, $n = 1, 2 \ldots$ of operators A and B be given, with eigenvalues a_n and b_n, respectively. Then A and B commute.

Proof:

Since $[A, B]\psi_n = (AB - BA)\psi_n = (a_n b_n - b_n a_n)\psi_n = 0$, it follows for arbitrary $\psi = \sum_n c_n \psi_n$ that

$$[A, B]\psi = 0$$

and thus

$$[A, B] = 0 \quad .$$

From Theorem 2 and Theorem 3 it follows that observables represented by commuting operators can simultaneously possess precise values, such that measurements of these quantities give unique results. Examples of commuting operators are x_1, x_2, x_3 or p_1, p_2, p_3 or x_1, p_2, p_3, but not x_1, p_1.

Definition 1. A complete set of eigenfunctions of the operator A is called a *basis* of A.

Definition 2. (Complete set of operators or equivalently complete set of observables.) The set of Hermitian operators $A, B, \ldots, M$ is called a *complete set* of operators if these operators all commute with each other and if the common set of eigenfunctions is no longer degenerate. These eigenfunctions can then be characterized by the corresponding eigenvalues $a, b, \ldots, m$: $\psi_{a,b,\ldots,m}$.

Remark: If for a given set of operators the eigenstates are still degenerate, there exists an additional symmetry of these operators and the generator of this symmetry operation also commutes with this set of operators.

Examples of complete sets of operators are:

-   for one-dimensional potentials: x or p,
-   for three-dimensional potentials: x, y, z or p_x, p_y, p_z,
-   for three-dimensional spherically symmetric potentials: x, y, z or p_x, p_y, p_z or $H, \boldsymbol{L}^2, L_z$,
-   for one-dimensional reflection symmetric potentials $(V(x) = V(-x))$: x or p or H, P.

The preceding theorems imply:

Theorem 4. If O is a function of the operators $A, B \ldots$ of a complete set, then the basis of the complete set is also a basis for O.

Theorem 5. An operator O that commutes with a complete set of operators is a function of these operators.

Since O commutes with the complete set of operators, it also has their basis

$$O\psi_{a,b\ldots} = o(a, b, \ldots)\psi_{a,b,\ldots} \quad .$$

Since the operator function $o(A, B, \ldots)$ possesses the same eigenvalues $o(a, b, \ldots)$, it thus follows that $O = o(A, B, \ldots)$, i.e., O is a function of the operators $A, B \ldots$.

We return again to the fact that commuting operators have common eigenfunctions. In such an eigenstate they simultaneously have precise, unique values, i.e., their eigenvalues. One refers to this by saying that "A and B are simultaneously measurable". In a measurement, the system is influenced, and in general the state is changed. If we consider an observable A with eigenfunctions ψ_n and eigenvalues a_n and if the state is given by $\psi = \sum c_n \psi_n$, then we measure the value a_n with the probability $|c_n|^2$. After this measurement, the state must have changed to ψ_n, because only then would any further measurement again give a_n. This is called "reduction of the wave function". We will analyze this situation in more detail in our discussion of the Stern–Gerlach experiment as an example of a measurement (Chap. 20).

If A and B do not commute, and if the system is in an eigenstate ψ_n of A, then a measurement of A produces the value a_n. A subsequent measurement of B will change the state of the system, because after an ideal measurement whose accuracy allows the unique determination of an eigenvalue of B, the state of the system will change to the corresponding eigenstate of B. For the results of a further measurement of A, only probabilistic statements can be made, which are determined by the expansion of the eigenstate of B thus obtained in terms of eigenfunctions of A. This explains the above expression "simultaneously measurable" for commuting operators. Only if A and B commute can sharp values be assigned to both of them simultaneously and measurements of the corresponding observables do not disturb (interfere with) each other. One expresses this also by saying that the two observables are compatible.

Problems

4.1 (a) Solve the differential equation (4.12), the condition for a minimal uncertainty product for the operators x and p.

(b) Why isn't the Gaussian wave packet of a free particle a minimal wave packet for finite times?

4.2 For a Gaussian wave packet, calculate the energy uncertainty ΔE and $\Delta E \Delta t$, where Δt is the characteristic time for the wave packet to pass through the point x.

5. Angular Momentum

For application to centrally symmetric potentials, we would now like to investigate properties of the angular momentum, which is also decisive in such problems in classical mechanics.

5.1 Commutation Relations, Rotations

The *orbital angular momentum operator* is defined in vector and component form by

$$L = x \times p = \frac{\hbar}{i} x \times \nabla \quad \text{or} \quad L_i = \varepsilon_{ijk} x_j p_k \quad . \tag{5.1}$$

In the second expression, the Einstein summation convention is employed: the expression is summed over repeated indices. ε_{ijk} is the completely anti-symmetric tensor of the third rank

$$\varepsilon_{ijk} = \begin{cases} 1 & \text{for even permutations of } (1\,2\,3) \\ -1 & \text{for odd permutations of } (1\,2\,3) \\ 0 & \text{otherwise} \end{cases} \quad .$$

The following commutation relations hold:

$$[L_i, L_j] = i\hbar \varepsilon_{ijk} L_k \quad , \tag{5.2a}$$
$$[L_i, x_j] = i\hbar \varepsilon_{ijk} x_k \quad , \tag{5.2b}$$
$$[L_i, p_j] = i\hbar \varepsilon_{ijk} p_k \quad . \tag{5.2c}$$

One can show this by direct calculation. The similarity of the commutators (5.2a)–(5.2c) suggests a general reason for this structure, which we would now like to find. To this end, we first show:

Theorem. The angular momentum L is the generator of rotations.
By this one means that the unitary operator

$$U_{\delta\varphi} \equiv \exp\left\{\frac{i}{\hbar}\delta\varphi \cdot L\right\} \approx 1 + \frac{i}{\hbar}\delta\varphi \cdot L \tag{5.3}$$

(for small $\delta\varphi$) generates infinitesimal rotations.

Proof:

(i) The unitarity of the operator $U_{\delta\varphi}$ is seen immediately, since by the Hermiticity of $\boldsymbol{L}$

$$U_{\delta\varphi}^{\dagger} = \exp\left\{-\frac{i}{\hbar}\delta\varphi \cdot \boldsymbol{L}\right\} = U_{\delta\varphi}^{-1} \tag{5.4}$$

holds and thus the relation $U_{\delta\varphi}^{\dagger}U_{\delta\varphi} = U_{\delta\varphi}U_{\delta\varphi}^{\dagger} = 1$ is fulfilled.

(ii) One sees that the operator $U_{\delta\varphi}$ acts as a rotation (Fig. 5.1), by letting it act on an arbitrary state $\psi(\boldsymbol{x})$:

$$U_{\delta\varphi}\psi(\boldsymbol{x}) = \left(1 + \frac{i}{\hbar}\delta\varphi \cdot \left(\boldsymbol{x} \times \frac{\hbar}{i}\boldsymbol{\nabla}\right)\right)\psi(\boldsymbol{x}) = (1 + (\delta\varphi \times \boldsymbol{x}) \cdot \boldsymbol{\nabla})\psi(\boldsymbol{x})$$

$$= \psi(\boldsymbol{x} + \delta\varphi \times \boldsymbol{x})$$

to first order in $\delta\varphi$ (Taylor expansion). The transformation of $\psi(\boldsymbol{x})$ is thus

$$U_{\delta\varphi}\psi(\boldsymbol{x}) = \psi(\boldsymbol{x}') \tag{5.5}$$

with

$$\boldsymbol{x}' = \boldsymbol{x} + \delta\varphi \times \boldsymbol{x} \quad \text{or} \quad x_i' = x_i + \varepsilon_{ijk}\delta\varphi_j x_k \quad . \tag{5.6}$$

This in fact represents a rotation, which proves the assertion. We would like to illustrate this geometrically. For an arbitrary vector $\boldsymbol{v}$, the rotated vector is $\boldsymbol{v}' = \boldsymbol{v} + \delta\varphi \times \boldsymbol{v}$. One can view the rotation actively: The vector $\boldsymbol{v}$ is transformed by a rotation through $\delta\varphi$ into the vector $\boldsymbol{v}'$ (Fig. 5.2a), or passively: The coordinate system S' results from the rotation of S through $-\delta\varphi$ (Fig. 5.2b).

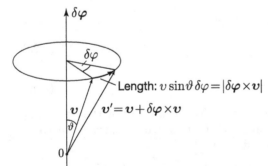

Fig. 5.1. Rotation about $\delta\phi$

We consider the rotation to be a passive transformation. The wave function $\psi(\boldsymbol{x})$ is transformed into the wave function $\psi(\boldsymbol{x}')$ which lies with respect to the coordinate system S' just as $\psi(\boldsymbol{x})$ does with respect to S, as illustrated by the level curves in Fig. 5.2b.

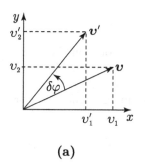 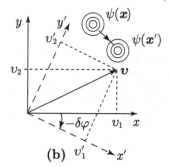

(a)   (b)

Fig. 5.2. $v_1' = v_1 - \delta\varphi v_2$, $v_2' = v_2 + \delta\varphi v_1$ (a) Active rotation; (b) passive rotation

How do the operators look in the rotated system? We begin with $A\psi(\boldsymbol{x}) = \varphi(\boldsymbol{x})$. "Inserting" $U^\dagger U = 1$ and multiplying the equation by $U \equiv U_{\delta\varphi}$ on the left, we obtain

$$UAU^\dagger(U\psi(\boldsymbol{x})) = U\varphi(\boldsymbol{x}) \quad UAU^\dagger\psi(\boldsymbol{x}') = \varphi(\boldsymbol{x}') \quad .$$

Thus,

$$A' = UAU^\dagger \tag{5.7a}$$

is the operator in the rotated coordinate system, since it has the same action on the transformed state $\psi(\boldsymbol{x}')$ as A on $\psi(\boldsymbol{x})$. Expanding (5.7a) up to first order in $\delta\varphi$, it follows immediately that

$$A' = A + \frac{\mathrm{i}}{\hbar}\delta\varphi_l[L_l, A] \quad . \tag{5.7b}$$

A few special cases are of general interest.

(i) Let A be a scalar (rotationally invariant) operator. Then $A' = A$, and it follows from (5.7b) that

$$[L_i, A] = 0 \quad \text{for} \quad i = 1, 2, 3 \quad . \tag{5.8}$$

Example: rotationally invariant Hamiltonian $H = \boldsymbol{p}^2/2m + V(r)$, $\boldsymbol{L}^2, \boldsymbol{p}^2$, etc.

(ii) Let $\boldsymbol{v}$ be a vector operator. The quantity $\boldsymbol{v}$ transforms as a vector under rotations according to $\boldsymbol{v}' = \boldsymbol{v} + \delta\boldsymbol{\varphi} \times \boldsymbol{v}$. Setting this equal to (5.7b) component by component

$$v_j + \varepsilon_{jlk}\delta\varphi_l v_k = v_j + \frac{\mathrm{i}}{\hbar}\delta\varphi_l[L_l, v_j]$$

shows

$$[L_i, v_j] = \mathrm{i}\hbar\varepsilon_{ijk}v_k \quad . \tag{5.9}$$

Examples: $\boldsymbol{L}, \boldsymbol{x}, \boldsymbol{p}$.

With this, (5.2) is proven in general.

Remarks:

(i)   The unitary operator $U_{\varphi e} = e^{i\varphi e \cdot L}$ generates a finite rotation about the unit vector e through the angle φ. This follows by decomposition into a sequence of infinitesimal rotations about e.

(ii) Momentum and translations: We note that the momentum operator $p = (\hbar/i)\nabla$ is the generator of translations

$$e^{ia \cdot p/\hbar}\psi(x) = e^{a \cdot \nabla}\psi(x) = \psi(x + a) \quad . \tag{5.10}$$

The middle part of Eq. (5.10) represents the Taylor expansion of $\psi(x+a)$ around x in compact form. The wave function $\psi(x)$ is shifted by $-a$. In the coordinate system S' which is shifted by $-a$ with respect to S, i.e., $x'_i = x_i + a_i$, the wave function $\psi(x + a)$ has the same position as $\psi(x)$ in S. The generators of symmetry operations are summarized in Table 8.1.

5.2 Eigenvalues of Angular Momentum Operators

The derivations of this section require only the algebraic properties of the angular momentum (the commutation relations (5.2a)) and thus hold not only for the orbital angular momentum, but for any angular momentum (spin, total angular momentum). Because different components of the angular momentum do not commute, they cannot be simultaneously diagonalized; thus, there is no common basis for all three components. However, since L^2 is a scalar, then according to (5.8) one has

$$[L^2, L_i] = 0 \quad \text{for} \quad i = 1, 2, 3 \quad , \tag{5.11}$$

and thus we can diagonalize L^2 and one component of L. We would now like to determine the eigenvalues of the common eigenfunctions of L^2 and L_z.

We first define two new operators

$$L_\pm = L_x \pm iL_y \tag{5.12}$$

with the following properties:

$$(L_\pm)^\dagger = L_\mp \quad , \tag{5.13a}$$

$$[L_z, L_\pm] = i\hbar L_y \pm \hbar L_x = \pm \hbar L_\pm \quad , \tag{5.13b}$$

$$[L_+, L_-] = -2i[L_x, L_y] = 2\hbar L_z \quad , \tag{5.13c}$$

$$[L^2, L_\pm] = 0 \quad , \tag{5.13d}$$

because of (5.11),

$$L_+L_- = L_x^2 + L_y^2 - i[L_x, L_y] = L_x^2 + L_y^2 + \hbar L_z \quad ,$$

and therefore

$$\boldsymbol{L}^2 = L_x^2 + L_y^2 + L_z^2 = L_+L_- - \hbar L_z + L_z^2 = L_-L_+ + \hbar L_z + L_z^2. \quad (5.13e)$$

Now let ψ_{l_z} be an eigenfunction of L_z:

$$L_z\psi_{l_z} = l_z\psi_{l_z} \quad .$$

With (5.13b) we find

$$L_z L_\pm \psi_{l_z} = L_\pm L_z \psi_{l_z} \pm \hbar L_\pm \psi_{l_z} \quad ,$$

and thus

$$L_z(L_\pm \psi_{l_z}) = (l_z \pm \hbar)L_\pm \psi_{l_z} \quad .$$

This means that if ψ_{l_z} is an eigenfunction of L_z with eigenvalue l_z, then $L_\pm \psi_{l_z}$ is an eigenfunction of L_z with eigenvalue $l_z \pm \hbar$ or, put more briefly, $L_\pm$ raises (lowers) the eigenvalue l_z by $\hbar$.

From now on we will denote the eigenfunctions of $\boldsymbol{L}^2$ and L_z by ψ_{lm}, where

$$\boldsymbol{L}^2\psi_{lm} = \hbar^2 l(l+1)\psi_{lm} \quad , \quad l \geq 0 \quad ,$$

$$L_z\psi_{lm} = \hbar m\psi_{lm} \quad . \quad (5.14)$$

Evidently, we can represent any eigenvalue of the positive semidefinite operator $\boldsymbol{L}^2$ and of the operator L_z in this form. We know already that $L_\pm$ raises (lowers) m by 1. From (5.13d) it further follows that

$$\boldsymbol{L}^2(L_\pm\psi_{lm}) = L_\pm\boldsymbol{L}^2\psi_{lm} = \hbar^2 l(l+1)(L_\pm\psi_{lm}) \quad .$$

$L_\pm\psi_{lm}$ is thus an eigenfunction of $\boldsymbol{L}^2$ with the *same* eigenvalue as ψ_{lm}.

Further information can be obtained from the normalization using (5.13a), (5.13e) and (5.14),

$$\begin{aligned} (L_\pm\psi_{lm}, L_\pm\psi_{lm}) &= (\psi_{lm}, L_\mp L_\pm\psi_{lm}) = (\psi_{lm}, (\boldsymbol{L}^2 - L_z^2 \mp \hbar L_z)\psi_{lm}) \\ &= \hbar^2(l(l+1) - m^2 \mp m) \quad , \end{aligned}$$

where we assume that ψ_{lm} is normalized. From this it follows that

$$L_\pm\psi_{lm} = \hbar\sqrt{l(l+1) - m(m \pm 1)}\psi_{l,m\pm1} \quad , \quad (5.15)$$

where the phase factor is set to unity. Since the norm is non-negative,

$$(L_\pm\psi_{lm}, L_\pm\psi_{lm}) = \hbar^2(l(l+1) - m(m \pm 1)) \geq 0 \quad .$$

Therefore, the restrictions

for $m > 0$: $l(l+1) \geq m(m+1)$,

for $m < 0$: $l(l+1) \geq m(m-1) = (-m)(-m+1) = |m|(|m|+1)$,

follow, that is $l(l+1) \geq |m|(|m|+1)$ and thus $|m| \leq l$.

Now let l be a fixed value and M the maximal corresponding m. In order that $L_+\psi_{lM}$ not be an eigenfunction with the larger eigenvalue $M+1$, the condition $L_+\psi_{lM} = 0$ must hold. From the normalization equation above, it follows immediately that $l(l+1) = M(M+1)$ and therefore $M = l$. The case where μ is the minimal m is handled analogously. In order that $L_-\psi_{l\mu}$ not be an eigenfunction with smaller eigenvalue $\mu-1$, $L_-\psi_{l\mu} = 0$ must hold and thus $\mu = -l$. Now one can obtain all the values of m recursively:

$$L_-\psi_{ll} \sim \psi_{l,l-1} \quad , \quad (L_-)^2\psi_{ll} \sim \psi_{l,l-2} \quad , \quad \text{etc.},$$

with eigenvalues $m = l, l-1$, etc. In order for this to give $-l$, the equation $l - k = -l$ must hold, where k is a non-negative integer. With this it follows that $l = k/2$. Summarizing, using only the algebraic structure of the commutation relations for angular momentum, we now obtain the following result for the eigenvalue spectrum:

$$l = 0, 1, 2, 3, \dots \quad \text{or} \quad l = \tfrac{1}{2}, \tfrac{3}{2}, \tfrac{5}{2}, \dots \tag{5.16}$$

with the corresponding values for m

$$m = l, l-1, \dots -l+1, -l \quad .$$

The angular momentum eigenvalues l are either integral or half integral, and the eigenvalues m range in integer steps from l to $-l$.

5.3 Orbital Angular Momentum in Polar Coordinates

We now leave these general considerations and determine the particular eigenvalues and eigenfunctions of the orbital angular momentum. Because of the intimate connection with rotations it is advantageous to transform to spherical polar coordinates (Fig. 5.3).

With $\boldsymbol{x} = r\boldsymbol{e}_r$ and

$$\boldsymbol{\nabla} = \boldsymbol{e}_r \frac{\partial}{\partial r} + \boldsymbol{e}_\vartheta \frac{1}{r}\frac{\partial}{\partial \vartheta} + \boldsymbol{e}_\varphi \frac{1}{r\sin\vartheta}\frac{\partial}{\partial\varphi} \tag{5.17}$$

one obtains after a simple calculation

$$L_x = \frac{\hbar}{i}\left(-\sin\varphi\frac{\partial}{\partial\vartheta} - \cos\varphi\cot\vartheta\frac{\partial}{\partial\varphi}\right) \quad , \tag{5.18a}$$

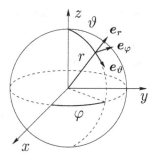

Fig. 5.3. Polar coordinates

$$L_y = \frac{\hbar}{i}\left(\cos\varphi\frac{\partial}{\partial\vartheta} - \sin\varphi\cot\vartheta\frac{\partial}{\partial\varphi}\right) \quad , \tag{5.18b}$$

$$L_z = \frac{\hbar}{i}\frac{\partial}{\partial\varphi} \quad , \tag{5.18c}$$

$$L_\pm = \hbar e^{\pm i\varphi}\left(\pm\frac{\partial}{\partial\vartheta} + i\cot\vartheta\frac{\partial}{\partial\varphi}\right) \quad , \tag{5.18d}$$

$$\mathbf{L}^2 = -\hbar^2\left[\frac{1}{\sin\vartheta}\frac{\partial}{\partial\vartheta}\left(\sin\vartheta\frac{\partial}{\partial\vartheta}\right) + \frac{1}{\sin^2\vartheta}\frac{\partial^2}{\partial\varphi^2}\right] \quad . \tag{5.18e}$$

The eigenvalue equations (5.14) are thus as follows:

$$\left[\frac{1}{\sin^2\vartheta}\frac{\partial^2}{\partial\varphi^2} + \frac{1}{\sin\vartheta}\frac{\partial}{\partial\vartheta}\left(\sin\vartheta\frac{\partial}{\partial\vartheta}\right)\right]\psi_{lm} = -l(l+1)\psi_{lm} \tag{5.19a}$$

and

$$\frac{\partial}{\partial\varphi}\psi_{lm} = im\psi_{lm} \quad . \tag{5.19b}$$

For ψ_{lm} we use the separation ansatz

$$\psi_{lm}(\vartheta,\varphi) = \Phi(\varphi)\,\Theta(\vartheta) \quad . \tag{5.20a}$$

From (5.19b) it follows immediately that

$$\Phi(\varphi) = e^{im\varphi} \quad . \tag{5.20b}$$

The continuity of the wave function demands $\Phi(\varphi + 2\pi) = \Phi(\varphi)$; for this reason, m must be an integer and, as a result, l as well, thus:

$$l = 0, 1, 2, 3, \ldots \quad ; \quad m = -l, -l+1, \ldots, 0, \ldots, l-1, l \quad . \tag{5.21}$$

Substituting $\psi_{lm} = \exp\{im\varphi\}\Theta(\vartheta)$ into (5.19a) we obtain the differential equation

$$\left[\frac{1}{\sin\vartheta}\frac{\partial}{\partial\vartheta}\left(\sin\vartheta\frac{\partial}{\partial\vartheta}\right) - \frac{m^2}{\sin^2\vartheta} + l(l+1)\right]\Theta(\vartheta) = 0 \quad , \tag{5.20c}$$

whose solutions are well known functions of mathematical physics, so that one finally gets for the eigenfunctions of the simultaneously diagonalizable operators L^2 and L_z of the orbital angular momentum in the polar representation the *spherical harmonics* whose definition and most important properties are summarized here:

$$\psi_{lm}(\vartheta, \varphi) = Y_{lm}(\vartheta, \varphi) = (-1)^{(m+|m|)/2} P_{l|m|}(\cos \vartheta) e^{im\varphi}$$

$$\times \left[\frac{2l+1}{4\pi} \frac{(l-|m|)!}{(l+|m|)!} \right]^{1/2} . \tag{5.22}$$

Here, the *associated Legendre functions*:

$$P_{lm}(\xi) = (1-\xi^2)^{m/2} \frac{d^m}{d\xi^m} P_l(\xi) = \frac{(-1)^l}{2^l \, l!} \sin^m \vartheta \, \frac{d^{l+m}(\sin^{2l} \vartheta)}{d \cos^{l+m} \vartheta} \tag{5.23}$$

enter, where $m \geq 0$, $\xi = \cos \vartheta$, and where the *Legendre polynomials* are given by

$$P_l(\xi) = \frac{1}{2^l l!} \frac{d^l}{d\xi^l} (\xi^2 - 1)^l = \frac{(-1)^l}{2^l \, l!} \frac{d^l \sin^{2l} \vartheta}{d \cos^l \vartheta} . \tag{5.24}$$

The $P_l(\xi)$ are lth order polynomials in ξ, and the $P_{lm}(\xi)$ are thus $(l-m)$th order polynomials in ξ, multiplied by $(\sin \vartheta)^m$. They have $l-m$ nodes in the interval $-1 < \xi < 1$.

The Legendre polynomials fulfil the following recursion relations:

$$(l+1)P_{l+1} = (2l+1)\xi P_l - lP_{l-1} , \tag{5.25a}$$

$$(1-\xi^2)\frac{dP_l}{d\xi} = l(P_{l-1} - \xi P_l) . \tag{5.25b}$$

The lowest Legendre polynomials are

$$P_0 = 1, \quad P_1 = \xi, \quad P_2 = \tfrac{1}{2}(3\xi^2 - 1), \quad P_3 = \tfrac{1}{2}(5\xi^3 - 3\xi), \dots .$$

The associated Legendre functions satisfy the differential equation

$$\left[(1-\xi^2)\frac{d^2}{d\xi^2} - 2\xi \frac{d}{d\xi} + l(l+1) - \frac{m^2}{1-\xi^2} \right] P_{lm}(\xi) = 0 \tag{5.26}$$

and have the properties

$$P_{lm}(-\xi) = (-1)^{l+m} P_{lm}(\xi) , \tag{5.27}$$

$$\int_{-1}^{+1} d\xi \, P_{lm}(\xi) P_{l'm}(\xi) = \frac{2}{2l+1} \frac{(l+m)!}{(l-m)!} \delta_{ll'} \quad (m \geq 0) ; \tag{5.28}$$

in particular, one has

$$P_{l0}(\xi) = P_l(\xi) , \quad P_{ll}(\xi) = (2l-1)!! \, (1-\xi^2)^{l/2} . \tag{5.29}$$

(Double factorial is defined by $(2l-1)!! = (2l-1)(2l-3) \dots 1$.) From this, the following properties of the spherical harmonics can be derived:

Orthogonality:

$$\int_0^\pi d\vartheta \, \sin \vartheta \int_0^{2\pi} d\varphi \, Y_{lm}(\vartheta, \varphi)^* Y_{l'm'}(\vartheta, \varphi) = \delta_{ll'}\delta_{mm'} \quad . \tag{5.30}$$

Completeness:

$$\sum_{l=0}^\infty \sum_{m=-l}^{+l} Y_{lm}(\vartheta, \varphi) Y_{lm}(\vartheta', \varphi')^* = (\sin \vartheta)^{-1}\delta(\vartheta - \vartheta')\delta(\varphi - \varphi') \quad . \tag{5.31}$$

Addition theorem:

$$\sum_{m=-l}^l Y_{lm}(\vartheta, \varphi) Y_{lm}(\vartheta', \varphi')^* = \frac{2l+1}{4\pi} P_l(\cos \Theta) \quad \text{(see Fig. 5.4)}, \tag{5.32}$$

where $\cos \Theta = \cos \vartheta \cos \vartheta' + \sin \vartheta \sin \vartheta' \cos(\varphi - \varphi')$.

$$Y_{l,-m}(\vartheta, \varphi) = (-1)^m Y_{lm}(\vartheta, \varphi)^* \quad . \tag{5.33}$$

A few spherical harmonics explicitly:

$$Y_{00} = \frac{1}{\sqrt{4\pi}} \quad ; \quad Y_{10} = \sqrt{\frac{3}{4\pi}} \cos \vartheta \quad , \quad Y_{11} = -\sqrt{\frac{3}{8\pi}} \sin \vartheta \, e^{i\varphi} \quad ;$$

$$Y_{20} = \sqrt{\frac{5}{16\pi}}(3 \cos^2 \vartheta - 1) \quad ,$$

$$Y_{21} = -\sqrt{\frac{15}{8\pi}} \sin \vartheta \cos \vartheta \, e^{i\varphi} \quad , \quad Y_{22} = \sqrt{\frac{15}{32\pi}} \sin^2 \vartheta \, e^{2i\varphi} \quad . \tag{5.34}$$

The corresponding $Y_{l,-m}$ are obtained from the Y_{lm} with the help of (5.33).[1]

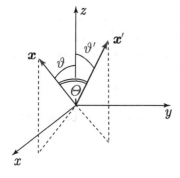

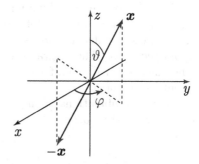

Fig. 5.4. Illustration concerning the addition theorem for spherical harmonics

Fig. 5.5. Reflection

---

[1] The angular momentum eigenfunctions can be determined also algebraically as shown in Appendix C.

What is the behavior of the $Y_{lm}(\vartheta, \varphi)$ under reflection (Fig. 5.5)? The parity operator P transforms the position vector $\boldsymbol{x}$ into $-\boldsymbol{x}$, or, acting on a state ψ, one has $P\psi(\boldsymbol{x}) = \psi(-\boldsymbol{x})$. The vector $-\boldsymbol{x}$ has polar coordinates $\pi - \vartheta, \varphi + \pi$.

Because of $\cos(\pi - \vartheta) = -\cos\vartheta$ we can write

$$PY_{lm}(\vartheta, \varphi) = Y_{lm}(\pi - \vartheta, \varphi + \pi) = e^{im\pi}(-1)^{l+|m|}Y_{lm}(\vartheta, \varphi) \quad , \qquad (5.35a)$$

or

$$PY_{lm}(\vartheta, \varphi) = (-1)^l Y_{lm}(\vartheta, \varphi) \quad . \qquad (5.35b)$$

The Y_{lm} are eigenfunctions of the parity operator with eigenvalues $(-1)^l$, i.e. Y_{lm} is even for even l and odd for odd l. By (5.30) and (5.31), the Y_{lm} form a complete orthonormal set of eigenfunctions of the orbital angular momentum operators $\boldsymbol{L}^2$ and L_z. The eigenvalues of $\boldsymbol{L}^2$ are $\hbar^2 l(l+1)$ with $l = 0, 1, 2, 3, \ldots$. For each fixed value of l, there exist $(2l+1)$ values $m = -l, -l+1, \ldots, l$ for the eigenvalue of the z-component L_z of orbital angular momentum. The components L_x and L_y are not diagonal in the states Y_{lm}. They have expectation value zero and the uncertainties ΔL_x and ΔL_y, respectively; for the states Y_{ll} for example, one has

$$\Delta L_x = \Delta L_y = \hbar\sqrt{l/2} \quad . \qquad (5.36)$$

The states Y_{ll} are concentrated mainly in the xy-plane. From (5.36) we see that the relative fluctuation $\Delta L_x/l$ decreases as $1/\sqrt{l}$ with increasing l.

States with orbital angular momentum quantum number $l = 0$ are known as "s-orbitals", states with $l = 1$ as "p-orbitals", with $l = 2$ as "d-orbitals" and with $l = 3$ as "f-orbitals".

So-called polar diagrams are often used, in which the value of $|Y_{lm}(\vartheta, \varphi)|^2 = (\Theta(\vartheta))^2$ is plotted as a function of angle ϑ (Fig. 5.6); this value can be read off from the radial distance of the curve from the origin of the diagram. Because of the φ-independence, such representations have rotational symmetry about the z-axis. The diagrams for m and $-m$ are identical (compare (5.33)). The functions $|Y_{ll}|^2$ (i.e., for maximal values of $m!$) are strongly concentrated about the xy-plane.

Linear combinations of the states Y_{lm} are also important. For example, one refers to (Fig. 5.7)

$$p_x = \frac{-1}{\sqrt{2}}(Y_{11} - Y_{1,-1}) = \sqrt{\frac{3}{4\pi}}\sin\vartheta\cos\varphi \quad ,$$

$$p_y = \frac{-1}{\sqrt{2}\,i}(Y_{11} + Y_{1,-1}) = \sqrt{\frac{3}{4\pi}}\sin\vartheta\sin\varphi \qquad (5.37)$$

as "p_x-orbitals", "p_y-orbitals", respectively. The "p_z-orbital" is identical to Y_{10}.

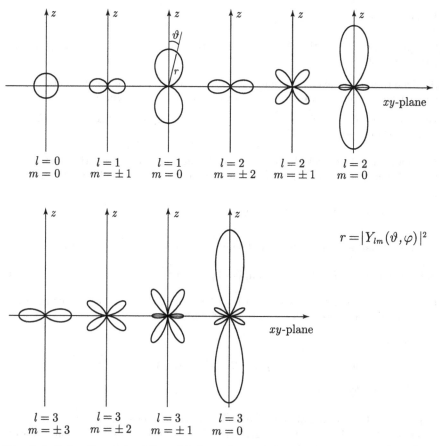

$$r = |Y_{lm}(\vartheta, \varphi)|^2$$

Fig. 5.6. Polar diagrams of the orbital angular momentum eigenfunctions Y_{lm} with $l = 0, 1, 2, 3$

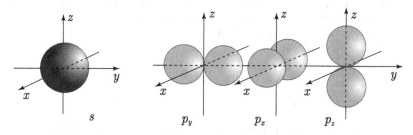

Fig. 5.7. Three-dimensional polar representations of the absolute value of the angular functions for the s-orbital and the three p-orbitals

Problems

5.1 (a) Show that for the eigenstates of L_z the expectation values of L_+, L_-, L_x, and L_y vanish.

(b) Determine the average square deviation $(\Delta L_i)^2$ for the components of the angular momentum operator in the states Y_{ll} and check the uncertainty relation.

(c) Show that for the eigenstates of L^2 and L_z the expression $(\Delta L_x)^2 + (\Delta L_y)^2$ is minimal if $m = \pm l$.

5.2 Show that the spherical harmonics satisfy the completeness relation (5.31). Use the addition theorem and the generating function for the Legendre polynomials

$$\frac{1}{\sqrt{1 - 2u \cos \vartheta + u^2}} = \sum_{l=0}^{\infty} u^l P_l (\cos \vartheta) \quad ,$$

and prove that your result has the defining properties of the $\delta^{(2)}$-function.

5.3 Show that $(\psi, L^2 \psi) = 0$ also implies $(\psi, L_k \psi) = 0$.

5.4 Show that the relation $[a \cdot L, b \cdot L] = i \hbar (a \times b) \cdot L$ holds under the assumption that a and b commute with each other and with L .

5.5 Show the following:

(a) For a vector operator $V(x, p)$, $[L^2, V] = 2i\hbar (V \times L - i\hbar V)$.

(b) $[L_i, p_j] = i\hbar \varepsilon_{ijk} p_k$.

5.6 (a) Find the eigenfunction ψ of L^2 and L_x with eigenvalues $2\hbar^2$ and $\hbar$, respectively. (Hint: Represent the eigenfunctions of L^2 and L_z of interest in terms of Cartesian coordinates, and determine ψ by a rotation through $\pi/2$.)

(b) Express ψ as a linear combination of eigenfunctions of L^2 and L_z.

5.7 Show that the eigenvalues l of the orbital angular momentum are non-negative integers by showing that the eigenvalues of L_z are integers.
Hint: Express by means of the transformation

$$x_i = \sqrt{\frac{\hbar}{2m\omega}} \left(a_i + a_i^\dagger \right) \quad ,$$

$$p_i = -i \sqrt{\frac{\hbar m \omega}{2}} \left(a_i - a_i^\dagger \right)$$

(with arbitrary m and ω) the operator L_z in terms of creation and annihilation operators. Bring L_z by means of a linear transformation to new annihilation operators b_1 and b_2 into the form

$$L_z = \hbar \left(b_2^\dagger b_2 - b_1^\dagger b_1 \right) \quad .$$

(See also Appendix C.)

6. The Central Potential I

In this chapter, we will consider motion in central potentials. We first reduce the time independent Schrödinger equation to a one-dimensional (radial) problem. We then determine the bound states for the most important
case of an attractive Coulomb potential. Finally, we transform the two-body
problem into a one-body problem with a potential, so that our treatment of
motion in a Coulomb potential also covers the nonrelativistic hydrogen atom.

6.1 Spherical Coordinates

We now study three-dimensional motion in a central-force field, which is
characterized by the fact that the potential energy $V(r)$ only depends on the
distance $r = |\boldsymbol{x}|$ from the origin. The Hamiltonian takes the form

$$H = \frac{1}{2m}\boldsymbol{p}^2 + V(r) \quad .$$

(6.1)

In *classical mechanics*,

$$\boldsymbol{L}^2 = (\boldsymbol{x} \times \boldsymbol{p})^2 = \boldsymbol{x}^2\boldsymbol{p}^2 - (\boldsymbol{x} \cdot \boldsymbol{p})^2 \quad ,$$

and thus

$$\boldsymbol{p}^2 = \frac{\boldsymbol{L}^2}{r^2} + p_r^2$$

holds, if $p_r \equiv (\boldsymbol{x}/r) \cdot \boldsymbol{p}$ designates the radial component of the momentum.

On the other hand, in *quantum mechanics* the noncommutativity of
position and momentum operators must be kept in mind

$$\boldsymbol{L}^2 = \boldsymbol{x}^2\boldsymbol{p}^2 - (\boldsymbol{x} \cdot \boldsymbol{p})^2 + i\hbar\boldsymbol{x} \cdot \boldsymbol{p} \quad .$$

(6.2)

Remark: This follows from

$$
\begin{aligned}
\boldsymbol{L}^2 &= \sum_{i=1}^{3} L_i^2 = \varepsilon_{ijk}x_j p_k \varepsilon_{irs}x_r p_s = x_j p_k x_j p_k - x_j p_k x_k p_j \\
&= \boldsymbol{x}^2\boldsymbol{p}^2 - i\hbar\boldsymbol{x} \cdot \boldsymbol{p} - x_j x_k p_k p_j + 3i\hbar x_j p_j \\
&= \boldsymbol{x}^2\boldsymbol{p}^2 - i\hbar\boldsymbol{x} \cdot \boldsymbol{p} - x_j p_j x_k p_k - i\hbar x_j p_j + 3i\hbar x_j p_j \quad ,
\end{aligned}
$$

where we have used $\varepsilon_{ijk}\varepsilon_{irs} = \delta_{jr}\delta_{ks} - \delta_{js}\delta_{kr}$, $p_k x_j = x_j p_k - i\hbar\delta_{jk}$, and $p_k x_k = x_k p_k - 3i\hbar$ in the intermediate steps.

Because of the rotational symmetry of the potential, one transforms to spherical coordinates. Since by (5.17) the projection of the momentum operator on the position operator is given by

$$x \cdot p = \frac{\hbar}{i} x \cdot \nabla = \frac{\hbar}{i} r \frac{\partial}{\partial r} \quad , \tag{6.3}$$

it follows from (6.2) that

$$p^2 = \frac{1}{r^2} L^2 - \frac{\hbar^2}{r^2} \left(\left(r\frac{\partial}{\partial r} \right)^2 + r\frac{\partial}{\partial r} \right) \quad . \tag{6.4}$$

If one considers the action on a function $f(r)$, one easily sees that the following expressions for the differential operator in the second term are equivalent:

$$\frac{1}{r}\frac{\partial}{\partial r} r \frac{\partial}{\partial r} + \frac{1}{r}\frac{\partial}{\partial r} = \frac{\partial^2}{\partial r^2} + \frac{2}{r}\frac{\partial}{\partial r} = \left(\frac{1}{r}\frac{\partial}{\partial r} r \right)^2 = \frac{1}{r}\frac{\partial}{\partial r}\frac{\partial}{\partial r} r \quad . \tag{6.5}$$

With this, (6.4) can be brought into the form

$$p^2 = -\hbar^2 \left(\frac{\partial^2}{\partial r^2} + \frac{2}{r}\frac{\partial}{\partial r} \right) + \frac{1}{r^2} L^2 = p_r^2 + \frac{1}{r^2} L^2 \quad , \tag{6.4'}$$

where, in contrast to the expression $(\hbar/i)\partial/\partial r$ that one might naively expect, the radial component of the momentum operator is given by

$$p_r = \frac{\hbar}{i}\frac{1}{r}\frac{\partial}{\partial r} r = \frac{\hbar}{i}\left(\frac{\partial}{\partial r} + \frac{1}{r} \right) \quad . \tag{6.6}$$

The operators p_r and r obey the commutation relation

$$[r, p_r] = \left[r, \frac{\hbar}{i}\left(\frac{\partial}{\partial r} + \frac{1}{r} \right) \right] = i\hbar \quad ; \tag{6.6'}$$

p_r is Hermitian.

Proof of Hermiticity

$$(i) \quad \int_0^\infty dr\, r^2 \varphi^* \frac{\hbar}{i}\frac{1}{r}\psi = -\int_0^\infty dr\, r^2 \left(\frac{\hbar}{i}\frac{1}{r}\varphi \right)^* \psi \quad .$$

$$(ii) \quad \int_0^\infty dr\, r^2 \varphi^* \frac{\hbar}{i}\frac{\partial}{\partial r}\psi = \frac{\hbar}{i} r^2 \varphi^* \psi \Big|_0^\infty - \frac{\hbar}{i}\int_0^\infty dr \left(\frac{\partial}{\partial r} r^2 \varphi^* \right) \psi$$

$$= \int_0^\infty dr\, r^2 \left(\frac{\hbar}{i}\frac{\partial}{\partial r}\varphi \right)^* \psi + 2 \cdot \int_0^\infty dr\, r^2 \left(\frac{\hbar}{i}\frac{1}{r}\varphi \right)^* \psi \quad .$$

The sum (i) + (ii) implies the Hermiticity of p_r; $(\hbar/i)\partial/\partial r$ by itself would not be Hermitian. In (ii), the boundary conditions of p. 122 were used.

If we substitute (6.4') into the Hamiltonian (6.1), we obtain for the time independent Schrödinger equation

$$\left[-\frac{\hbar^2}{2m} \left(\frac{\partial^2}{\partial r^2} + \frac{2}{r}\frac{\partial}{\partial r} \right) + \frac{L^2}{2mr^2} + V(r) \right] \psi(r,\vartheta,\varphi) = E\psi(r,\vartheta,\varphi). \quad (6.7)$$

Since we already know the eigenfunctions of L^2 we introduce the separation ansatz

$$\psi(r,\vartheta,\varphi) = R(r)\,Y_{lm}(\vartheta,\varphi) \quad . \qquad (6.8)$$

If this is substituted into the differential equation (6.7), then using (5.14) and (5.21) a differential equation for the radial part $R(r)$ immediately follows:

$$\left[-\frac{\hbar^2}{2m} \left(\frac{\partial^2}{\partial r^2} + \frac{2}{r}\frac{\partial}{\partial r} \right) + \frac{\hbar^2 l(l+1)}{2mr^2} + V(r) \right] R(r) = ER(r) \quad . \qquad (6.9)$$

In the separation ansatz for the wave function, we encounter a general principle. Since H as a scalar is invariant under rotations, i.e., possesses rotational symmetry, then according to (5.9), $[H, L] = [H, L^2] = 0$, which means that H, L_z, and L^2 are simultaneously diagonalizable. As in the present situation in which the conservation of angular momentum L follows from the rotational symmetry of the Hamiltonian, a continuous symmetry always leads to a corresponding conservation law.

In any case, we now would like to continue the investigation of (6.9), which, although it depends only on one coordinate, does not yet take the form of a one-dimensional Schrödinger equation. In order to bring it into this form, we substitute

$$R(r) = u(r)/r \quad , \qquad (6.10)$$

and note that by (6.5) one has

$$\left(\frac{\partial^2}{\partial r^2} + \frac{2}{r}\frac{\partial}{\partial r} \right) R(r) = \left(\frac{1}{r}\frac{\partial}{\partial r}r \right)^2 \frac{u}{r} = \frac{1}{r}\frac{\partial^2}{\partial r^2}u \quad ,$$

and therefore

$$\left[-\frac{\hbar^2}{2m}\frac{d^2}{dr^2} + \frac{\hbar^2 l(l+1)}{2mr^2} + V(r) \right] u(r) = Eu(r) \quad . \qquad (6.11)$$

Thus, the central potential problem has been reduced to a one-dimensional problem; (6.11) is a one-dimensional Schrödinger equation with the effective potential (Fig. 6.1)

$$V_{\text{eff}}(r) = V(r) + \frac{\hbar^2 l(l+1)}{2mr^2} \quad . \qquad (6.12)$$

This is in complete analogy to classical mechanics, in which the central potential is also modified by a repulsive centrifugal term (second term in (6.12)). We now must determine boundary and normalization conditions for $u(r)$.

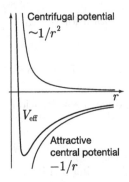

Centrifugal potential
$\sim 1/r^2$

V_{eff}

r

Attractive
central potential
$-1/r$

Fig. 6.1. The effective potential of the radial Schrödinger equation for an attractive Coulomb potential

1. From normalizability

$$\int d^3x |\psi(\boldsymbol{x})|^2 = \int_0^\infty dr\, r^2 \frac{1}{r^2} |u(r)|^2 < \infty$$

it follows for bound states that

$$\lim_{r \to \infty} |u(r)| \leq \frac{a}{r^{(1/2+\varepsilon)}} \quad \text{with} \quad \varepsilon > 0 \quad .$$

Thus, $u(r)$ must decrease for large r more strongly with r than $1/\sqrt{r}$.

2. Behavior for $r \to 0$: For $V(r) \neq \delta^{(3)}(\boldsymbol{x})$, the boundary condition $u(0) = 0$ must further hold, since otherwise $\Delta\psi = \nabla^2 u(0)/r \sim \delta^{(3)}(\boldsymbol{x})\, u(0)$, in contradiction to (6.11).

6.2 Bound States in Three Dimensions

We now pose the question of what general statements can be made concerning the existence of bound states. Let us first consider the orbital angular momentum $l = 0$, where one has $V_{\text{eff}}(r) = V(r)$. According to (3.58), the boundary condition $u(0) = 0$ means that the equivalent one-dimensional problem has the potential (Fig. 6.2)

$$V_1(x) = \begin{cases} V(x) & x > 0 \\ \infty & x < 0 \end{cases} .$$

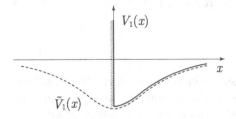

$V_1(x)$

x

$\tilde{V}_1(x)$

Fig. 6.2. The effective one-dimensional potential for $l = 0, V_1(x)$. The eigenfunctions are identical to the odd eigenfunctions of $\tilde{V}_1(x)$ (*dashed*)

As a reference, we also define the symmetric potential $\tilde{V}_1(x) \equiv V(|x|)$. We know from Chap. 4 that symmetric, one-dimensional problems always possess at least one even bound state. However, an odd bound state exists only if the potential reaches a minimum strength. The ground state of the potential $V_1(x)$ coincides, for positive x, with the first excited state, i.e., the lowest odd bound state of $\tilde{V}_1(x)$. Therefore, $V(r)$ must exceed a minimum strength, in order that in three dimensions a bound state exist. The centrifugal term is increasingly repulsive with increasing l. If a potential has no bound state for $l = 0$, then it certainly has none for $l > 0$.

At this point, we would like to gain further information about the behavior of $u(r)$ from a consideration of the radial Schrödinger equation (6.11) in *limiting cases*, before we finally turn to the Coulomb problem in the next section.

Limit $r \to 0$

For potentials like the Coulomb potential or the square-well potential, the centrifugal term dominates for decreasing r compared to $(V(r) - E)$, so that (6.11) becomes

$$\left[-\frac{\hbar^2}{2m} \frac{d^2}{dr^2} + \frac{\hbar^2 l(l+1)}{2mr^2} \right] u(r) = 0 \quad .$$

This differential equation of second order has the general solution

$$u(r) = A\, r^{l+1} + B\, r^{-l} \quad .$$

Because of the boundary condition $u(0) = 0$, the B term is not permitted, so that in the limit $r \to 0$, $u(r) \to a_0 r^{l+1}$, and one can in general take

$$u(r) = r^{l+1}(a_0 + a_1 r + \ldots) \quad . \tag{6.13a}$$

Remark: For potentials $V(r) \sim 1/r$ and $V(r) \sim r^0$ (Coulomb potential, square well), $r = 0$ is a "regular singular point" of the differential equation (6.11).[1]

Limit $r \to \infty$

In this case, V_{eff} is negligible and (6.11) approaches

$$-\frac{\hbar^2}{2m} \frac{d^2}{dr^2} u = Eu \quad .$$

For bound states $(E < 0)$ one obtains the solutions $\exp\{\pm \kappa r\}$ with

$$\kappa = \frac{1}{\hbar} \sqrt{2m(-E)} \quad ,$$

---

[1] See for instance P. Dennery and A. Krzywicki: *Mathematics for Physicists* (Harper & Row, New York 1967) or E.T. Whittaker and G.N. Watson: *A Course of Modern Analysis* (Cambridge University Press 1963)

where because of the normalization condition only the exponentially decreasing solution is relevant for $u(r)$, i.e.,

$$u(r) = Ce^{-\kappa r} \quad \text{for} \quad r \to \infty \quad . \tag{6.13b}$$

This result suggests the introduction of the dimensionless variable $\varrho = \kappa r$ in the differential equation (6.11)

$$\left[\frac{d^2}{d\varrho^2} - \frac{l(l+1)}{\varrho^2} - \frac{V(\varrho/\kappa)}{|E|} - 1\right] u(\varrho) = 0 \quad . \tag{6.14}$$

6.3 The Coulomb Potential

We now make our investigations concrete by considering an electron in the field of an atomic nucleus. For this purpose we choose for $V(r)$ the Coulomb potential

$$V(r) = -\frac{e_0^2 Z}{r} \quad , \tag{6.15}$$

$e_0 = 4.803 \times 10^{-10}$ esu (elementary charge). We define

$$\varrho_0 = \frac{e_0^2 Z \kappa}{|E|} = \sqrt{\frac{2m}{|E|}} \frac{Z e_0^2}{\hbar} \quad , \tag{6.16}$$

and thus

$$\frac{V}{|E|} = -\frac{\varrho_0}{\varrho} \quad .$$

Equation (6.14) then becomes

$$\left[\frac{d^2}{d\varrho^2} - \frac{l(l+1)}{\varrho^2} + \frac{\varrho_0}{\varrho} - 1\right] u(\varrho) = 0 \quad . \tag{6.17}$$

For $u(\varrho)$ we now make a substitution which takes into account the asymptotic behavior derived in (6.13a) and (6.13b):

$$u(\varrho) = \varrho^{l+1} e^{-\varrho} w(\varrho) \quad . \tag{6.18}$$

Substituting into (6.17), we thus obtain the following second-order differential equation for $w(\varrho)$:

$$\varrho \frac{d^2 w}{d\varrho^2} + 2(l+1-\varrho)\frac{dw}{d\varrho} + (\varrho_0 - 2(l+1))w = 0 \quad . \tag{6.19}$$

For $w(\varrho)$ we introduce the power series expansion

$$w(\varrho) = \sum_{k=0}^{\infty} a_k \varrho^k \quad , \tag{6.20}$$

and find from (6.19)

$$\sum_{k=0}^{\infty} a_k [k(k-1)\varrho^{k-1} + 2(l+1)k\varrho^{k-1}$$

$$- 2k\varrho^k + (\varrho_0 - 2(l+1))\varrho^k] = 0 \quad . \tag{6.21}$$

The coefficients of each power of ϱ must vanish, and thus for ϱ^k

$$[(k+1)k + 2(l+1)(k+1)]a_{k+1} + [-2k + (\varrho_0 - 2(l+1))]a_k = 0 \quad .$$

This leads to a recursion relation, which allows calculation of the next higher coefficient a_{k+1} from a_k :

$$a_{k+1} = \frac{2(k+l+1) - \varrho_0}{(k+1)(k+2l+2)} a_k \quad . \tag{6.22}$$

For convergence, the ratio of successive coefficients in the limit of large k is crucial. Equation (6.22) gives

$$\frac{a_{k+1}}{a_k} \rightarrow \frac{2}{k} \quad \text{for} \quad k \rightarrow \infty \quad .$$

If this is compared to the exponential series

$$e^{2\varrho} = \sum_{k=0}^{\infty} \frac{1}{k!} (2\varrho)^k \quad ,$$

in which the ratio of successive coefficients also satisfies

$$\frac{2^{k+1}/(k+1)!}{2^k/k!} = \frac{2}{k+1} \approx \frac{2}{k} \quad ,$$

one sees that $w(\varrho)$ behaves like $e^{2\varrho}$. In order that (6.18) not lead to $u(r) \sim e^{\varrho} = e^{\kappa r}$ for large r, the series must terminate. If the series terminates after the Nth term, then $w(\varrho)$ is a polynomial of Nth order. The termination condition $a_{N+1} = a_{N+2} = \ldots = 0$ leads, using (6.22), to

$$\varrho_0 = 2(N+l+1) \quad , \quad N = 0, 1, 2, \ldots \quad . \tag{6.23}$$

The quantity N is called the *"radial quantum number"*. The energy eigenvalues of the bound states of the Coulomb potential result from (6.23) and (6.16):

$$E = -\frac{2mZ^2 e_0^4}{\varrho_0^2 \hbar^2} = -\frac{mZ^2 e_0^4}{2\hbar^2(N+l+1)^2} \quad . \tag{6.24}$$

Table 6.1. Quantum numbers and energy eigenvalues

$n = 1$	$l = 0$ (s-orbital)	$m = 0$	E_1 (1-fold)
$n = 2$	$l = 0$ (s)	$m = 0$	$\left.\rule{0pt}{20pt}\right\}$ E_2 (4-fold)
	$l = 1$ (p)	$m = -1, 0, 1$	
$n = 3$	$l = 0$ (s)	$m = 0$	
	$l = 1$ (p)	$m = -1, 0, 1$	$\left.\rule{0pt}{30pt}\right\}$ E_3 (9-fold)
	$l = 2$ (d)	$m = -2, -1, 0, 1, 2$	
$n = 4$	$l = 0$ (s)	$m = 0$	
	$l = 1$ (p)	$m = -1, 0, 1$	$\left.\rule{0pt}{40pt}\right\}$ E_4 (16-fold)
	$l = 2$ (d)	$m = -2, -1, 0, 1, 2$	
	$l = 3$ (f)	$m = -3, -2, -1, 0, 1, 2, 3$	

If we introduce the *"principal quantum number"* n,

$$n = N + l + 1 \quad , \tag{6.25}$$

we obtain

$$E_n = -\frac{mZ^2 e_0^4}{2\hbar^2 n^2} \quad , \quad n = 1, 2, 3, 4, \ldots \quad . \tag{6.24'}$$

The energy eigenvalues E_n depend only on the combination $n = N + l + 1$. For a given, fixed n, the angular momentum quantum numbers $l = 0, 1, 2, \ldots, n-1$ are possible. If one further remembers that for each l there are $2l+1$ distinct values of m, then by

$$\sum_{l=0}^{n-1} (2l + 1) = 2\frac{n(n - 1)}{2} + n = n^2$$

the energy eigenvalue E_n is n^2-fold degenerate.

Table 6.1 gives an overview of the assignment of quantum numbers to the individual energy eigenvalues.

In (6.24) we determined the energy eigenvalues for the Coulomb potential and, using the recursion relation (6.22), we could also calculate the corresponding wave functions. Instead of pursuing this line further, we would like to establish a connection between the radial Schrödinger equation (6.19) with the eigenvalues we have found and a differential equation which has been thoroughly investigated in mathematics, and we will make use of its known solutions. For this purpose, we multiply (6.19) by 1/2 and, using (6.23) and (6.25), we substitute $\varrho_0 = 2n$, with the result

$$(2\varrho)\frac{d^2 w}{d(2\varrho)^2} + ((2l + 1) + 1 - (2\varrho))\frac{dw}{d(2\varrho)} + ((n + l) - (2l + 1))w = 0 \quad . \tag{6.26}$$

This is the Laguerre differential equation. In order to see this, and for completeness, we insert a short overview of the most important (for our purposes) properties of the *Laguerre polynomials*.

The *Laguerre polynomials*[2] enter as coefficients in the power series expansion of the following *generating function*:

$$\frac{1}{1-s} \exp\left\{-x\frac{s}{1-s}\right\} = \sum_{r=0}^{\infty} L_r(x)\frac{s^r}{r!} \quad . \tag{6.27}$$

Partial differentiation of this relation with respect to x gives

$$-\frac{s}{(1-s)^2} \exp\left\{-x\frac{s}{1-s}\right\} = \sum L'_r(x)\frac{s^r}{r!} \quad ,$$

or, upon substitution of the power series (6.27),

$$-\sum L_r\frac{s^{r+1}}{r!} = \sum L'_r(x)\frac{1}{r!}(s^r - s^{r+1}) \quad .$$

Finally by comparison of the coefficients of s^r,

$$-\frac{1}{(r-1)!}L_{r-1} = L'_r\frac{1}{r!} - L'_{r-1}\frac{1}{(r-1)!} \quad ,$$

one obtains the following recursion relation:

(i) $L'_r = r\left(L'_{r-1} - L_{r-1}\right)$. $\tag{6.28}$

A second recursion relation results in complete analogy by partial differentiation with respect to s:

(ii) $L_{r+1} = (2r+1-x)L_r - r^2 L_{r-1}$. $\tag{6.29}$

From the recursion relations, we note (without proof) that the *Laguerre differential equation*

$$x L''_r + (1-x) L'_r + r L_r = 0 \tag{6.30}$$

follows. The *associated Laguerre polynomials* are given by the definition

$$L^s_r = \frac{d^s}{dx^s} L_r(x) \quad . \tag{6.31}$$

They can also be represented in the form

$$L^s_r = \frac{d^s}{dx^s} e^x \frac{d^r}{dx^r} e^{-x} x^r \tag{6.32a}$$

and are polynomials of order $r - s$ with $r - s$ distinct real zeros. The explicit representation of these polynomials is

$$L^s_r(x) = \sum_{k=0}^{r-s}(-1)^{k+s}\frac{(r!)^2}{k!\,(k+s)!\,(r-k-s)!}\,x^k \quad . \tag{6.32b}$$

---

[2] When using the literature for Laguerre polynomials one should be aware of the fact that differing definitions and notations are in use.

In addition, they obey the normalization relation

$$\int_0^\infty dx\, x^{s+1} e^{-x} [L_r^s(x)]^2 = \frac{(2r-s+1)\,(r!)^3}{(r-s)!} \quad . \tag{6.33}$$

If (6.30) is differentiated s times with respect to x using the product rule, one gets

$$x\,L_r^{s''} + s\,L_r^{s'} + L_r^{s'} - x\,L_r^{s'} - s\,L_r^{s} + r\,L_r^{s} = 0 \quad ,$$

that is, the following differential equation for the associated Laguerre polynomials:

$$x\,L_r^{s''} + (s+1-x)\,L_r^{s'} + (r-s)\,L_r^{s} = 0 \quad . \tag{6.34}$$

Now we see by comparison of (6.26) and (6.34) that the $w(\varrho)$ are solutions of the differential equation for the associated Laguerre polynomials with $s = 2l+1$, $r = n+l$, that is,

$$w(\varrho) = A\,L_{n+l}^{2l+1}(2\varrho) \quad , \quad \varrho = \kappa r \quad . \tag{6.35}$$

One can easily convince oneself that the coefficients (6.32b) of the polynomial $L_{n+l}^{2l+1}(2\varrho)$ satisfy the recursion relation (6.22) with (6.23) and (6.25).

Let us summarize: With (6.8), (6.10), (6.18), (6.35) and corresponding normalization according to (6.33), the bound stationary states of the *Coulomb potential* become

$$\psi_{nlm}(r, \vartheta, \varphi, t) = e^{-itE_n/\hbar}\, R_{nl}(r) Y_{lm}(\vartheta, \varphi) \quad , \tag{6.36}$$

where

$$R_{nl}(r) = \frac{u(r)}{r} = -\left[\frac{(n-l-1)!(2\kappa)^3}{2n((n+l)!)^3}\right]^{1/2} (2\kappa r)^l\, e^{-\kappa r}\, L_{n+l}^{2l+1}(2\kappa r) \tag{6.37}$$

and

$$\kappa = \frac{\sqrt{2m|E|}}{\hbar} = \frac{mZe_0^2}{\hbar^2 n} \quad ,$$

and thus

$$\kappa = \frac{Z}{na} \quad , \tag{6.38}$$

provided the *Bohr radius* is defined by

$$a = \frac{\hbar^2}{me_0^2} = 0.529 \times 10^{-8}\, \text{cm} \quad . \tag{6.39}$$

According to (6.24'), the corresponding energy eigenvalues are

$$E_n = -\frac{mZ^2e_0^4}{2\hbar^2 n^2} = -\frac{(Ze_0)^2}{2an^2} = -\frac{mc^2}{2}\,\alpha^2\,\frac{Z^2}{n^2} \quad . \tag{6.24''}$$

Here, $mc^2 = 0.510\,98\,\mathrm{MeV}$ stands for the rest energy of the electron and

$$\alpha = \frac{e_0^2}{\hbar c} = \frac{1}{137.037} \tag{6.40}$$

is the *Sommerfeld fine-structure constant*.

From this, one obtains for the binding energy in the ground state of the hydrogen atom ($Z = 1$), which is equivalent to the negative of the ionization energy of the hydrogen atom,

$$E_1(Z = 1) = -13.6\,\mathrm{eV} = -1\,\mathrm{Ry}\ \text{(Rydberg)}\quad . \tag{6.41}$$

Since $\psi_{nlm}(r, \vartheta, \varphi, t)$ was normalized both in the angular and in the radial part, the orthonormality relation

$$\int d^3x\,\psi^*_{nlm}\psi_{n'l'm'} = \delta_{nn'}\delta_{ll'}\delta_{mm'} \tag{6.42}$$

holds.

Remarks:

(i)   Due to the properties of the associated Laguerre polynomials, the radial part R_{nl} has $N = n - l - 1$ positive zeros (nodes).
(ii) R_{nl} does not depend on the quantum number m.
(iii) $|\psi_{nlm}(r, \vartheta, \varphi, t)|^2 r^2 dr\, d\Omega$ expresses the position probability in dr and $d\Omega$. The radial position probability is obtained by angular integration. It is the probability of finding the particle in the interval dr at a distance r from the origin. Because of the normalization of the angular part, $Y_{lm}(\Omega)$, the radial position probability is given by $|R_{nl}(r)|^2 r^2 dr$.

We give the lowest radial wave functions:

$n = 1,\ l = 0$ ("K-shell, s-orbital"):

$$R_{10}(r) = 2\left(\frac{Z}{a}\right)^{3/2} e^{-Zr/a}$$

$n = 2,\ l = 0$ ("L-shell, s-orbital"):

$$R_{20}(r) = 2\left(\frac{Z}{2a}\right)^{3/2}\left(1 - \frac{Zr}{2a}\right) e^{-Zr/2a}$$

$l = 1$ ("L-shell, p-orbital"):

$$R_{21}(r) = \frac{1}{\sqrt{3}}\left(\frac{Z}{2a}\right)^{3/2}\frac{Zr}{a} e^{-Zr/2a} \tag{6.43}$$

$n = 3$, $l = 0$ ("M-shell, s-orbital"):

$$R_{30}(r) = 2 \left(\frac{Z}{3a}\right)^{3/2} \left[1 - \frac{2Zr}{3a} + \frac{2(Zr)^2}{27a^2}\right] e^{-Zr/3a}$$

$l = 1$ ("M-shell, p-orbital"):

$$R_{31}(r) = \frac{4\sqrt{2}}{3} \left(\frac{Z}{3a}\right)^{3/2} \frac{Zr}{a} \left(1 - \frac{Zr}{6a}\right) e^{-Zr/3a}$$

$l = 2$ ("M-shell, d-orbital"):

$$R_{32}(r) = \frac{2\sqrt{2}}{27\sqrt{5}} \left(\frac{Z}{3a}\right)^{3/2} \left(\frac{Zr}{a}\right)^2 e^{-Zr/3a}.$$

One recognizes that the states with $l = 0$ are finite at the origin, whereas the states with $l \geq 1$ vanish there. The functions $R_{nl}(r)$ and the radial probability densities $r^2 R_{nl}^2$ are shown graphically in Fig. 6.3.

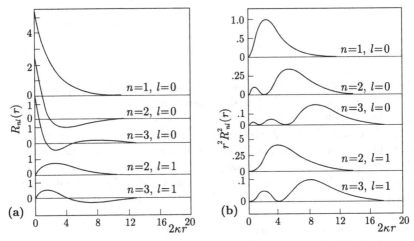

Fig. 6.3a,b. The radial wave function $R_{nl}(r)$ for the attractive Coulomb potential ($Z = 1$). (a) The radial wave function R_{nl}. (b) The radial probability density $r^2 R_{nl}^2$

The energy levels can be represented conveniently in the form of an *energy level diagram*. This is shown for hydrogen in Fig. 6.4.

Atomic spectra arise due to transitions between the discrete levels; in the hydrogen atom, the energy change in a transition from the initial state m into the final state n is given by

$$\hbar\omega_{mn} = E_m - E_n = 1\,\mathrm{Ry} \left(-\frac{1}{m^2} + \frac{1}{n^2}\right) . \tag{6.44}$$

We have thus found the quantum mechanical basis for the combination principle formulated by Ritz in 1905. The best known spectral series, characterized by the particular final state (n) of the transition, are given for the hydrogen atom in Table 6.2.

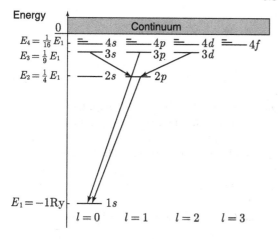

Energy

$E_4 = \frac{1}{16}E_1$ $E_3 = \frac{1}{9}E_1$ $E_2 = \frac{1}{4}E_1$

$E_1 = -1\,\mathrm{Ry}$ 1s

$l=0$ $l=1$ $l=2$ $l=3$

Fig. 6.4. The energy level diagram for the Coulomb potential

Table 6.2. Spectral series of the H-atom

Series	Final state	Orbitals of transition	Characteristic wavelength
Lyman	$n = 1$	$p \to s$	Ultraviolet about 100 nm
Balmer	$n = 2$	$s \to p$	Ultraviolet up to visible (400–600 nm)
Paschen	$n = 3$	$d \to p$	
Bracket	$n = 4$	$f \to d$	Infrared
Pfund	$n = 5$		(1000–7000 nm)

The energy eigenvalues of the Coulomb Hamiltonian exhibit, as we saw earlier, multiple degeneracy in the quantum numbers l and m. To what degree does this depend on the special features of the physical situation investigated here?

(a) For central potentials, i.e., whenever $V = V(r)$, the energy eigenvalues are always independent of m.

To this end, we consider a rotationally symmetric Hamiltonian, that is $[H, L_i] = 0$ for $i = 1, 2, 3$. In this case, $H, \boldsymbol{L}^2$, and L_z are simultaneously diagonalizable, i.e., they have common eigenfunctions, and in addition one has $[L_\pm, H] = 0$. Let $\psi_{.lm}$ be an eigenfunction[3] of $H : H\psi_{.lm} = E\psi_{.lm}$. We now operate on this equation with L_+.

$$HL_+\psi_{.lm} = L_+H\psi_{.lm} = EL_+\psi_{.lm} \quad ,$$

---

[3] The dot refers to further quantum numbers of H, e.g. n in the case of the Coulomb potential.

or, because

$$L_+\psi_{.lm} \propto \psi_{.lm+1} \quad ,$$

$$H\psi_{.lm+1} = E\psi_{.lm+1} \quad .$$

However, this implies that if $\psi_{.lm}$ is an eigenfunction with the eigenvalue E then this also holds for $\psi_{.lm+1}$. Thus, as a consequence of rotational symmetry, the energy eigenvalues are independent of m, as one sees in elementary fashion on the basis of the m-independence of the radial Schrödinger equation (6.9).

(b) The degeneracy of the energy eigenvalues with respect to l represents a specific property of the $1/r$-potential.

The reason for this is that for the Coulomb potential the Lenz vector

$$\boldsymbol{A} = \frac{1}{2m}(\boldsymbol{p} \times \boldsymbol{L} - \boldsymbol{L} \times \boldsymbol{p}) - \frac{e_0^2}{r}\boldsymbol{x}$$

is an additional conserved quantity ($\boldsymbol{A}\cdot\boldsymbol{L} = \boldsymbol{L}\cdot\boldsymbol{A} = 0; [\boldsymbol{A}, H] = 0$). Classically the implication of this is that the major axis of the elliptical orbit is spatially fixed.

The degree of degeneracy of the energy eigenvalues was determined earlier as n^2; if one additionally takes into account the double occupancy of each state by two electrons of opposite spin, this becomes $2\,n^2$. In line with spectroscopic notation, we will later designate the $2\,(2l+1)$ states with fixed values of n and l as a "shell".

The position expectation values and uncertainties in the eigenstates of the Coulomb Hamiltonian are also of interest. One obtains for the expectation value of the radial distance in the state ψ_{nlm}, using the recursion relation for the Laguerre polynomials,

$$\langle r \rangle_{nl} = \int d^3x\, \psi_{nlm}^* r \psi_{nlm} = \frac{1}{2}\frac{a}{Z}(3n^2 - l(l+1)) \quad . \tag{6.45}$$

In particular, $\langle r \rangle_{n,n-1} = (a/Z)n\,(n+1/2)$.

Analogously,

$$\langle r^2 \rangle_{n,n-1} = \frac{a^2}{Z^2}n^2(n+1)\left(n+\frac{1}{2}\right) \quad . \tag{6.46}$$

For the state $\psi_{n,n-1,m}$ this gives a radial uncertainty of

$$\Delta r = \sqrt{\langle r^2 \rangle_{n,n-1} - (\langle r \rangle_{n,n-1})^2} = \frac{an}{Z}\sqrt{\frac{1}{2}\left(n+\frac{1}{2}\right)}$$

and thus a relative fluctuation of

$$\frac{\Delta r}{\langle r \rangle_{n,n-1}} = \frac{1}{\sqrt{2n+1}} \ . \tag{6.47}$$

One sees that for $n \to \infty$, this fraction approaches 0, i.e., the orbitals are very well defined for high energies.

The states with angular momentum quantum number $l = 0$ are proportional to

$$\psi_{n00} \propto e^{-\kappa r} L_n^1(2\kappa r) \ .$$

They have $n - 1$ nodes and are non vanishing at $r = 0$ ($\psi_{n00}(0, \vartheta, \varphi) \neq 0$). Additionally, the states ψ_{n00} are independent of angle, i.e., spherically symmetric, because of $Y_{00}(\vartheta, \varphi) = 1/\sqrt{4\pi}$, whereas classically motion with orbital angular momentum $L = 0$ runs along a straight line through the center.

In contrast, according to (6.37), the states with maximal orbital angular momentum, i.e. $l = n - 1$, have the radial dependence

$$R_{n,n-1}(r) = \frac{(2\kappa)^{3/2}}{((2n)!)^{1/2}}(2\kappa r)^{n-1}e^{-\kappa r}, \quad \text{(Fig. 6.5)}.$$

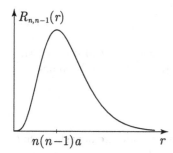

Fig. 6.5. A radial eigenfunction with maximal l, shown here for $n = 5$

The maximum of the radial probability density

$$p(r) = \int d\Omega r^2 |\psi_{n,n-1,m}|^2 = C\,r^{2n}\,e^{-2\kappa r}$$

is found by differentiation

$$\frac{dp(r)}{dr} \propto r^{2n}e^{-2\kappa r}\left(\frac{2n}{r} - 2\kappa\right) \ ,$$

to be at the position $r_0 = n/\kappa = n^2a/Z$; this value grows with increasing n. In addition, we calculated in (6.47) the relative fluctuation, which falls off with increasing n. Classically, for a given energy, the orbits with maximal orbital angular momentum L are circular. By superposition of the states $\psi_{n,n-1,n-1}$ for large n, one can construct wave packets which represent exactly the classical circular orbits and fulfill *Kepler's third law* (radius $\propto n_0^2$, orbital period $\propto n_0^3$; see the end of this section).

Whereas L^2 and L_z assume well defined values in the eigenstates ψ_{nlm}, this is not true of the angular momentum components L_x and L_y. Because of the symmetry properties of the ψ_{nlm} in angular coordinates, one finds for the expectation values

$$\langle L_{x,y} \rangle_{l,m} = (\psi_{nlm}, L_{x,y}\psi_{nlm}) = (Y_{lm}, L_{x,y}Y_{lm}) = 0 \quad . \tag{6.48}$$

With $L_x^2 + L_y^2 = L^2 - L_z^2$ one further obtains

$$\langle L_x^2 \rangle_{l,m} = \langle L_y^2 \rangle_{l,m} = (Y_{lm}, L_x^2 Y_{lm})$$
$$= \tfrac{1}{2}\left(Y_{lm}, \left(L^2 - L_z^2\right)Y_{lm}\right) = \tfrac{1}{2}\hbar^2[l(l+1) - m^2]$$

and hence

$$\Delta L_x = \Delta L_y = \sqrt{\langle L_y^2 \rangle_{l,m} - \langle L_y \rangle_{l,m}^2} = \hbar\sqrt{\frac{l(l+1) - m^2}{2}} \quad . \tag{6.49}$$

The uncertainties therefore do not vanish here, but they are minimal for $|m| = l$

$$\Delta L_x \big|_{l,\pm l} = \Delta L_y \big|_{l,\pm l} = \hbar\sqrt{l/2} \quad , \tag{6.50}$$

that is, as expected, in the states with the strongest z-orientation of the orbital angular momentum. For large l, the relative fluctuation becomes

$$\frac{\Delta L_x}{\langle L^2 \rangle^{1/2}} \sim l^{-1/2} \quad .$$

In order to get a feeling for atomic orders of magnitude, we now show that we can obtain the ground state energy and the Bohr radius by minimizing the energy while taking into account the uncertainty relation. With the help of the uncertainty relation, one can estimate the ground state energy without further knowledge of what the state looks like. If in the Hamiltonian

$$H = \frac{p^2}{2m} - \frac{e_0^2 Z}{r}$$

p is replaced by Δp and r by Δr, then with the help of the uncertainty relation ($\Delta p \Delta r \approx \hbar$) the result is

$$E = \frac{\hbar^2}{2m(\Delta r)^2} - \frac{e_0^2 Z}{\Delta r} \quad .$$

Differentiating,

$$\frac{dE}{d(\Delta r)} = -\frac{\hbar^2}{m(\Delta r)^3} + \frac{e_0^2 Z}{(\Delta r)^2} \quad ,$$

one determines the position of the energy minimum to be

$$(\Delta r)_0 = \frac{\hbar^2}{m e_0^2 Z} = \frac{a}{Z} \quad ,$$

and then the minimal energy compatible with the uncertainty relation takes the value

$$E_{\min} = \frac{\hbar^2}{2m(\Delta r)_0^2} - \frac{e_0^2 Z}{(\Delta r)_0} = -\frac{mZ^2 e_0^4}{2\hbar^2} \quad ,$$

which (remarkably) is in complete agreement with the ground state energy calculated exactly.

We add a few remarks concerning the *characteristic lengths* that arise in the atomic domain and their connection with the fine-structure constant $\alpha = e_0^2/\hbar c$. These are:

(i)   the *Compton wavelength*

$$\lambda_c = \frac{\hbar}{mc} \quad ,$$

which entered in Chap. 1 in the scattering of γ-quanta from electrons, and which also can be interpreted as the de Broglie wavelength of highly relativistic electrons;

(ii) the *Bohr radius* (6.39) as the characteristic extension of atomic electron states

$$a = \frac{\hbar^2}{m e_0^2} = \frac{1}{\alpha} \lambda_c \quad ;$$

(iii) the typical wavelength $2\pi\lambda$ of the light emitted in quantum transitions. At a transition energy of the order $\Delta E = e_0^2/a$, corresponding to a frequency of $\omega = \Delta E/\hbar$, one gets

$$\lambda = \frac{1}{k} = \frac{c}{\omega} = \frac{c\hbar}{\Delta E} = \frac{c\hbar a}{e_0^2} = \frac{a}{\alpha} \quad ;$$

(iv) the *classical electron radius* defined by $e_0^2/r_e = mc^2$; thus

$$r_e = \frac{e_0^2}{mc^2} = \alpha \lambda_c \quad .$$

For these four quantities, the sequence of ratios can be given as

$$r_e : \lambda_c : a : \lambda = \alpha^3 : \alpha^2 : \alpha : 1 \quad .$$

From the uncertainty relation, the typical velocity of the electron in the atom is $v = \Delta p/m = Z\hbar/am = Z\alpha c$.

In our treatment of the *Coulomb potential* in the above sections, we have occasionally referred to the *hydrogen atom*. To what extent is this justified? Indeed, it turns out that in very accurate experiments, deviations were discovered which result from the following effects, which will be treated in later chapters.

(i)   First of all, the hydrogen atom is a two-body system consisting of a nucleus of mass m_N and an electron. In the next section we will show that the transformation to center-of-mass and relative coordinates leads to a one-body problem with the reduced mass

$$\mu = \frac{m\,m_N}{m + m_N} = \frac{m}{1 + m/m_N} \quad .$$

   In hydrogen, this gives a correction in the fourth decimal place, that is, $1 + m_e/m_N = 1.00054463$, whereas for example in positronium the change is considerable: $m_N = m_{e^+} = m_e$, and therefore $\mu = m_e/2$.

(ii) Relativistic effects give rise to the so-called "fine structure" and are of order α^2 in comparison to the original levels. These include the relativistic correction to the electron mass, the Darwin term and the spin–orbit coupling (Chap. 12).

(iii) The Lamb shift, a quantum electrodynamic phenomenon, is of order $\alpha^3 \ln \alpha$ in comparison to the original energy levels.

(iv) The hyperfine structure, which results from the interaction between the electron and the nuclear spin, is smaller by a factor of about $m_e/m_N \approx 1/2000$ than the fine structure.

Time dependence: *circular Keplerian orbits.* We now discuss the *classical limit* mentioned on p. 133 more precisely and in particular we derive Kepler's third law of planetary motion from quantum theory. Considering our experience with the classical limit of the harmonic oscillator at the end of Sect. 3.1, we expect that we have to construct a superposition of states with large quantum numbers. We restrict consideration to circular orbits. In classical mechanics, circular orbits have the maximal allowed angular momentum corresponding to a given energy. This means that we must superimpose states $\psi_{n,n-1,n-1}(r, \vartheta, \varphi)$, since $l = n - 1$ is the maximal angular momentum quantum number to n. Moreover, we know from (5.36) that the orbital plane is best defined for $m = l$, and therefore we have set $m = n-1$. We thus arrive at the following superposition:

$$\psi(r, \vartheta, \varphi, t) = \sum_n c'_n\, \psi_{n,n-1,n-1}(r, \vartheta, \varphi)\, e^{-itE_n/\hbar} \quad . \tag{6.51}$$

Suppose that the coefficients c'_n have their maximum at some large value n_0 and are concentrated about this value. We write

$$n = n_0 + \varepsilon$$

and expand the quantities occurring in (6.51) in ε, which according to the above assumption is small compared to n_0. In particular, we obtain for the energy eigenvalues (6.24') of the hydrogen atom ($Z = 1$ and $\mathrm{Ry} = me_0^4/2\hbar^2$)

$$E_n = -\mathrm{Ry}\frac{1}{n^2} = -\mathrm{Ry}\left(\frac{1}{n_0^2} - \frac{2\varepsilon}{n_0^3}\right) + O\left(\varepsilon^2\right)$$

and for the wave packet

$$\psi(r, \vartheta, \varphi, t) = \sum_\varepsilon c_n' \frac{1}{\sqrt{\pi n!}\, n^n a^{3/2}} \left(-\frac{r}{a}\sin\vartheta\right)^{n-1} e^{-r/na}$$

$$\times \exp\left\{i\left(n_0 + \varepsilon - 1\right)\varphi + i\frac{t\mathrm{Ry}}{\hbar}\left(\frac{1}{n_0^2} - \frac{2\varepsilon}{n_0^3}\right)\right\} \,, \tag{6.51'}$$

where we have used (C.9). From (6.51') it is again clear that for large n_0 the wave packet is completely concentrated within $\vartheta = \pi/2$. We now absorb all the normalization constants into a new expansion coefficient c_ε and get for the radial and angular dependence of the wave packet in the orbital plane ($\vartheta = \pi/2$)

$$\psi\left(r, \frac{\pi}{2}, \varphi, t\right) \propto \sum_\varepsilon c_\varepsilon\, r^{n-1}\, e^{-r/na} \exp\left\{i\varepsilon\left(\varphi - \frac{2\,\mathrm{Ry}}{\hbar n_0^3}t\right)\right\} \,. \tag{6.52}$$

By appropriate choice of c_ε, e.g., a Gaussian, one obtains a wave packet having the form $\psi(r, \pi/2, \varphi, t) = g(\varphi - t\nu)$ with respect to azimuthal angular dependence and circulating with the frequency $\nu = 2\,\mathrm{Ry}/\hbar n_0^3$. Concerning the permissibility of superimposing energy eigenstates with distinct principal quantum numbers, we note that

$$\langle r \rangle_{n,n-1} = \frac{a}{Z}n\left(n + \frac{1}{2}\right) \,, \tag{6.53}$$

which yields in the limit of large n

$$\frac{\langle r \rangle_{n+1,n} - \langle r \rangle_{n,n-1}}{\langle r \rangle_{n,n-1}} = \frac{2n + 3/2}{n^2 + n/2} \approx \frac{2}{n} \to 0 \,. \tag{6.53'}$$

For large n_0 and $\varepsilon \ll n_0$ the relative uncertainty of the orbital radius is vanishingly small.

We can summarize the result as follows. We obtain wave packets moving on circular orbits and sharply concentrated with respect to both r and φ. The orbital radius is $\langle r \rangle = an_0^2$, and the orbital period is $\nu = 2\,\mathrm{Ry}/\hbar n_0^3$. Hence, we have derived *Kepler's third law* from the classical limit of quantum mechanics. This result is of interest for Rydberg states of the hydrogen atom. Of course, it is also instructive to go from the Coulomb interaction to the gravitational interaction by $e_0^2 \to MmG$ (G is the gravitational constant, M and m are the mass of celestial bodies such as the sun and a planet). One can apply the result to planetary motion and compute, for example, n_0 for the earth or for a satellite orbiting the earth and estimate the aforementioned uncertainties. We have restricted consideration here to circular orbits; the calculation of quantum mechanical wave packets in elliptical orbits turns out to be more complicated.[4]

---

[4] M. Nauenberg: Phys Rev. A **40**, 1133 (1989); J.-C. Gay, D. Delande, A. Bommier: Phys. Rev. A **39**, 6587 (1989).

6.4 The Two-Body Problem

If we consider the motion of an electron about the atomic nucleus or that of the earth about the sun, then (forgetting for the moment about other electrons or planets) we are faced with a two-body problem. In general, the Hamiltonian of a two-body problem involves the sum of the kinetic energies of both particles and their interaction, which is assumed to depend only on the separation vector:

$$H = \frac{p_1^2}{2m_1} + \frac{p_2^2}{2m_2} + V(\boldsymbol{x}_1 - \boldsymbol{x}_2) \quad . \tag{6.54}$$

Since the potential here depends only on the relative separation, one introduces relative and center-of-mass coordinates in the form

$$\boldsymbol{x}_r = \boldsymbol{x}_1 - \boldsymbol{x}_2 \quad ; \quad \boldsymbol{p}_r = \mu(\boldsymbol{v}_1 - \boldsymbol{v}_2) = \frac{m_2\boldsymbol{p}_1 - m_1\boldsymbol{p}_2}{m_1 + m_2}$$

$$\boldsymbol{x}_{cm} = \frac{m_1\boldsymbol{x}_1 + m_2\boldsymbol{x}_2}{m_1 + m_2} \quad ; \quad \boldsymbol{p}_{cm} = \boldsymbol{p}_1 + \boldsymbol{p}_2 \tag{6.55}$$

with reduced mass

$$\mu = \frac{m_1 m_2}{m_1 + m_2} \tag{6.56a}$$

and total mass

$$M = m_1 + m_2 \quad . \tag{6.56b}$$

Since $[x_{\nu i}, p_{\mu j}] = i\hbar\delta_{ij}\delta_{\nu\mu}$ ($\nu, \mu = 1, 2$ particle index; $i, j = 1, 2, 3$ Cartesian component index), the canonical commutation relations also hold for $\boldsymbol{p}_r$, $\boldsymbol{p}_{cm}$ and $\boldsymbol{x}_r$, $\boldsymbol{x}_{cm}$:

$$[x_{ri}, p_{rj}] = i\hbar\delta_{ij} = [x_{cmi}, p_{cmj}] \quad , \tag{6.57}$$

while the remaining commutators of these operators vanish. It follows from this property that the momentum operators $\boldsymbol{p}_r$ and $\boldsymbol{p}_{cm}$ satisfy

$$\boldsymbol{p}_r = \frac{\hbar}{i}\boldsymbol{\nabla}_r \quad , \quad \boldsymbol{p}_{cm} = \frac{\hbar}{i}\boldsymbol{\nabla}_{cm} \quad . \tag{6.58}$$

As in classical mechanics, one has

$$\frac{p_1^2}{2m_1} + \frac{p_2^2}{2m_2} = \frac{p_r^2}{2\mu} + \frac{p_{cm}^2}{2M} \quad , \tag{6.59}$$

so that using (6.54) one obtains for the time independent Schrödinger equation for the two-body problem

$$\left[\frac{p_r^2}{2\mu} + \frac{p_{cm}^2}{2M} + V(\boldsymbol{x}_r)\right]\psi(\boldsymbol{x}_r, \boldsymbol{x}_{cm}) = \hat{E}\psi(\boldsymbol{x}_r, \boldsymbol{x}_{cm}) \quad . \tag{6.60}$$

Since the potential depends only on $\boldsymbol{x}_r$, the separation ansatz

$$\psi(\boldsymbol{x}_r, \boldsymbol{x}_{cm}) = e^{i\boldsymbol{k}_{cm} \cdot \boldsymbol{x}_{cm}} \psi(\boldsymbol{x}_r) \tag{6.61}$$

brings (6.60) into the form

$$\left[\frac{\boldsymbol{p}_r^2}{2\mu} + V(\boldsymbol{x}_r) \right] \psi(\boldsymbol{x}_r) = E\psi(\boldsymbol{x}_r) \tag{6.62}$$

with the energy eigenvalues $E = \hat{E} - \hbar^2 k_{cm}^2/2M$ or

$$\hat{E} = E + \frac{\hbar^2 k_{cm}^2}{2M} \quad . \tag{6.63}$$

By the introduction of relative and center-of-mass coordinates, the two-body problem has thus been successfully reduced to a one-body problem: (6.62) corresponds precisely to the situation discussed earlier, which was represented by the Hamiltonian (6.1). Here, we have separated off the free motion of the center-of-mass with energy $E_{cm} = \hbar^2 k_{cm}^2/2M$ and wave function

$$\psi_{cm}(\boldsymbol{x}_{cm}, t) = \exp\left\{ -\frac{i}{\hbar} E_{cm} t + i\boldsymbol{k}_{cm} \cdot \boldsymbol{x}_{cm} \right\} \quad .$$

Aside from the trivial term E_{cm} in the energy eigenvalues $\hat{E}$, a modification of the earlier one-particle energy eigenvalues E also enters, in that the reduced mass μ replaces the particle mass m in (6.62). With this modification, we can therefore apply the results of Sect. 6.3 to the nonrelativistic hydrogen atom.

Similarly, one may start from the time dependent Schrödinger equation

$$i\hbar \frac{\partial}{\partial t} \psi(\boldsymbol{x}_r, \boldsymbol{x}_{cm}, t) = H\psi(\boldsymbol{x}_r, \boldsymbol{x}_{cm}, t) \tag{6.64}$$

and introduce the separation ansatz

$$\psi(\boldsymbol{x}_r, \boldsymbol{x}_{cm}, t) = \psi_r(\boldsymbol{x}_r, t)\psi_{cm}(\boldsymbol{x}_{cm}, t) \quad , \tag{6.65}$$

where $\psi_{cm}(\boldsymbol{x}, t)$ is a solution of the center of mass Schrödinger equation

$$i\hbar \frac{\partial}{\partial t} \psi_{cm}(\boldsymbol{x}_{cm}, t) = -\frac{\hbar^2}{2M} \boldsymbol{\nabla}_{cm}^2 \psi_{cm}(\boldsymbol{x}_{cm}, t) \quad . \tag{6.66}$$

This leads to

$$i\hbar \frac{\partial}{\partial t} \psi_r(\boldsymbol{x}_r, t) = \left[\frac{\boldsymbol{p}_r^2}{2m} + V(\boldsymbol{x}_r) \right] \psi_r(\boldsymbol{x}_r, t) \quad . \tag{6.67}$$

for the relative motion.

Problems

6.1 To the Coulomb problem: Derive the differential equation for $w(\varrho)$, (6.19) starting from (6.17).

6.2 Show that for the associated Laguerre polynomials

$$L_r^s(x) = (d/dx)^s \, L_r(x) \quad ,$$
$$L_r(x) = e^x \, (d/dx)^r \, e^{-x} x^r \quad ,$$

the following relations hold:

(a) $\displaystyle L_r^s(x) = \sum_{k=0}^{r-s} (-1)^{k+s} \frac{[r!]^2 x^k}{k! \, (k+s)! \, (r-k-s)!} \quad ,$

(b) $\displaystyle \frac{(-1)^m \, e^{-xt/(1-t)}}{(1-t)^{m+1}} = \sum_{n=0}^{\infty} \frac{t^r}{(r+m)!} L_{r+m}^m(x) \quad .$

6.3 Prove the following recursion formula for the matrix elements of r^k

$$\langle r^k \rangle = \langle nl | r^k | nl \rangle$$

in the hydrogen atom:

$$\frac{(k+1)}{n^2} \langle r^k \rangle - (2k+1)a\langle r^{k-1} \rangle + \frac{k}{4} \left[(2l+1)^2 - k^2 \right] a^2 \langle r^{k-2} \rangle = 0 \quad .$$

Hint: It is easier to formulate the equation first for the dimensionless variable $\varrho = \kappa r$ with $\kappa = \sqrt{-2mE}/\hbar$ and, for the computation of the $\langle \varrho^k \rangle$, to begin with the differential equation

$$\left[\frac{d^2}{d\varrho^2} - \frac{l(l+1)}{\varrho^2} + \frac{2n}{\varrho} - 1 \right] u(\varrho) = 0 \quad .$$

Multiplying this equation by $\varrho^{k+1} u'_{nl}(\varrho)$ and $\varrho^k u$, respectively, and partially integrating, you can prove the claim.

6.4 Calculate the following expectation values for the hydrogen atom:

(a) $\langle nl | r^2 | nl \rangle$,

(b) $\langle nl | r | nl \rangle$,

(c) $\langle nl | r^{-1} | nl \rangle$,

(d) $\langle nl | r^{-2} | nl \rangle$,

(e) $\langle nl | r^{-3} | nl \rangle$,

(f) $\langle nlm | \delta^{(3)}(x) | nlm \rangle$.

Note concerning (a–e): Consider the recursion formula proven in the preceding exercise. You then need only compute one expectation value directly.

6.5 Determine the probability density of the momentum values for the ground state of the hydrogen atom.

6.6 Calculate the electrostatic potential of a charge distribution corresponding to the probability density of the hydrogen electron in the ground state. Be sure to take the nuclear potential into account, and use the Poisson equation.

6.7 The potential acting on an electron near the surface of liquid helium can be approximately described by the sum of an infinitely high potential threshold (repulsive action of the He-surface) and an attractive potential due to the mirror charge (problem of a point charge in front of a dielectric half-space)

$$
V(x) = \begin{cases} \infty & \text{for } x \leq 0 \\ -\dfrac{Ze^2}{x} & \text{for } x > 0 \end{cases} .
$$

Here, x is the coordinate of the electron perpendicular to the surface, and $Z = (\varepsilon - 1)/4(\varepsilon + 1)$, where for He $\varepsilon = 1.05723$.

(a) Determine in analogy to the H-atom the energy eigenstates and eigenvalues.

(b) Give the value of the ground state energy.

Hint: Separate variables with respect to the coordinate x and the coordinates parallel to the surface.

6.8 (a) Determine the quantum number n_0 corresponding to the motion of the earth about the sun.

(b) Determine the radial uncertainty.
Hint: Approximate the earth's orbit by a circular orbit.

6.9 (a) Show that the parity operator P is a Hermitian operator.

(b) Show that $P^2 = 1$ and $[H, P] = 0$ for a particle in a central potential.

(c) What are the eigenvalues of the parity operator for the eigenfunctions of the one-dimensional harmonic oscillator?

6.10 Express $\dfrac{-\hbar^2}{2m_1}\nabla_1^2 - \dfrac{\hbar^2}{2m_2}\nabla_2^2$ in terms of relative and center-of-mass coordinates.

6.11 Compute $(\Delta r)^2 (\Delta p_r)^2$ for the ground state of the H-atom, i.e., for $\psi_{1,0,0}$.

6.12 Compute $\langle z^2 \rangle$ and $\langle z^2 \rangle / \langle r \rangle^2$ for $\psi_{n,n-1,n-1}$.

6.13 The Hermitian vector operator $\boldsymbol{A}$ corresponding to the Lenz vector is

$$
\boldsymbol{A} = \frac{1}{2m}(\boldsymbol{p} \times \boldsymbol{L} - \boldsymbol{L} \times \boldsymbol{p}) - \frac{Ze^2}{r}\boldsymbol{r} .
$$

Show that $\boldsymbol{A}^\dagger = \boldsymbol{A}$ and that $\boldsymbol{A}$, as in classical theory, is a constant of motion and is normal to $\boldsymbol{L}$, i.e.,

$$
[\boldsymbol{A}, H] = 0 \quad , \qquad \boldsymbol{A} \cdot \boldsymbol{L} = \boldsymbol{L} \cdot \boldsymbol{A} = 0 \quad ,
$$

where H is the Hamiltonian of the hydrogen atom.

7. Motion in an Electromagnetic Field

7.1 The Hamiltonian

We now consider a particle of mass m and charge e in an electromagnetic field. The representation of the field by the vector potential $\boldsymbol{A}$ and the scalar potential Φ

$$\boldsymbol{E} = -\frac{1}{c}\frac{\partial \boldsymbol{A}}{\partial t} - \boldsymbol{\nabla}\Phi \quad ; \quad \boldsymbol{B} = \boldsymbol{\nabla} \times \boldsymbol{A} \tag{7.1}$$

and the classical Hamiltonian

$$H = \frac{1}{2m}\left(\boldsymbol{p} - \frac{e}{c}\boldsymbol{A}(\boldsymbol{x},t)\right)^2 + e\Phi(\boldsymbol{x},t) \tag{7.2}$$

is known from electrodynamics. By the correspondence principle (Sect. 2.5.1), the replacement of $\boldsymbol{p}$ by the momentum operator turns (7.2) into the Hamiltonian operator, and the time dependent Schrödinger equation takes the form

$$i\hbar\frac{\partial}{\partial t}\psi = \left[\frac{1}{2m}\left(\frac{\hbar}{i}\boldsymbol{\nabla} - \frac{e}{c}\boldsymbol{A}\right)^2 + e\Phi\right]\psi \quad . \tag{7.3}$$

Expanding the square on the right-hand side of the Schrödinger equation, one obtains for the mixed term

$$-\frac{\hbar e}{2imc}(\boldsymbol{\nabla}\cdot\boldsymbol{A} + \boldsymbol{A}\cdot\boldsymbol{\nabla})\psi = \frac{ie\hbar}{mc}\boldsymbol{A}\cdot\boldsymbol{\nabla}\psi \quad ,$$

where the Coulomb gauge condition[1] $\boldsymbol{\nabla}\cdot\boldsymbol{A} = 0$ has been imposed. The result in this gauge is

$$i\hbar\frac{\partial}{\partial t}\psi = -\frac{\hbar^2}{2m}\boldsymbol{\nabla}^2\psi + \frac{ie\hbar}{mc}\boldsymbol{A}\cdot\boldsymbol{\nabla}\psi + \frac{e^2}{2mc^2}\boldsymbol{A}^2\psi + e\Phi\psi \quad . \tag{7.3$'$}$$

---

[1] We recall that in the Coulomb gauge the wave equations take the form

$$\boldsymbol{\nabla}^2\Phi = -4\pi\varrho \quad \Box\boldsymbol{A} = \frac{4\pi}{c}\boldsymbol{j} - \frac{1}{c}\boldsymbol{\nabla}\frac{\partial}{\partial t}\Phi = \frac{4\pi}{c}\boldsymbol{j}_{\text{trans}} \quad ,$$

where

7.2 Constant Magnetic Field B

For a constant magnetic field B, the vector potential can be written as

$$A = -\tfrac{1}{2}[x \times B] \quad , \tag{7.4a}$$

since

$$(\text{curl } A)_i = \varepsilon_{ijk}\partial_j(-\tfrac{1}{2}\varepsilon_{kst}x_s B_t) = \varepsilon_{ijk}\varepsilon_{kjt}(-\tfrac{1}{2})B_t = B_i \quad . \tag{7.4b}$$

Let us now consider more carefully in this special case the terms in (7.3′) that depend on the vector potential. The second term in the Schrödinger equation (7.3′) has the form

$$\frac{i\hbar e}{mc}A \cdot \nabla\psi = \frac{i\hbar e}{mc}(-\tfrac{1}{2})(x \times B) \cdot \nabla\psi$$

$$= \frac{i\hbar e}{2mc}(x \times \nabla) \cdot B\psi = -\frac{e}{2mc}L \cdot B\psi \quad ,$$

where L is the orbital angular momentum operator. With (7.4), the third term in the Schrödinger equation (7.3′) can also be written in the form

$$\frac{e^2}{2mc^2}A^2\psi = \frac{e^2}{8mc^2}(x \times B)^2\psi = \frac{e^2}{8mc^2}(x^2 B^2 - (x \cdot B)^2)\psi$$

$$= \frac{e^2 B^2}{8mc^2}(x^2 + y^2)\psi \quad ,$$

where without loss of generality $B \parallel e_z$ has been imposed. The second term $(-e/2mc)L \cdot B$ contributes to *paramagnetism*, whereas the third term $e^2 B^2(x^2 + y^2)/8mc^2$ is responsible for *diamagnetism*.

We now compare the order of magnitude of these two terms for electrons in atoms:

$$\frac{(e^2/8mc^2)\langle x^2 + y^2\rangle B^2}{|-(e/2mc)\langle L_z\rangle B|} \simeq \frac{e_0}{4c}\frac{a^2 B^2}{\hbar B} = \frac{e_0^2}{4\hbar c}\frac{B}{e_0/a^2}$$

$$= 1.1 \times 10^{-10} B \text{ (in gauss)} \quad .$$

Here, $\langle x^2 + y^2\rangle \sim a^2$, involving the Bohr radius, and $\langle L_z\rangle \sim \hbar$ have been used.

$$\Box = \frac{1}{c^2}\frac{\partial^2}{\partial t^2} - \nabla^2 \quad \text{(D'Alembert operator)}$$

and

$$j_{\text{trans}} = \frac{1}{4\pi}\nabla \times \nabla \times \int d^3x' \frac{j}{|x - x'|} \quad .$$

As has been done up to now, all calculations are performed in the cgs system.

The order of magnitude is the fine-structure constant times the ratio of B to atomic electric field strengths. Experimentally, fields of about 10^5 G are achievable. Thus, the quadratic term in $\boldsymbol{A}$ is negligible whenever $\langle L_z \rangle \neq 0$. Therefore, under laboratory conditions, diamagnetic effects are smaller than paramagnetic effects for electrons bound in atoms. However, there do exist situations in which the diamagnetic and paramagnetic terms can be of comparable magnitude. This is the case for metal electrons, and in particular for free electrons a comparison of the susceptibilities for Landau diamagnetism and Pauli paramagnetism yields

$$\chi_{\text{Landau}} = -\tfrac{1}{3}\chi_{\text{Pauli}} \quad .$$

The diamagnetic term is also important in the treatment of matter under conditions such as those prevailing on the surfaces of neutron stars: there, fields up to 10^{12} gauss occur, which leads to a considerable change in the atomic structure. (See also the end of Sect. 14.1.)

Finally, we would like to compare the paramagnetic term with the Coulomb energy:

$$\frac{(-e/2mc)\langle L_z \rangle B}{e^2/a} \cong \frac{(-e/2mc)\hbar B}{e^2/a} = \frac{a^2 \alpha B}{2e_0}$$
$$= 2 \times 10^{-10} B \ (\text{in gauss}) \quad ,$$

i.e., the change in the energy levels due to laboratory fields is very small.

7.3 The Normal Zeeman Effect

The estimates of Sect. 7.2 suggest that we take as a Hamiltonian for the hydrogen atom in a constant magnetic field

$$H = H_0 - \frac{e}{2mc} B L_z \quad , \tag{7.5}$$

where the magnetic field is chosen parallel to the z-axis. Here, H_0 is the Coulomb Hamiltonian $[(-\hbar^2/2m)\boldsymbol{\nabla}^2 - e_0^2/r]$, whose eigenfunctions ψ_{nlm_l} we know.[2] How does the total Hamiltonian (7.5) act on ψ_{nlm_l}?

---

[2] To avoid confusion with the particle mass we use m_l here for the eigenvalue of $L_z/\hbar$ instead of m as in previous chapters. Because of (7.7) one calls m_l also the magnetic quantum number.

$$H\psi_{nlm_l} = \left(-\frac{\mathrm{Ry}}{n^2} - \frac{e\hbar B}{2mc}m_l\right)\psi_{nlm_l} \quad. \tag{7.6}$$

Thus, the Coulomb wave functions are also eigenfunctions of H with the energy eigenvalues

$$E_{nlm_l} = -\frac{\mathrm{Ry}}{n^2} + \hbar\omega_{\mathrm{L}}\,m_l \quad, \tag{7.7}$$

where

$$\omega_{\mathrm{L}} = -\frac{eB}{2mc} = \frac{e_0 B}{2mc} \tag{7.8}$$

is the *Larmor frequency*. The magnetic field removes the $(2l+1)$-fold degeneracy of the energy levels. By (7.7), a level with angular momentum quantum number l is split into $(2l+1)$ equidistant levels. The magnitude of the splitting is

$$\frac{e_0 \hbar B}{2mc} = 13.6\,\mathrm{eV} \times (4 \times 10^{-10}\, B\ \text{(in gauss)}) \quad,$$

and it is independent of l.

This splitting leads to additional transitions, which, however, are restricted by the selection rule $\Delta m_l = -1, 0, 1$. This equidistant splitting caused by the magnetic field is called the "normal" Zeeman effect (Fig. 7.1). Indeed, in the hydrogen atom the splitting is completely different, and in fact there is an even number of levels, as if the angular momentum were half integral. This will lead us in Chap. 9 to an additional kind of angular momentum, the spin.

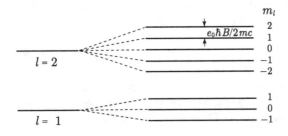

Fig. 7.1. Level splitting in the "normal" Zeeman effect

In many-electron atoms, L refers to the total angular momentum and l to the corresponding quantum number. For historical reasons, one speaks of the "normal" Zeeman effect if the splitting is determined by the orbital angular momentum alone and is given by (7.7). In fact, as in the hydrogen atom, the situation is usually more complicated and cannot be explained purely by the orbital angular momentum (the "anomalous" Zeeman effect). In Chap. 14 this will be discussed in more detail.

We note in passing that, according to the general definition of the magnetic moment $\boldsymbol{\mu} \equiv -\partial H/\partial \boldsymbol{B}$, it follows for the paramagnetic part that

$$\boldsymbol{\mu} = \frac{e}{2mc}\boldsymbol{L} = -\mu_{\mathrm{B}}\boldsymbol{L}/\hbar, \tag{7.9}$$

where

$$\mu_{\mathrm{B}} = \frac{e_0\hbar}{2mc} = 0.927 \times 10^{-20}\,\mathrm{erg/G} \tag{7.10}$$

is the Bohr magneton. The paramagnetic term leads to the following contribution to atomic magnetism:

$$|\langle\boldsymbol{\mu}\rangle| = \mu_{\mathrm{B}}|\langle\boldsymbol{L}\rangle|/\hbar \approx \mu_{\mathrm{B}} \quad .$$

If the paramagnetic contribution is zero, the diamagnetic part becomes noticeable:

$$\langle\boldsymbol{\mu}\rangle = -\frac{e^2\boldsymbol{B}}{4mc^2}\langle x^2 + y^2\rangle \approx -\frac{e^2\boldsymbol{B}}{6mc^2}a^2 \quad .$$

The paramagnetic part must be augmented by the spin contribution (Chap. 9).

7.4 Canonical and Kinetic Momentum, Gauge Transformation

7.4.1 Canonical and Kinetic Momentum

In Appendix B, the concept of canonical and kinetic momentum from classical mechanics is reviewed. The quantity $\boldsymbol{p}$ is the *canonical momentum*, whereas

$$m\dot{\boldsymbol{x}} = \boldsymbol{p} - \frac{e}{c}\boldsymbol{A} \tag{7.11}$$

is the *kinetic momentum*. The corresponding operators satisfy the following commutation relations:

$$[x_i, p_j] = i\hbar\delta_{ij} \quad , \quad [x_i, x_j] = [p_i, p_j] = 0 \quad , \tag{7.12a,b}$$

$$[x_i, m\dot{x}_j] = i\hbar\delta_{ij} \quad , \quad [m\dot{x}_i, m\dot{x}_j] = i\hbar\frac{e}{c}\varepsilon_{ijk}B_k \tag{7.13a,b}$$

with $-A_{i,j} + A_{j,i} = \varepsilon_{ijk}B_k$ and $A_{i,j} = \partial A_i/\partial x_j$.

The fact that the components of the kinetic momentum do not commute among themselves has important consequences for motion in a magnetic field.

7.4.2 Change of the Wave Function
Under a Gauge Transformation

The Lorentz force (B.3) depends only on $\boldsymbol{B}$ (see Appendix B), whereas the Schrödinger equation (7.3) contains the vector potential $\boldsymbol{A}$. The question thus immediately arises as to whether the wave function ψ depends on the gauge and whether charged particles react to $\boldsymbol{B}$ or $\boldsymbol{A}$ (possibly only to its gauge invariant part).

To answer this question, we first study the influence of a *gauge transformation*

$$\boldsymbol{A} \to \boldsymbol{A}' = \boldsymbol{A} + \boldsymbol{\nabla}\Lambda \qquad \Phi \to \Phi' = \Phi - \frac{1}{c}\frac{\partial}{\partial t}\Lambda \quad , \tag{7.14}$$

where $\Lambda(\boldsymbol{x},t)$ is a scalar function. The Schrödinger equation in the first gauge takes the form

$$\left[\frac{1}{2m}\left(\frac{\hbar}{i}\boldsymbol{\nabla} - \frac{e}{c}\boldsymbol{A}(\boldsymbol{x},t)\right)^2 + e\Phi(\boldsymbol{x},t)\right]\psi(\boldsymbol{x},t) = i\hbar\frac{\partial}{\partial t}\psi(\boldsymbol{x},t) \quad . \tag{7.15}$$

We would like to show now that the wave function $\psi'(\boldsymbol{x},t)$ in the second gauge, characterized by the primed potential, is

$$\psi'(\boldsymbol{x},t) = \exp\left\{\frac{ie}{\hbar c}\Lambda(\boldsymbol{x},t)\right\}\psi(\boldsymbol{x},t) \quad . \tag{7.16}$$

Multiplication of the Schrödinger equation (7.15) from the left by the factor $\exp\left\{(ie/\hbar c)\Lambda(\boldsymbol{x},t)\right\}$ and using the identity

$$e^{f(y)}\frac{\partial}{\partial y} = \left(\frac{\partial}{\partial y} - \frac{\partial f}{\partial y}\right)e^{f(y)} \tag{7.17}$$

twice, leads to

$$\left[\frac{1}{2m}\left(\frac{\hbar}{i}\boldsymbol{\nabla} - \frac{e}{c}\boldsymbol{A} - \frac{\hbar}{i}\frac{ie}{\hbar c}\boldsymbol{\nabla}\Lambda\right)^2 + e\Phi\right]\exp\left\{\frac{ie\Lambda}{\hbar c}\right\}\psi$$

$$= i\hbar\left(\frac{\partial}{\partial t} - \frac{ie}{\hbar c}\frac{\partial\Lambda}{\partial t}\right)\exp\left\{\frac{ie\Lambda}{\hbar c}\right\}\psi \quad .$$

This is evidently identical to

$$\left[\frac{1}{2m}\left(\frac{\hbar}{i}\boldsymbol{\nabla} - \frac{e}{c}\boldsymbol{A}'\right)^2 + e\Phi'\right]\psi' = i\hbar\frac{\partial}{\partial t}\psi' \quad , \tag{7.18}$$

the Schrödinger equation with the primed potentials. The gauge transformation introduces an additional space and time dependent phase factor into the wave function. However, the change in gauge has no observable physical consequences, since $|\psi|^2$ does not change.[3] Matrix elements of $\boldsymbol{x}$ and $m\dot{\boldsymbol{x}}$ and functions thereof remain unchanged.

---

[3] However, the energy eigenvalues of extended stationary states may depend on $\boldsymbol{A}$. This can be seen for free particles in the presence of a uniform vector potential

7.5 The Aharonov–Bohm Effect

7.5.1 The Wave Function in a Region Free of Magnetic Fields

We now consider the motion of an electron in the presence of a time independent magnetic field $\boldsymbol{B}(\boldsymbol{x})$. Let this field vanish within some region of space

$$\boldsymbol{B} = \text{curl } \boldsymbol{A} = 0 \quad , \tag{7.19}$$

as is the case outside of an infinitely long coil (Fig. 7.2). In this region, $\boldsymbol{A}$ can be expressed as the gradient of a scalar field Λ :

$$\boldsymbol{A} = \boldsymbol{\nabla}\Lambda \quad , \tag{7.20a}$$

$$\Lambda(\boldsymbol{x}) = \int_{\boldsymbol{x}_0}^{\boldsymbol{x}} d\boldsymbol{s} \cdot \boldsymbol{A}(\boldsymbol{s}) \quad . \tag{7.20b}$$

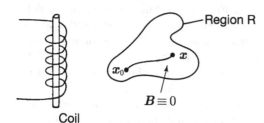

Fig. 7.2. The field of an infinitely long coil is zero on the exterior; see (7.19) and (7.20)

Here, $\boldsymbol{x}_0$ is an arbitrary initial point in the field-free region. The wave function in this region can be found either from

$$\frac{1}{2m}\left(\frac{\hbar}{\mathrm{i}}\boldsymbol{\nabla} - \frac{e}{c}\boldsymbol{A}\right)^2 \psi + V\psi = \mathrm{i}\hbar\frac{\partial}{\partial t}\psi \quad , \tag{7.21}$$

or from the gauge transformed equation in which the vector potential

$$\boldsymbol{A}' = \boldsymbol{A} + \boldsymbol{\nabla}(-\Lambda) = 0$$

does not appear, that is

$$\frac{1}{2m}\left(\frac{\hbar}{\mathrm{i}}\boldsymbol{\nabla}\right)^2 \psi' + V\psi' = \mathrm{i}\hbar\frac{\partial}{\partial t}\psi' \quad . \tag{7.22}$$

$\boldsymbol{A}$ obeying periodic boundary conditions – a problem equivalent to Problems 7.4 and 7.5. Wave packets constructed from such stationary states with and without $\boldsymbol{A}$ are related by a combined gauge and Galilei transformation (Problem 16.8).

If we replace Λ in (7.16) by $-\Lambda$, we obtain the following connection between these two wave functions:

$$\psi = \psi' \, \exp\left\{\frac{ie}{\hbar c}\Lambda\right\} = \psi' \, \exp\left\{\frac{ie}{\hbar c}\int_{x_0}^{x} ds \cdot A(s)\right\} , \qquad (7.23)$$

where ψ' is the wave function in the potential V with field $B \equiv 0$ (identically zero) in all space. (Since no electric field acts, we have set $\Phi = 0$. The potential $V(x)$ is intended to represent influences which are not electric in origin.) At this point we recall from electrodynamics that $\int_{x_0}^{x} ds \cdot A(s)$ of Eq. (7.20b) depends only on the end points but not on the paths, as long as the loop formed by a pair of different paths does not enclose a magnetic flux.

7.5.2 The Aharonov–Bohm Interference Experiment

We would now like to investigate whether an electron moving only in regions of nonvanishing $A(x)$ but vanishing $B(x)$ feels anything of the existence of the magnetic field in the inaccessible region. To this end, we now consider the interference experiment shown in Fig. 7.3 in which the magnetic field is restricted to the interior of the "infinitely" long coil, perpendicular to the plane of Fig. 7.3 and depicted by a circle. It is assumed that the coil is screened by a wall of the double slit device and thus the electrons are restricted to the region $B = 0$. The lack of penetration of ψ into the field region can be represented formally by an infinitely high potential barrier.

In order to find the solution as a function of the field, we determine first the solutions with only one slit open at a time and then linearly superimpose them. Let $\psi_{1,B}(x)$ be the wave function when only slit 1 is open. By (7.23), it can be obtained from the field-free wave function $\psi_{1,0}$ and becomes

$$\psi_{1,B}(x) = \psi_{1,0}(x) \, \exp\left\{\frac{ie}{\hbar c}\int_{1} ds \cdot A(s)\right\} , \qquad (7.24a)$$

where the line integral runs from the source through slit 1 to x. Similarly, for the wave function when only slit 2 is open, we have

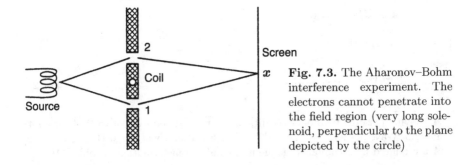

Fig. 7.3. The Aharonov–Bohm interference experiment. The electrons cannot penetrate into the field region (very long solenoid, perpendicular to the plane depicted by the circle)

$$\psi_{2,B}(\boldsymbol{x}) = \psi_{2,0}(\boldsymbol{x}) \exp\left\{\frac{ie}{\hbar c}\int_2 d\boldsymbol{s}\cdot\boldsymbol{A}(\boldsymbol{s})\right\} \quad, \tag{7.24b}$$

where now the line integral runs from the source to $\boldsymbol{x}$ through slit 2. If both slits are open, we superimpose (7.24a) and (7.24b) to obtain

$$\psi_B(\boldsymbol{x}) = \psi_{1,0}(\boldsymbol{x}) \exp\left\{\frac{ie}{\hbar c}\int_1 d\boldsymbol{s}\cdot\boldsymbol{A}(\boldsymbol{s})\right\} + \psi_{2,0}(\boldsymbol{x}) \exp\left\{\frac{ie}{\hbar c}\int_2 d\boldsymbol{s}\cdot\boldsymbol{A}(\boldsymbol{s})\right\} \quad .$$

The relative phase of the two terms, caused by the magnetic field, is related to

$$\int_1 d\boldsymbol{s}\cdot\boldsymbol{A}(\boldsymbol{s}) - \int_2 d\boldsymbol{s}\cdot\boldsymbol{A}(\boldsymbol{s}) = \oint d\boldsymbol{s}\cdot\boldsymbol{A}(\boldsymbol{s}) = \int d\boldsymbol{a}\cdot\operatorname{curl}\boldsymbol{A} = \Phi_B \quad ,$$

where $d\boldsymbol{a}$ is the area element and Φ_B is the magnetic flux. Thus, we have

$$\psi_B(\boldsymbol{x}) = \left(\psi_{1,0}(\boldsymbol{x})\exp\left\{\frac{ie}{\hbar c}\Phi_B\right\} + \psi_{2,0}(\boldsymbol{x})\right)\exp\left\{\frac{ie}{\hbar c}\int_2 d\boldsymbol{s}\cdot\boldsymbol{A}(\boldsymbol{s})\right\}. \tag{7.25}$$

The phase relation between ψ_1 and ψ_2 changes under a change in the enclosed magnetic flux Φ_B, and thus the interference pattern is also shifted.

Remark: If we consider in detail the cylindrical waves leaving the slits in (7.25),

$$\psi_{1,0} = \frac{e^{ikr_1}}{\sqrt{r_1}} \quad , \quad r_1 = |\boldsymbol{x} - \boldsymbol{x}_{\text{slit1}}| \quad ,$$

$$\psi_{2,0} = \frac{e^{ikr_2}}{\sqrt{r_2}} \quad , \quad r_2 = |\boldsymbol{x} - \boldsymbol{x}_{\text{slit2}}| \quad ,$$

the condition for constructive interference in the presence of a magnetic field is

$$kr_1 + \frac{e}{\hbar c}\Phi_B - kr_2 = 2\pi n \quad ,$$

where n is an integer, and from this

$$r_1 - r_2 = \frac{\lambda}{2\pi}\left(2\pi n - \frac{e\Phi_B}{\hbar c}\right) \tag{7.26}$$

follows.

The positions of the interference maxima are shifted due to the variation in Φ_B, although the electron does not penetrate into the region of finite magnetic field. This is known as the *Aharonov–Bohm effect*[4]. The unit of flux entering here is

$$2\Phi_0 = \frac{2\pi\hbar c}{e_0} = 4.135 \times 10^{-7}\,\mathrm{G\,cm^2} \quad . \tag{7.27}$$

---

[4] Y. Aharonov, D. Bohm: Phys. Rev. **115**, 485 (1959)

The first experimental demonstration using a magnetic one-domain iron whisker was given by Chambers[5].

A related phenomenon (Fig. 7.4) can be found in a SQUID (*superconducting quantum interference device*)[6]. The maximum current that can pass through two Josephson junctions in parallel,

$$I_{\max} = 2I_0 |\cos \frac{\pi \Phi_B}{\Phi_0}| \quad , \tag{7.28}$$

shows the interference pattern of Fig. 7.5. This results if one maximizes the sum of the Josephson currents

$$I = I_0 (\sin \gamma_A + \sin \gamma_B)$$

subject to the constraint[6]

$$\gamma_A - \gamma_B = \frac{2\pi \Phi_B}{\Phi_0} \quad .$$

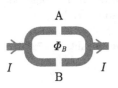

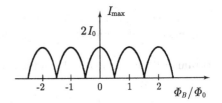

Fig. 7.4. The SQUID: Two Josephson contacts, A and B, connected in parallel with enclosed flux Φ_B

Fig. 7.5. Maximal current as a function of flux

The Aharonov–Bohm interference oscillations were also observed recently in normally conducting metal rings.

Summarizing: Classically, $\boldsymbol{E}$ and $\boldsymbol{B}$ are the physically relevant quantities, since they determine the Lorentz force. In regions where $\boldsymbol{E} = \boldsymbol{B} = 0$ the particle feels no force. The potentials $\boldsymbol{A}$ and Φ serve in classical physics only as auxiliary quantities. In quantum mechanics, $\boldsymbol{A}(\boldsymbol{x})$ is the fundamental physical field; however, the wave function always has the property that physical quantities and effects depend only on *gauge invariant* quantities.

---

[5] R.C. Chambers: Phys. Rev. Lett. **5**, 3 (1960); see also H. Börsch, H. Hamisch, K. Grohmann, D. Wohlleben: Z. Phys. **165**, 79 (1961)

[6] J.E. Zimmerman, J.E. Mercereau: Phys. Rev. Lett. **13**, 125 (1964); R.C. Jaklevic, J.E. Lambe, J.E. Mercereau, E.H. Silver: Phys. Rev. **140**, A 1628 (1965); M. Tinkham: *Introduction to Superconductivity* (McGraw-Hill, New York 1975) p. 202

7.6 Flux Quantization in Superconductors

Many metals and, as discovered recently, many oxidic semiconductors become superconducting below a critical temperature T_c characteristic for each particular substance. The electrons form Cooper pairs. We consider a type-I superconductor in the form of a hollow cylinder in an external magnetic field taken parallel to the axis of the cylinder. It is found experimentally (the Meissner effect) that the magnetic field is expelled from the superconductor and thus vanishes within it (except for a thin boundary layer). The doubly charged Cooper pairs thus move in a field-free region, and therefore the wave function (7.23) can be used to describe them. If the wave function of the Cooper pairs in the absence of a field is given by $\psi_0(\boldsymbol{x})$, then in the presence of a field it becomes according to (7.23)

$$\psi_B(\boldsymbol{x}) = \exp\left\{\frac{i2e}{\hbar c}\int_{\boldsymbol{x}_0}^{\boldsymbol{x}} d\boldsymbol{s}\cdot\boldsymbol{A}(\boldsymbol{s})\right\}\psi_0(\boldsymbol{x}) \quad . \tag{7.23'}$$

The vector potential in (7.23') has the property that within the superconductor curl $\boldsymbol{A} = 0$ (i.e., for any curve within the superconductor which can be shrunk to a point, $\oint d\boldsymbol{s}\cdot\boldsymbol{A}(\boldsymbol{s}) = 0$), whereas $\Phi_B = \int d\boldsymbol{a}\cdot\text{curl}\,\boldsymbol{A} = \oint d\boldsymbol{s}\cdot\boldsymbol{A}(\boldsymbol{s})$ gives the magnetic flux through the hollow cylinder (i.e., for curves encircling the cavity, $\oint d\boldsymbol{s}\cdot\boldsymbol{A}(\boldsymbol{s}) = \Phi_B$). A closed path about the cylinder starting at the point $\boldsymbol{x}_0$ (Fig. 7.6) gives

$$\psi_B(\boldsymbol{x}_0) = \psi_0(\boldsymbol{x}_0) = \exp\left\{\frac{i2e}{\hbar c}\oint d\boldsymbol{s}\cdot\boldsymbol{A}(\boldsymbol{s})\right\}\psi_0(\boldsymbol{x}_0) \quad .$$

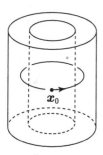

Fig. 7.6. Flux quantization

The requirement that the wave function $\psi_B(\boldsymbol{x})$ be single valued implies the quantization of the enclosed flux:

$$\Phi_B = \Phi_0 n \quad , \quad n = 0, \pm 1, \ldots \quad ,$$

$$\Phi_0 = \frac{\hbar c\pi}{e_0} = 2.07 \times 10^{-7}\,\text{G cm}^2 \text{ (the flux quantum)} \quad .$$

This quantization has also been observed experimentally[7]. The occurrence of twice the electronic charge in the quantization represents an important test of the existence of Cooper pairs, which are the basis of BCS (Bardeen–Cooper–Schrieffer) theory.

7.7 Free Electrons in a Magnetic Field

We now investigate free electrons in a magnetic field oriented in the x_3-direction. The vector potential (7.4a) has only components perpendicular to $\boldsymbol{B}$, so that the p_3-contribution to the kinetic energy is the same as that of free particles, and the Hamiltonian is given by

$$H = H_\perp + \frac{p_3^2}{2m} \quad . \tag{7.29a}$$

Expressed in terms of the components of the kinetic momentum (7.11), the transverse part of the Hamiltonian takes the form

$$H_\perp = \frac{m}{2}(\dot{x}_1^2 + \dot{x}_2^2) \quad . \tag{7.29b}$$

The second term in (7.29a) is diagonalized by $\exp\{ip_3 x_3/\hbar\}$, corresponding to free motion in the x_3-direction, which can be separated off, since p_3 commutes with the $\dot{x}_i$. We now turn to the transverse part, which contains the magnetic effects. For electrons, $e = -e_0$, and the commutation relations

$$[m\dot{x}_1, m\dot{x}_2] = i\hbar\frac{eB}{c} \quad , \quad [\dot{x}_1, \dot{x}_1] = [\dot{x}_2, \dot{x}_2] = 0 \quad , \tag{7.30}$$

suggest the introduction of

$$\pi_i = \frac{m\dot{x}_i}{\sqrt{e_0 B/c}} \quad . \tag{7.31}$$

Now, these operators satisfy the commutation relations

$$[\pi_2, \pi_1] = i\hbar \quad , \quad [\pi_1, \pi_1] = [\pi_2, \pi_2] = 0 \quad , \tag{7.32}$$

and, in analogy to position and momentum, they represent canonical variables with the Hamiltonian

$$H_\perp = \frac{1}{2}\frac{e_0 B}{c m}(\pi_1^2 + \pi_2^2) \quad . \tag{7.33}$$

According to the theory of the harmonic oscillator (Sect. 3.1) using

$$a = \frac{\pi_2 + i\pi_1}{\sqrt{2\hbar}} \quad , \tag{7.34}$$

---

[7] R. Doll, M. Näbauer: Phys. Rev. Lett. **7**, 51 (1961); B.S. Deaver, Jr., W.M. Fairbank: Phys. Rev. Lett. **7**, 43 (1961)

this can be brought into the standard form

$$H_\perp = \hbar\omega_c(a^\dagger a + \tfrac{1}{2}) \quad , \tag{7.35}$$

where

$$\omega_c = \frac{e_0 B}{mc} \tag{7.36}$$

is the cyclotron frequency. Consequently, the energy eigenvalues of (7.29b) are

$$E_n = \hbar\omega_c(n + \tfrac{1}{2}) \quad , \tag{7.37}$$

with $n = 0, 1, \ldots$. We have thus found the energy levels for free electrons in a homogeneous magnetic field – also known as *Landau levels*. These play an important role in solid state physics. The problem is not yet completely solved, since for example we have not yet determined the degeneracy and the wave function of our particles. Formally it is clear that, beginning with the four canonical operators x_1, x_2, p_1, p_2, we need two more operators, in addition to π_1, π_2 introduced above, for a complete characterization. In Sect. 8.6 we will show in the Heisenberg representation that these are given by

$$X = x - \frac{1}{\omega_c}\tau\dot{x} \quad ,$$

where

$$\tau = \begin{pmatrix} 0 & 1 \\ -1 & 0 \end{pmatrix} \quad .$$

In classical mechanics, X is the center of the circular orbits $(x - X)^2 = \dot{x}^2/\omega_c^2 = \text{const}$. In quantum mechanics, X_1 and X_2 are canonical variables and cannot simultaneously be specified with arbitrary accuracy. X is also referred to as the "guiding center".

Problems

7.1 Let the Hamiltonian be

$$H = \frac{1}{2m}\left(p - \frac{e}{c}A(x,t)\right)^2 + e\Phi(x,t) \quad .$$

Prove the continuity equation $(\partial/\partial t)\,\psi^*\psi + \boldsymbol{\nabla} \cdot \boldsymbol{j} = 0$, with

$$\boldsymbol{j} \equiv \frac{\hbar}{2mi}\left[\psi^*\boldsymbol{\nabla}\psi - (\boldsymbol{\nabla}\psi^*)\psi - \frac{2ie}{\hbar c}A(x,t)\psi^*\psi\right]$$

$$\equiv \frac{1}{2m}\left(\psi^*\left(\frac{\hbar}{i}\boldsymbol{\nabla} - \frac{e}{c}A(x,t)\right)\psi + \text{c.c.}\right) \quad .$$

7.2 Determine the energy eigenfunctions and eigenvalues for a charged particle in a homogeneous, constant magnetic field oriented along the z-axis.

Hint: Use the Coulomb gauge $\boldsymbol{A} = -\frac{1}{2}\boldsymbol{x} \times \boldsymbol{B}$. Transform to cylindrical coordinates; for the wave function, use the separation ansatz $\psi(\boldsymbol{x}) = R_m(\varrho)\, e^{im\varphi}\, e^{ikz}$. Justify and use the ansatz $R(\varrho) = \varrho^{|m|}\, e^{-c\varrho^2}\, w(\varrho)$ with $c > 0$, an appropriate constant. The substitution $y = \varrho^2$ in the differential equation for $w(\varrho)$ leads to the Laguerre differential equation.

Compare the result with Sect. 8.6.

7.3 Consider a particle with mass m and charge e in a homogeneous electromagnetic field $\boldsymbol{B} = (0,0,B)$, $\boldsymbol{E} = (E,0,0)$ with $|E| < |B|$. Take the gauge $\boldsymbol{A} = (0, Bx, 0)$. Determine the eigenfunctions and eigenvalues for the Hamiltonian

$$H = \frac{1}{2m}\left(\boldsymbol{p} - \frac{e}{c}\boldsymbol{A}\right)^2 - eEx \quad .$$

In the case $E = 0$, discuss also the degeneracy of the energy levels.

7.4 Consider a plane rotator of radius a described in cylindrical coordinates (r, φ, z) by the Hamiltonian

$$H_0 = \frac{p_0^2}{2m}$$

with

$$p_0 = \frac{\hbar}{i}\frac{1}{a}\frac{\partial}{\partial\varphi} \quad .$$

A plane rotator is a particle of mass m moving on a circumference of radius a.

(a) Determine the energy eigenfunctions and eigenvalues, imposing periodic boundary conditions.

(b) Calculate the wave functions and energy levels for the plane rotator in the presence of a vector potential

$$\boldsymbol{A} = \boldsymbol{e}_\varphi \begin{cases} \dfrac{B}{2}r & \text{for } r \leq r_0 \\[2mm] \dfrac{(B/2)r_0^2}{r} & \text{for } r \geq r_0 \end{cases} \quad ,$$

where $r_0 < a$.

(c) Now consider the vector potential

$$\boldsymbol{A}' = \boldsymbol{A} + \nabla\chi \quad .$$

For the particular choice

$$\chi = -\frac{Br_0^2}{2}\varphi \quad ,$$

calculate the vector potential $\boldsymbol{A}'$ and the magnetic field $\boldsymbol{B}'$. Pay careful attention to the behavior of $\boldsymbol{B}'$ at the origin. Does the transformation $\boldsymbol{A} \to \boldsymbol{A}'$ represent a gauge transformation?

(d) Find the wave functions and the energy eigenvalues for A .

(e) Compare the results of (b) and (d) and explain why the Aharonov–Bohm effect is not contradicted by the transformation from A to A'.

7.5 Consider again the plane rotator

$$H_0 = \frac{p_0^2}{2m} \quad , \quad p_0 = \frac{\hbar}{ia} \frac{\partial}{\partial \varphi} \quad .$$

Start from the stationary states obeying periodic boundary conditions. Investigate the transformation

$$p_0 \;\to\; p_s = e^{-is(\varphi)/\hbar} \, p_0 \, e^{is(\varphi)/\hbar} \quad , \quad \text{etc.}$$

and calculate p_s, H_s, and the new wave functions. Show that the latter in general no longer obey periodic boundary conditions.

8. Operators, Matrices, State Vectors

8.1 Matrices, Vectors, and Unitary Transformations

In the following sections, we write all relations, to the extent that a specification is necessary at all, for one spatial dimension. All formulae can of course be rewritten in three dimensions with the replacements $x \to \mathbf{x}$ and $dx \to d^3x$.

Let a complete orthonormal set of functions $\{\psi_n(x)\}$ and an arbitrary operator A be given. Then the matrix representation of A in the basis $\{\psi_n(x)\}$ is defined as

$$A_{nm} = (\psi_n, A\psi_m) \quad . \tag{8.1}$$

The matrix A_{nm} may be finite or, more generally, infinite dimensional, depending on the number of basis functions $\psi_n(x)$. We now list a few properties of the matrix A_{nm}.

(i)   If A is Hermitian, $A^\dagger = A$, then A_{nm} is a Hermitian matrix, i.e.,

$$A^*_{nm} = A_{mn} \quad . \tag{8.2}$$

   Proof:

$$A^*_{nm} = (\psi_n, A\psi_m)^* = (A\psi_m, \psi_n) = (\psi_m, A\psi_n) = A_{mn} \quad .$$

(ii)   If $\{\psi_n(x)\}$ is a basis of A, then

$$A_{nm} = a_n \delta_{nm} \tag{8.3}$$

   holds, where the a_n are the eigenvalues of A corresponding to $\psi_n(x)$, i.e.,

$$A\psi_n(x) = a_n \psi_n(x) \quad .$$

Since $\{\psi_n(x)\}$ is a complete, orthonormal system, any arbitrary state $\psi(x)$ can be represented as

$$\psi(x) = \sum_n c_n \psi_n(x) \tag{8.4a}$$

with

$$c_n = (\psi_n, \psi) \quad . \tag{8.4b}$$

We now consider a second complete, orthonormal system $\{\psi'_n(x)\}$. One can also represent the operator A and the state $\psi(x)$ in this new basis system:

$$A'_{nm} = (\psi'_n, A\psi'_m) \quad , \tag{8.2'}$$

$$\psi(x) = \sum_n c'_n \psi'_n(x) \tag{8.3'}$$

with

$$c'_n = (\psi'_n, \psi) \quad . \tag{8.4'}$$

Our goal is now to find the relationship between these two representations. For this, we first remark that the $\psi'_n(x)$ can be expanded in terms of the system of functions $\{\psi_n(x)\}$ according to

$$\psi'_n(x) = \sum_m S_{mn} \psi_m(x) \tag{8.5a}$$

with

$$S_{mn} = (\psi_m, \psi'_n) = \int dx\, \psi_m^*(x)\psi'_n(x) \quad . \tag{8.5b}$$

(iii) The transformation matrix S_{nm} is unitary, i.e.,

$$SS^\dagger = S^\dagger S = \mathbb{1} \quad , \tag{8.6}$$

($\mathbb{1}$ = unit matrix) or equivalently

$$\sum_n S_{mn} S_{m'n}^* = \sum_n S_{nm}^* S_{nm'} = \delta_{mm'} \quad . \tag{8.6'}$$

Proof:

$$\sum_n S_{mn} S_{m'n}^* = \sum_n \int dx \int dy\, \psi_m^*(x)\psi'_n(x)\psi_{m'}(y)\psi_n^{'*}(y)$$

$$= \int dx \int dy\, \delta(x-y)\psi_{m'}(y)\psi_m^*(x) = \delta_{mm'} \quad ,$$

where the completeness relation for the $\{\psi'_n(x)\}$ and orthonormality of the $\{\psi_n(x)\}$ have been used. Analogously, one can show that $\sum_n S_{nm}^* S_{nm'} = \delta_{mm'}$ by utilizing the completeness relation for $\{\psi_n(x)\}$ and the orthonormality of the $\{\psi'_n(x)\}$.

Thus, the following transformation laws result:

(iv) $$c'_n = \sum_m (S^\dagger)_{nm} c_m \quad , \tag{8.7a}$$

and, since S is unitary,

$$c_m = \sum_n S_{mn} c'_n \quad .$$ (8.7b)

In matrix notation, these equations take the form

$$\begin{pmatrix} c'_1 \\ c'_2 \\ \vdots \end{pmatrix} = S^\dagger \begin{pmatrix} c_1 \\ c_2 \\ \vdots \end{pmatrix} \quad ; \quad \begin{pmatrix} c_1 \\ c_2 \\ \vdots \end{pmatrix} = S \begin{pmatrix} c'_1 \\ c'_2 \\ \vdots \end{pmatrix} \quad .$$

Proof:

$$c'_n = (\psi'_n, \psi) = \sum_m (S_{mn}\psi_m, \psi) = \sum_m S^*_{mn}(\psi_m, \psi)$$

$$= \sum_m S^*_{mn} c_m = \sum_m (S^\dagger)_{nm} c_m \quad .$$

For the matrix representation of the operators, we have

(v) $$A'_{nm} = \sum_{l,k} S^*_{ln} A_{lk} S_{km}$$ (8.8a)

or in matrix notation

$$A' = S^\dagger A S \quad .$$ (8.8b)

Proof:

$$A'_{nm} = (\psi'_n, A\psi'_m) = \sum_{l,k} S^*_{ln}(\psi_l, A\psi_k) S_{km} = \sum_{l,k} S^*_{ln} A_{lk} S_{km}$$

$$= \sum_{l,k} (S^\dagger)_{nl} A_{lk} S_{km} \quad ,$$

where (8.5a) has been used.

Summarizing: Operators can be represented by matrices and states by vectors. The representations in different basis systems are related by unitary transformations.

To illustrate this, we give three examples:

1. *Energy eigenfunctions of the harmonic oscillator.* In Sect. 3.1, we found for the one-dimensional harmonic oscillator the complete set of energy eigenfunctions

$$\psi_n(x) = (2^n n! \sqrt{\pi} x_0)^{-1/2} \exp\left\{ -\frac{1}{2}\left(\frac{x}{x_0}\right)^2 \right\} H_n(x/x_0)$$ (8.9)

with $(\psi_n, \psi_{n'}) = \delta_{nn'}$. In this basis, the position operator x (for example) takes the form (see (3.5a))

$$x_{nm} = \frac{x_0}{\sqrt{2}} \left\{ \sqrt{n}\, \delta_{n,m+1} + \sqrt{n+1}\, \delta_{n,m-1} \right\} \quad .$$ (8.10)

2. *Momentum eigenfunctions.* The eigenfunctions ψ_p of the momentum operator $(\hbar/\mathrm{i})\,\partial/\partial x$ are

$$\psi_p = \frac{1}{\sqrt{2\pi\hbar}}\mathrm{e}^{\mathrm{i}px/\hbar} \quad , \tag{8.11}$$

$$\frac{\hbar}{\mathrm{i}}\frac{\partial}{\partial x}\psi_p = p\psi_p \quad .$$

The eigenvalue spectrum of the momentum operator is continuous, and the orthonormality relation in this case is

$$(\psi_p, \psi_{p'}) = \int \frac{dx}{2\pi\hbar}\mathrm{e}^{\mathrm{i}(p'-p)x/\hbar} = \delta(p-p') \quad .$$

3. *Position eigenfunctions.* In Sect. 2.9, we found the position eigenfunctions

$$\psi_\xi(x) = \delta(x - \xi) \tag{8.12a}$$

which satisfy

$$x\psi_\xi = \xi\psi_\xi \quad . \tag{8.12b}$$

The spectrum is once again continuous, and one has

$$(\psi_\xi, \psi_{\xi'}) = \int dx\,\delta(x-\xi)\delta(x-\xi') = \delta(\xi - \xi') \quad .$$

Remark:

As always, the index on the wave functions indicates the eigenvalue as well as the operator to which it belongs.

We now calculate the matrix representation of a few important operators with respect to the momentum and position eigenfunctions, using the following notation:

$$A_{pp'} = (\psi_p, A\psi_{p'}) \quad , \quad A_{\xi\xi'} = (\psi_\xi, A\psi_{\xi'}) \quad . \tag{8.13}$$

With this, one obtains

$$x_{\xi\xi'} = \int dx\,\delta(x-\xi)x\delta(x-\xi') = \xi\delta(\xi - \xi') \quad , \tag{8.14a}$$

$$p_{\xi\xi'} = \int dx\,\delta(x-\xi)\frac{\hbar}{\mathrm{i}}\frac{\partial}{\partial x}\delta(x-\xi') = \frac{\hbar}{\mathrm{i}}\frac{\partial}{\partial \xi}\delta(\xi - \xi') \quad , \tag{8.14b}$$

$$\begin{aligned}
x_{pp'} &= \int dx\frac{\mathrm{e}^{-\mathrm{i}px/\hbar}}{\sqrt{2\pi\hbar}}x\frac{\mathrm{e}^{\mathrm{i}p'x/\hbar}}{\sqrt{2\pi\hbar}} \\
&= -\frac{\hbar}{\mathrm{i}}\frac{\partial}{\partial p}\int dx\frac{\mathrm{e}^{\mathrm{i}(p'-p)x/\hbar}}{2\pi\hbar} = -\frac{\hbar}{\mathrm{i}}\frac{\partial}{\partial p}\delta(p-p') \quad ,
\end{aligned} \tag{8.14c}$$

$$p_{pp'} = \int dx \frac{e^{-ipx/\hbar}}{\sqrt{2\pi\hbar}} \frac{\hbar}{i} \frac{\partial}{\partial x} \frac{e^{ip'x/\hbar}}{\sqrt{2\pi\hbar}} = p'\delta(p-p') = p\delta(p-p') \quad , \qquad (8.14d)$$

$$(p^2)_{\xi\xi'} = -\hbar^2 \frac{\partial^2}{\partial\xi^2} \delta(\xi-\xi') \quad , \qquad (8.14e)$$

$$V(x)_{\xi\xi'} = V(\xi)\delta(\xi-\xi') \quad . \qquad (8.14f)$$

Functions of position are diagonal in the coordinate representation, functions of momentum in the momentum representation. Note the similarity between (8.14a) and (8.14d) as well as between (8.14b) and (8.14c).

To conclude this section, we discuss the expansion of an arbitrary state $\psi(x)$ in momentum and position eigenfunctions.

(i)   Expansion in position eigenfunctions

$$\psi(x) = \int d\xi\, c_\xi \psi_\xi(x) = \int d\xi\, \psi(\xi)\psi_\xi(x) \quad . \qquad (8.15a)$$

We know from (2.100) that the expansion coefficient here is precisely the wave function

$$c_\xi = (\psi_\xi, \psi) = \psi(\xi) \quad . \qquad (8.15b)$$

(ii)   Expansion in momentum eigenfunctions

$$\psi(x) = \int dp\, c_p \psi_p(x) \quad . \qquad (8.16a)$$

Comparison with (2.5),

$$\psi(x) = \int \frac{dp}{2\pi\hbar} \varphi(p) e^{ipx/\hbar} \quad ,$$

shows that

$$c_p = (\psi_p, \psi) = \frac{1}{\sqrt{2\pi\hbar}} \varphi(p) \quad . \qquad (8.16b)$$

Up to the factor $(2\pi\hbar)^{-1/2}$, c_p is the Fourier transform $\varphi(p)$ of the wave function. The probability of finding the value p in a momentum measurement is given by $|c_p|^2$.

As emphasized in Sect. 2.9, in our axiomatic system we can dispense with the intermediate hypothesis (which was convenient in the didactic development) that $|\psi(x)|^2$ and $|\varphi(p)|^2$ are respectively the probability density for position and momentum. This is a consequence of the form of the position and momentum operators and the general results for the probability densities of observables in Sect. 2.9.

8.2 State Vectors and Dirac Notation

It is often useful and more transparent to formulate a theory independently of a particular basis system. This can be illustrated in the case of three-dimensional vectors $\boldsymbol{v}$ in the space $\mathbb{R}^3$ (generally $\mathbb{R}^n$).

One can characterize a vector $\boldsymbol{v} \in \mathbb{R}^3$ in some basis $\{\boldsymbol{e}_i\}$ or in another basis $\{\boldsymbol{e}'_i\}$ of the infinitely many bases rotated relative to the first one by its projections onto the respective coordinate axes (Fig. 8.1):

$$\boldsymbol{v} = \sum_i v_i \boldsymbol{e}_i \quad , \quad v_i = \boldsymbol{e}_i \cdot \boldsymbol{v} \quad ,$$
$$\boldsymbol{v} = \sum_i v'_i \boldsymbol{e}'_i \quad , \quad v'_i = \boldsymbol{e}'_i \cdot \boldsymbol{v} \quad . \tag{8.17}$$

The vector $\boldsymbol{v}$ itself does not change; only its components with respect to the coordinate axes change under the transformation from the basis $\{\boldsymbol{e}_i\}$ to the basis $\{\boldsymbol{e}'_i\}$ according to the transformation law

$$v'_i = \sum_j v_j \boldsymbol{e}'_i \cdot \boldsymbol{e}_j = \sum_j D_{ij} v_j \quad . \tag{8.18}$$

The matrix $S^\dagger$ of Sect. 8.1 corresponds here to the matrix D. Instead of characterizing a vector $\boldsymbol{v}$ by its components v_i with respect to a particular coordinate system, it is often more convenient to use the coordinate independent vector notation $\boldsymbol{v}$.

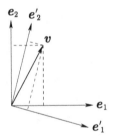

Fig. 8.1. Coordinate transformation

Analogously, in quantum mechanics one can represent the state $\psi(x)$ in various basis systems:

$$\psi(x) = \sum_n c_n \psi_n(x) = \int d\xi \, c_\xi \psi_\xi(x) = \int dp \, c_p \psi_p(x) = \dots \quad . \tag{8.19}$$

These representations are called the energy, coordinate, and momentum representations (and so on), where the respective basis system is formed by the energy eigenfunctions ψ_n, the position eigenfunctions ψ_ξ, the momentum eigenfunctions ψ_p, etc. The expansion coefficients c_n, c_ξ, c_p, etc. characterize the state ψ equally well. Instead of giving one or another set of infinitely many components, we introduce a vector notation for the state (Dirac notation) which is independent of the basis:

$$\psi(x) \to |\psi\rangle \quad . \tag{8.20}$$

Here, $|\psi\rangle$ is a vector in an infinite-dimensional space. For particular states we will use the abbreviated notation

$$\psi_p(x) \to |p\rangle \quad , \quad \psi_\xi(x) \to |\xi\rangle \quad ,$$
$$\psi_n(x) \to (|\psi_n\rangle) \to |n\rangle \quad , \quad \psi_a \to |a\rangle \quad . \tag{8.21}$$

For a general (arbitrary) state, we will write for brevity $|\ \rangle$. The sum of state vectors and multiplication by complex numbers is defined by

$$\psi_c = \alpha\psi_a + \beta\psi_b \to |c\rangle = \alpha|a\rangle + \beta|b\rangle \quad . \tag{8.22}$$

In particular, from this it follows for multiplication by 1 that

$$1\,\psi = \psi \to 1\,|\psi\rangle = |\psi\rangle$$

and for the addition of the zero element

$$\psi + 0 = \psi \to |\psi\rangle + 0 = |\psi\rangle \quad .$$

Since the wave functions $\psi(x)$ form a linear space, the vectors $|\ \rangle$ also form a *linear space*.

Remark: We collect here the mathematical definitions of a *linear space*. A linear space S is defined by the following properties[1]:

1a)   If $|a\rangle$, $|b\rangle \in S$, then also the sum $|a\rangle + |b\rangle \in S$.
 b)   For $|a\rangle \in S$, $\alpha \in \mathbb{C}$: $\alpha|a\rangle \in S$.
 c)   There exists a zero element 0 with $|a\rangle + 0 = |a\rangle$.
 d)   For each $|a\rangle \in S$, the inverse element $|a'\rangle \in S$ exists, with $|a\rangle + |a'\rangle = 0$.

2)   For all $|a\rangle, |b\rangle, |c\rangle \in S$ and complex numbers $\alpha, \beta \in \mathbb{C}$, one has the following properties of addition and multiplication:
 a) $|a\rangle + |b\rangle = |b\rangle + |a\rangle$ (Commutativity)
       $(|a\rangle + |b\rangle) + |c\rangle = |a\rangle + (|b\rangle + |c\rangle)$ (Associativity)
 b) $1\,|a\rangle = |a\rangle$
 c) $\alpha\,(\beta|a\rangle) = (\alpha\,\beta)|a\rangle$ (Associativity)
       $(\alpha + \beta)|a\rangle = \alpha|a\rangle + \beta|a\rangle$ (Distributivity)
       $\alpha(|a\rangle + |b\rangle) = \alpha|a\rangle + \alpha|b\rangle$ (Distributivity)

---

[1] P. Dennery, A. Krzywicki: *Mathematics for Physicists* (Harper & Row, New York 1967) p. 104

The *scalar product* of two state vectors $|a\rangle$ and $|b\rangle$ is introduced by the definition

$$\langle a|b\rangle = (\psi_a, \psi_b) \quad . \tag{8.23}$$

The dual vector space

Corresponding to the assignment

$$\psi \rightarrow \begin{pmatrix} c_1 \\ c_2 \\ \vdots \end{pmatrix} \rightarrow |\psi\rangle \quad , \tag{8.24a}$$

we assign

$$\psi^* \rightarrow (c_1^*, c_2^*, \dots) \rightarrow \langle\psi| \quad . \tag{8.24b}$$

The vector $\langle\psi|$ is called "the dual vector to $|\psi\rangle$". The space of $\langle\psi|$ is called the dual space[2]. The sum in dual space results from the mapping

$$\psi_c^* = \alpha^*\psi_a^* + \beta^*\psi_b^* \rightarrow \langle c| = \alpha^*\langle a| + \beta^*\langle b| \quad . \tag{8.25}$$

The dual vectors also form a linear space.

We now define the *product* of the dual vector $\langle a|$ with the vector $|b\rangle$. We denote it by $\langle a\|b\rangle$, usually abbreviated $\langle a|b\rangle$, and define it by

$$\langle a|b\rangle = (\psi_a, \psi_b) \quad . \tag{8.23'}$$

The quantity $\langle a|b\rangle$ can be read as the product of $\langle a|$ and $|b\rangle$ or as the scalar product of $|a\rangle$ and $|b\rangle$.

Properties of the product (scalar product)

$$\langle a|b\rangle^* = \langle b|a\rangle \quad , \tag{8.26a}$$

$$\langle a|a\rangle \geq 0 \quad , \quad 0 \text{ if and only if } |a\rangle = 0 \quad , \tag{8.26b}$$

$$|\langle a|b\rangle|^2 \leq \langle a|a\rangle\langle b|b\rangle \quad \text{(Schwarz inequality)} \quad . \tag{8.26c}$$

For $|c\rangle = \alpha|a\rangle + \beta|b\rangle$ one has

$$\langle d|c\rangle = \alpha\langle d|a\rangle + \beta\langle d|b\rangle \quad , \tag{8.26d}$$

$$\langle c|d\rangle = \alpha^*\langle a|d\rangle + \beta^*\langle b|d\rangle \quad . \tag{8.26e}$$

The product is linear in both the right and the left factor.

From the word "bracket" and the notation $\langle a|b\rangle$ for the product, the expressions bra-vector and ket-vector have been introduced, standing for $\langle|$ and $|\rangle$, respectively (*Dirac notation*).

---

[2] To be mathematically precise, one introduces the dual vectors by means of the linear functionals $\psi^* \rightarrow f_\psi(\varphi) = (\psi, \varphi)$. These functionals are linear: $f_\psi(\alpha\varphi_a + \beta\varphi_b) = \alpha f_\psi(\varphi_a) + \beta f_\psi(\varphi_b)$ and form a linear space: $\alpha^*\psi^* + \gamma^*\varrho^* \rightarrow f_{\alpha\psi+\gamma\varrho} = \alpha^* f_\psi + \gamma^* f_\varrho$. With the notation $f_\psi \equiv \langle\psi|$ the product $f_\psi(\varphi) \equiv \langle\psi|\varphi\rangle$ results.

A few *important scalar products* are

$$\langle n|m \rangle = \delta_{nm} \quad , \quad \langle \xi|\xi' \rangle = \delta(\xi - \xi') \quad , \quad \langle p|p' \rangle = \delta(p - p') \quad , \tag{8.27}$$

$$\langle \xi|\psi \rangle = \psi(\xi) \quad , \quad \langle \xi|p \rangle = \frac{e^{ip\xi/\hbar}}{\sqrt{2\pi\hbar}} \quad , \quad \langle p|\xi \rangle = \frac{e^{-ip\xi/\hbar}}{\sqrt{2\pi\hbar}} \quad . \tag{8.28}$$

The *expansion of a state* in terms of basis vectors is represented in Dirac notation as follows:

$$\begin{aligned} |\psi\rangle &= \sum c_n |n\rangle \quad , \qquad c_n = \langle n|\psi \rangle \\ &= \int d\xi \, \psi(\xi) |\xi\rangle \quad , \quad \psi(\xi) = \langle \xi|\psi \rangle \\ &= \int dp \, c_p |p\rangle \quad , \qquad c_p = \langle p|\psi \rangle \quad . \end{aligned} \tag{8.29}$$

Operators

We now need to introduce operators acting on the states of our general state space. Given an operator A on the space of wave functions (in coordinate space), let it transform the wave function ψ_a into ψ_b, i.e.,

$$A\psi_a = \psi_b \quad . \tag{8.30a}$$

We then write this in basis independent notation as

$$A|a\rangle = |b\rangle \quad . \tag{8.30b}$$

The action of the operator A in the vector space is obtained from its action in the coordinate representation (8.30a) as follows:

$$A|a\rangle = \sum_n |n\rangle\langle n|b\rangle = \sum_n |n\rangle(\psi_n, A\psi_a) \quad , \tag{8.31}$$

where $|n\rangle$ is an arbitrary basis.

Projection operators

An important class of operators are the projection operators, characterized by the property

$$P^2 = P \quad . \tag{8.32}$$

The projection operator which projects an arbitrary state $|\psi\rangle$ onto the normalized state $|a\rangle$ ($\langle a|a\rangle = 1$) is denoted P_a:

$$P_a|\psi\rangle = \langle a|\psi\rangle|a\rangle = |a\rangle\langle a|\psi\rangle \quad .$$

Therefore, we can represent P_a by

$$P_a = |a\rangle\langle a| \quad . \tag{8.33}$$

P_a is indeed a projection operator, since $P_a^2 = |a\rangle\langle a|a\rangle\langle a| = |a\rangle\langle a| = P_a$. We call $|n\rangle$ a *complete, orthonormal set,* if the $\psi_n(x)$ form a complete, orthonormal set. For a complete, orthonormal set, one has

$$\langle n|m\rangle = \delta_{nm} \quad , \tag{8.34a}$$

$$\sum_n |n\rangle\langle n| = \mathbb{1} \quad , \tag{8.34b}$$

where $\mathbb{1}$ is the unit operator. The orthogonality relation follows from the definition of the scalar product and the orthogonality of the ψ_n, while the completeness relation follows from the possibility of expanding an arbitrary state $|\psi\rangle$:

$$|\psi\rangle = \sum_n |n\rangle(\psi_n, \psi) = \sum_n |n\rangle\langle n|\psi\rangle = \left(\sum_n |n\rangle\langle n|\right)|\psi\rangle \quad .$$

From the completeness relation (8.34b), one can immediately rederive that of the wave functions:

$$\sum_n \psi_n(x)\psi_n^*(x') = \langle x|\left(\sum_n |n\rangle\langle n|\right)|x'\rangle = \langle x|x'\rangle = \delta(x - x') \quad .$$

In the continuum case, one has in analogy

$$\int d\xi\, |\xi\rangle\langle\xi| = \mathbb{1} \quad , \quad \int dp\, |p\rangle\langle p| = \mathbb{1} \quad . \tag{8.35}$$

Using the completeness relation (8.34b) twice, we can represent an arbitrary operator A by its matrix elements:

$$A = \sum_{n,m} |n\rangle\langle n|A|m\rangle\langle m| = \sum_{n,m}\langle n|A|m\rangle|n\rangle\langle m| = \sum_{n,m} A_{nm}|n\rangle\langle m| . \tag{8.36}$$

Up to now, in a matrix element of the type $\langle c|A|a\rangle$, the action of A is only defined to the right. We define $\langle c|A$ by

$$\langle c|A = \sum_n \langle c|A|n\rangle\langle n| \quad . \tag{8.37}$$

Thus, the action of an operator to the left derives from its action to the right. Together with the completeness relation, this implies

$$(\langle c|A)|a\rangle = \langle c|(A|a\rangle) \quad . \tag{8.38}$$

Definition of the adjoint operator

The adjoint operator $B^\dagger$ to B is defined by

$$\langle a|B|b\rangle^* = \langle b|B^\dagger|a\rangle \quad . \tag{8.39}$$

It then follows from the equation

$$|d\rangle = B|c\rangle = \sum_n |n\rangle\langle n|B|c\rangle \tag{8.40a}$$

that for bra vectors

$$\langle d| = \sum_n \langle n|\langle n|B|c\rangle^* = \sum_n \langle n|\langle c|B^\dagger|n\rangle = \langle c|B^\dagger \quad . \tag{8.40b}$$

Example: For the annihilation and creation operators a and $a^\dagger$ of the harmonic oscillator, one has

$$a|n\rangle = \sqrt{n}\,|n-1\rangle \text{ and } a^\dagger|n\rangle = \sqrt{n+1}\,|n+1\rangle \quad .$$

From the preceding relation, it follows that

$$\langle n|a = \sqrt{n+1}\,\langle n+1| \quad ,$$

i.e., the annihilation operator a acts to the left in the same way as the creation operator acts to the right.

Definition. The operator A is called Hermitian if $A^\dagger = A$.

8.3 The Axioms of Quantum Mechanics

The foundations of quantum mechanics may be summarized in the following axioms:

I.   The state of a system is described by the state vector $|\psi\rangle$.
II. The observables are represented by Hermitian operators A, with functions of observables being represented by the corresponding functions of the operators.
III. The expectation value of the observable A – represented by the operator A – is given by $\langle A\rangle = \langle \psi|A|\psi\rangle$.
IV. The time evolution is determined by the Schrödinger equation[3]

$$i\hbar\frac{\partial}{\partial t}|\psi,t\rangle = H|\psi,t\rangle \quad . \tag{8.41}$$

V.   If in a measurement of A the value a_n is found, then the state of the system changes to the corresponding eigenstate $|n\rangle$.[4]

From II and III and Sect. 2.9, it follows that if a system is in the state

$$|\psi\rangle = \sum_n c_n|n\rangle \text{ with } c_n = \langle n|\psi\rangle \quad ,$$

---

[3] We could also use the ordinary time derivative here, since in the state $|\psi,t\rangle$ the time t is the only independent variable occurring.
[4] This is called reduction of the state. See Sect. 20.3, in particular 20.3.4.

where $|n\rangle$ are the eigenstates of A, $A|n\rangle = a_n|n\rangle$, then $|c_n|^2$ expresses the probability that, in a measurement of A, the eigenvalue a_n will be found.[5] This also implies the probability interpretation of the wave function $\psi(x)$.

For *stationary states*

$$|\psi_n, t\rangle = \mathrm{e}^{-\mathrm{i}E_n t/\hbar}|\psi_n\rangle \tag{8.42a}$$

(8.41) yields the *time independent Schrödinger equation* for $|\psi_n\rangle$,

$$H|\psi_n\rangle = E_n|\psi_n\rangle \quad . \tag{8.42b}$$

If at time zero the state is $|\psi\rangle$, then, as in (2.77), the expansion in terms of stationary states reads

$$|\psi, t\rangle = \sum_n \langle\psi_n|\psi\rangle \mathrm{e}^{-\mathrm{i}\,E_n t/\hbar}|\psi_n\rangle \quad . \tag{8.43}$$

We now derive various representations of the Schrödinger equation from the basis independent version (8.41).

8.3.1 Coordinate Representation

The spatial wave function is obtained by projection onto $|x\rangle$ (8.28) as

$$\langle x|\psi, t\rangle = \psi(x, t) \quad .$$

The position eigenvector $\langle x|$ is time independent. We multiply the Schrödinger equation (8.41) by this on the left,

$$\begin{aligned}
\mathrm{i}\hbar\frac{\partial}{\partial t}\langle x|\psi, t\rangle &= \int dx' \langle x|H|x'\rangle\langle x'|\psi, t\rangle \\
&= \int dx' \left[-\frac{\hbar^2}{2m}\frac{\partial^2}{\partial x^2}\delta(x - x') + V(x)\delta(x - x')\right]\psi(x', t) \quad ,
\end{aligned}$$

where we have used (8.14e) and (8.14f); thus

$$\mathrm{i}\hbar\frac{\partial}{\partial t}\psi(x, t) = \left[-\frac{\hbar^2}{2m}\frac{\partial^2}{\partial x^2} + V(x)\right]\psi(x, t) \quad . \tag{8.44}$$

This is our well known Schrödinger equation for the wave function.

---

[5] Here we use the same symbol A for the operator as for the observable. (See also Remark (i), Sect. 2.9.4.)

8.3.2 Momentum Representation

With

$$\langle p|\psi, t\rangle = c_p(t) = \varphi(p, t)/\sqrt{2\pi\hbar}$$

it follows that

$$i\hbar\frac{\partial}{\partial t}c_p(t) = \int dp' \langle p|H|p'\rangle\langle p'|\psi, t\rangle \quad .$$

Using (8.14d) and (8.14f),

$$\langle p|H|p'\rangle = \frac{p^2}{2m}\delta(p - p') + \int dx\,dx' \langle p|x\rangle\langle x|V(x)|x'\rangle\langle x'|p'\rangle$$

$$= \frac{p^2}{2m}\delta(p - p') + \int dx\,V(x)e^{-i(p-p')x/\hbar}/2\pi\hbar \quad ,$$

and introducing the Fourier transform

$$\tilde{V}(q) = \int dx\,e^{-iqx/\hbar}V(x)$$

of the potential, we obtain the Schrödinger equation in the momentum representation,

$$i\hbar\frac{\partial}{\partial t}\varphi(p, t) = \frac{p^2}{2m}\varphi(p, t) + \int\frac{dp'}{2\pi\hbar}\tilde{V}(p - p')\varphi(p', t) \quad . \tag{8.45}$$

This is in general an integro-differential equation. We can also write the potential term using

$$e^{-ipx/\hbar}V(x) = V\left(-\frac{\hbar}{i}\frac{\partial}{\partial p}\right)e^{-ipx/\hbar}$$

in the form

$$\int dx\int dp'e^{-i(p-p')x/\hbar}V(x)\varphi(p', t)/2\pi\hbar$$

$$= V\left(-\frac{\hbar}{i}\frac{\partial}{\partial p}\right)\int\frac{dx\,dp'}{2\pi\hbar}e^{-i(p-p')x/\hbar}\varphi(p', t)$$

$$= V\left(-\frac{\hbar}{i}\frac{\partial}{\partial p}\right)\varphi(p, t) \quad ,$$

and we obtain from (8.45)

$$i\hbar\frac{\partial}{\partial t}\varphi(p, t) = \left[\frac{p^2}{2m} + V\left(-\frac{\hbar}{i}\frac{\partial}{\partial p}\right)\right]\varphi(p, t) \quad . \tag{8.46}$$

8.3.3 Representation in Terms of a Discrete Basis System

Projecting the state onto a discrete basis system, we obtain

$$\langle n|\psi,t\rangle = c_n(t) \quad ,$$

$$i\hbar\frac{\partial}{\partial t}\langle n|\psi,t\rangle = \sum_{m'}\langle n|H|m'\rangle\langle m'|\psi,t\rangle \quad .$$

The Schrödinger equation then consists of the following linear system of equations:

$$i\hbar\frac{d}{dt}c_n(t) = \sum_{m'}H_{nm'}c_{m'}(t) \quad . \tag{8.47}$$

For a basis with a *discrete* and a *continuous* spectrum, $\sum_{m'}$ should be replaced by $\underset{m'}{S}$.

8.4 Multidimensional Systems and Many-Particle Systems

Up to now, only one-dimensional systems have been treated in this chapter. The state space was spanned by the position eigenvectors $|\xi\rangle$, and an arbitrary state was given by

$$|\psi\rangle = \int d\xi\,\psi(\xi)|\xi\rangle \quad .$$

In three-dimensional systems, we have for the three Cartesian components the basis vectors

$$|\xi_x\rangle \quad , \quad |\xi_y\rangle \quad , \quad |\xi_z\rangle \quad . \tag{8.48}$$

A state characterizing a particle located at the position $\boldsymbol{\xi}$ is given by the direct product of these basis vectors

$$|\boldsymbol{\xi}\rangle = |\xi_x\rangle|\xi_y\rangle|\xi_z\rangle \quad . \tag{8.49}$$

We have

$$x_i|\boldsymbol{\xi}\rangle = \xi_i|\boldsymbol{\xi}\rangle \quad , \quad i = x, y, z \quad ,$$

$$\langle\boldsymbol{\xi}|\boldsymbol{\xi}'\rangle = \delta^{(3)}(\boldsymbol{\xi}-\boldsymbol{\xi}') \quad . \tag{8.50}$$

A general state is given by

$$|\psi\rangle = \int d^3\xi\,\psi(\boldsymbol{\xi})|\boldsymbol{\xi}\rangle \quad . \tag{8.51}$$

If $|\xi_i\rangle$ $i = 1, 2, \ldots, N$ are one-particle position eigenstates, one can then write the N-particle state as

$$|\xi_1, \xi_2, \ldots, \xi_N\rangle = |\xi_1\rangle|\xi_2\rangle \ldots |\xi_N\rangle \quad . \tag{8.52}$$

The corresponding orthonormality relation is

$$\langle\xi_1, \xi_2 \ldots \xi_N|\xi_1', \xi_2' \ldots \xi_N'\rangle = \delta^{(3)}(\xi_1 - \xi_1') \ldots \delta^{(3)}(\xi_N - \xi_N') \quad .$$

Arbitrary states are obtained by superposition of the position eigenstates (8.52). It is thus clear that the formulae of the previous sections of this chapter also hold for higher spatial dimensions and many-particle systems, where ξ, x etc. stand for the whole set of variables.

Remark: Additional points concerning the direct product: Let $v_i^{(1)}$, $i = 1, \ldots N_1$ be elements of an N_1-dimensional vector space and $v_j^{(2)}$, $j = 1, \ldots N_2$ be elements of an N_2-dimensional vector space, then their direct product $v_i^{(1)} v_j^{(2)}$ spans an $N_1 N_2$-dimensional space. In the state space 1 of the states $|1\rangle$ let operators A_1 act, and in the state space 2 of the states $|2\rangle$ let operators A_2 act. In the direct product space of the states $|1, 2\rangle = |1\rangle|2\rangle$ the operators $A_1 \otimes A_2$ then act, where

$$\langle 1, 2|A_1 \otimes A_2|1', 2'\rangle = \langle 1|A_1|1'\rangle\langle 2|A_2|2'\rangle \quad .$$

This corresponds to the definition of the direct product of matrices. The operator A_1 corresponds to $A_1 \otimes I_2$ in the product space, and the operator A_2 corresponds to $I_1 \otimes A_2$ in the product space, where I_i is the unit operator in the state space i. One has $A_1 B_1 \otimes C_2 D_2 = (A_1 \otimes C_2)(B_1 \otimes D_2)$.

8.5 The Schrödinger, Heisenberg and Interaction Representations

8.5.1 The Schrödinger Representation

We assume that the Hamiltonian H is time independent. The generalization to a time dependent H will be treated in Chap. 16. The time evolution of the state vector $|\psi, t\rangle$ is then determined by the Schrödinger equation

$$i\hbar\frac{\partial}{\partial t}|\psi, t\rangle = H|\psi, t\rangle \quad , \tag{8.53}$$

which has the formal solution

$$|\psi, t\rangle = e^{-iHt/\hbar}|\psi, 0\rangle \quad . \tag{8.54}$$

This representation of quantum theory, which we have employed up until now, is called the *Schrödinger representation,* or *Schrödinger picture.* The state vectors depend on time, while the operators corresponding to physical observables are independent of time, aside from explicit time dependence. In particular, operators such as x, p, L, etc. are time independent.

8.5.2 The Heisenberg Representation

In the *Heisenberg picture*, on the other hand, the operators obey an equation of motion. Starting with operators A in the Schrödinger picture, one defines the Heisenberg operators

$$A_H = e^{iHt/\hbar} A\, e^{-iHt/\hbar} \quad . \tag{8.55}$$

By an elementary calculation, one obtains from this definition the Heisenberg equation for the operator A_H

$$\frac{d}{dt} A_H = \frac{i}{\hbar}[H, A_H] + \frac{\partial}{\partial t} A_H \quad . \tag{8.56}$$

The last term $\partial A_H/\partial t$ enters only if the operator A depends explicitly on time, as is the case for example in the presence of an external field which is periodic in time. In the last term,

$$e^{iHt/\hbar} \left(\frac{\partial}{\partial t} A(x, p, \dots, t) \right) e^{-iHt/\hbar}$$
$$= \frac{\partial}{\partial t} A(x_H(t), p_H(t), \dots, t) = \frac{\partial A_H}{\partial t} \tag{8.57}$$

has been used. Here, we have expanded the operator function $\acute{A}$ in a power series (with the powers written as products) and inserted

$$1 = \exp\left\{ \frac{-i}{\hbar} Ht \right\} \exp\left\{ \frac{i}{\hbar} Ht \right\}$$

between successive factors.

The Heisenberg state vector is defined by

$$|\psi\rangle_H = e^{iHt/\hbar} |\psi, t\rangle \quad . \tag{8.58}$$

Comparison with (8.54) shows that $|\psi, t\rangle_H$ is identical to the initial value of the Schrödinger state vector $|\psi, 0\rangle$ and is therefore time independent, which, incidentally, can also be seen from

$$\frac{\partial}{\partial t} |\psi\rangle_H = \frac{i}{\hbar} H e^{iHt/\hbar} |\psi, t\rangle + e^{iHt/\hbar} \frac{1}{i\hbar} H |\psi, t\rangle = 0 \quad . \tag{8.59}$$

These two descriptions are related by a unitary transformation and of course give the same physical results. Thus, the expectation value of the observable A is given by

$$\langle \psi, t | A | \psi, t \rangle = \langle \psi |_H A_H(t) | \psi \rangle_H \quad ,$$

where again $\exp\{-iHt/\hbar\} \exp\{iHt/\hbar\} = 1$ has been inserted. The time dependence of the expectation value comes from the state vector in the Schrödinger picture and from the operator in the Heisenberg picture.

We can also write the Heisenberg equation of motion (8.56) in the form

$$\frac{d}{dt}A_{\mathrm{H}} = \frac{i}{\hbar}[H_{\mathrm{H}}, A_{\mathrm{H}}] + \frac{\partial}{\partial t}A_{\mathrm{H}} \quad , \tag{8.56'}$$

since evidently

$$H_{\mathrm{H}} = e^{iHt/\hbar}H\,e^{-iHt/\hbar} = H \quad . \tag{8.60}$$

If we additionally make use of the fact that H is a function of x, p, etc., it then also follows that

$$H_{\mathrm{H}} = H(x,p) = H(x_{\mathrm{H}}(t), p_{\mathrm{H}}(t)) \quad . \tag{8.61}$$

Let us consider the example of the one-dimensional harmonic oscillator,

$$H = \frac{1}{2m}p^2 + \frac{m\omega^2}{2}x^2 \quad .$$

Here, the equations of motion are

$$\dot{x}_{\mathrm{H}} = \frac{i}{\hbar}[H_{\mathrm{H}}, x_{\mathrm{H}}] = \frac{1}{m}p_{\mathrm{H}}$$

$$\dot{p}_{\mathrm{H}} = \frac{i}{\hbar}[H_{\mathrm{H}}, p_{\mathrm{H}}] = -m\omega^2 x_{\mathrm{H}} \quad , \tag{8.62}$$

analogous in structure to the classical ones.

Conservation laws

As in classical mechanics, if appropriate symmetries are present, then conservation laws hold for the Hamiltonian, the angular momentum, and the momentum. In Table 8.1, these conservation laws are listed together with the pertinent symmetry operations, see also Eqs. (5.5), (5.10) and (8.54).

The angular momentum and the momentum of N particles are defined by

$$\boldsymbol{L}_{\mathrm{H}} = \sum_{n=1}^{N}\boldsymbol{x}_{n\mathrm{H}} \times \boldsymbol{p}_{n\mathrm{H}} \quad , \quad \boldsymbol{P}_{\mathrm{H}} = \sum_{n=1}^{N}\boldsymbol{p}_{n\mathrm{H}} \quad . \tag{8.63}$$

For translationally invariant systems, the theorem regarding uniform motion of the center of mass holds in the form

$$\boldsymbol{R}_{\mathrm{H}}(t) = \frac{1}{M}\boldsymbol{P}_{\mathrm{H}}\,t + \boldsymbol{R}_{\mathrm{H}}(0) \quad , \tag{8.64}$$

where

$$\boldsymbol{R}_{\mathrm{H}}(t) = \frac{1}{M}\sum_{n=1}^{N}m_n\boldsymbol{x}_{n\mathrm{H}}(t)$$

Table 8.1. Conserved quantities and generators of the symmetry operations

Conserved quantity			Generator of
H	$\dfrac{d}{dt}H_{\mathrm{H}} = \dfrac{i}{\hbar}[H_{\mathrm{H}}, H_{\mathrm{H}}] = 0$	for time-independent Hamiltonian	time translation $\mathrm{e}^{-iHt/\hbar}$
L	$\dfrac{d}{dt}L_{\mathrm{H}} = \dfrac{i}{\hbar}[H_{\mathrm{H}}, L_{\mathrm{H}}] = 0$	for rotationally invariant problem	rotation $\mathrm{e}^{i\boldsymbol{\varphi}\cdot\boldsymbol{L}/\hbar}$
P	$\dfrac{d}{dt}P_{\mathrm{H}} = \dfrac{i}{\hbar}[H_{\mathrm{H}}, P_{\mathrm{H}}] = 0$	for translationally invariant problem	translation $\mathrm{e}^{i\boldsymbol{a}\cdot\boldsymbol{P}/\hbar}$

and

$$M = \sum_{n=1}^{N} m_n$$

are the center-of-mass operator and the total mass, respectively.

8.5.3 The Interaction Picture (or Dirac Representation)

For problems whose Hamiltonian

$$H = H_0 + V(t) \tag{8.65}$$

can be separated into a time independent part H_0 and a (possibly time dependent) perturbation $V(t)$, the interaction picture is convenient as a point of departure for time dependent perturbation theory (Sect. 16.3).

The state vectors and the operators in the interaction picture are defined by

$$|\psi, t\rangle_{\mathrm{I}} = \mathrm{e}^{iH_0t/\hbar}|\psi, t\rangle \tag{8.66}$$

and

$$A_{\mathrm{I}}(t) = \mathrm{e}^{iH_0t/\hbar}A(t)\mathrm{e}^{-iH_0t/\hbar} \quad . \tag{8.67}$$

These satisfy the following equations of motion:

$$i\hbar\frac{\partial}{\partial t}|\psi, t\rangle_{\mathrm{I}} = V_{\mathrm{I}}(t)|\psi, t\rangle_{\mathrm{I}} \tag{8.68}$$

and

$$\frac{d}{dt}A_{\mathrm{I}}(t) = \frac{i}{\hbar}[H_0, A_{\mathrm{I}}(t)] + \frac{\partial}{\partial t}A_{\mathrm{I}}(t) \quad . \tag{8.69}$$

The interaction picture lies in a certain sense between the Schrödinger and Heisenberg pictures, since both the state vectors and the operators are time dependent. The state vectors evolve due to the perturbative part of the Hamiltonian, the operators due to the free part H_0.

8.6 The Motion of a Free Electron in a Magnetic Field

As an example of the use of the Heisenberg representation, and in order to complete our discussion of free motion in a magnetic field, we take up where we left off in Sect. 7.7. We omit the index H. The Hamiltonian

$$H = \frac{1}{2m} \left(\boldsymbol{p} - \frac{e}{c} \boldsymbol{A}(\boldsymbol{x}) \right)^2 \quad , \tag{8.70}$$

together with the commutation relations (7.12a,b) and (7.13a,b), leads to the equations of motion

$$m\dot{\boldsymbol{x}} = \frac{\mathrm{i}}{\hbar}[H, m\boldsymbol{x}] = \boldsymbol{p} - \frac{e}{c} \boldsymbol{A}(\boldsymbol{x}) \quad , \tag{8.71}$$

$$m\ddot{\boldsymbol{x}} = \frac{\mathrm{i}}{\hbar}[H, m\dot{\boldsymbol{x}}] = \frac{e}{c} \dot{\boldsymbol{x}} \times \boldsymbol{B} \quad . \tag{8.72}$$

From (8.71), it is now clear from a quantum theoretical point of view why we referred to this quantity as the kinetic momentum.

We take $\boldsymbol{B}$ to point in the z-direction, $\boldsymbol{B} = (0, 0, B)$, and consider only the motion perpendicular to $\boldsymbol{B}$. Then both equations of motion

$$m\ddot{x}_1 = \frac{eB}{c}\dot{x}_2 \quad , \quad m\ddot{x}_2 = -\frac{eB}{c}\dot{x}_1 \tag{8.73}$$

can be written compactly in the form

$$\ddot{x} = -\omega_{\mathrm{c}}\tau\dot{x} \quad , \tag{8.73'}$$

where

$$x \equiv \begin{pmatrix} x_1 \\ x_2 \end{pmatrix} \quad \text{and} \quad \tau = \begin{pmatrix} 0 & 1 \\ -1 & 0 \end{pmatrix} \quad , \quad \omega_{\mathrm{c}} = \frac{B(-e)}{mc} = \frac{Be_0}{mc} \quad . \tag{8.74}$$

The first integral in (8.73') becomes, because of $\tau^2 = -\mathbb{1}$,

$$\dot{x}(t) = \mathrm{e}^{-\omega_{\mathrm{c}}\tau t}\dot{x}(0) = (\cos \omega_{\mathrm{c}}t - \tau \sin \omega_{\mathrm{c}}t)\dot{x}(0) \quad , \tag{8.75a}$$

and the solution is then

$$x(t) = X + \omega_{\mathrm{c}}^{-1}\tau \mathrm{e}^{-\omega_{\mathrm{c}}\tau t}\dot{x}(0) \quad . \tag{8.75b}$$

From (8.75a) one sees that $\dot{x}(t)^2 = \dot{x}(0)^2$, which is a consequence of energy conservation. The solution is formally identical to the classical one, except that the integration constant X is an operator. In the classical case, $X = \binom{X_1}{X_2}$ is the center of circular motion

$$(x(t) - X)^2 = \omega_c^{-2} \dot{x}(0)^2 \quad . \tag{8.76}$$

We will shortly see that X_1 and X_2 do not commute, and therefore the center of the orbit cannot be defined quantum mechanically with arbitrary precision. If we work in the gauge

$$A = \frac{B}{2} \binom{-x_2}{x_1}, \tag{8.77}$$

using (8.71), (8.75a), and (8.75b), the operators $\dot{x}$ and X become

$$m\dot{x} = \binom{p_1}{p_2} - \frac{eB}{2c} \binom{-x_2}{x_1} = \binom{p_1}{p_2} + \frac{m\omega_c}{2} \binom{-x_2}{x_1} \quad , \tag{8.78}$$

$$X = x - \omega_c^{-1} \tau \dot{x} = \binom{x_1}{x_2} - \frac{\tau}{m\omega_c} \left[\binom{p_1}{p_2} + \frac{m\omega_c}{2} \binom{-x_2}{x_1} \right]$$

$$= \frac{1}{2} \binom{x_1}{x_2} - \frac{1}{m\omega_c} \binom{p_2}{-p_1} \quad . \tag{8.79}$$

Besides the commutation relations familiar from (7.30),

$$[m\dot{x}_1, m\dot{x}_2] = \frac{i\hbar eB}{c} \quad , \tag{8.80a}$$

one also has

$$[X_1, X_2] = \frac{i\hbar}{m\omega_c} \quad , \quad [X_i, \dot{x}_j] = 0 \quad . \tag{8.80b}$$

Thus, instead of the canonical variables x_1, p_1 and x_2, p_2, one can also use $\dot{x}_2, \dot{x}_1$ and X_1, X_2. This led us in Sect. 7.7 to the introduction of

$$a = \frac{\pi_2 + i\pi_1}{\sqrt{2\hbar}} = \frac{m(\dot{x}_2 + i\dot{x}_1)}{\sqrt{2\hbar\omega_c m}} \quad . \tag{8.81}$$

We then are again confronted with the harmonic oscillator

$$H = \hbar\omega_c \left(a^\dagger a + \tfrac{1}{2} \right) \tag{8.82}$$

with eigenstates

$$|0\rangle, \dots, |n\rangle = \frac{1}{\sqrt{n!}} (a^\dagger)^n |0\rangle, \dots \quad . \tag{8.83}$$

In order to calculate a explicitly, we introduce the definitions

$$x_\pm = \frac{x_2 \pm ix_1}{\sqrt{2}} \quad , \quad p_\pm = \frac{p_2 \mp ip_1}{\sqrt{2}} \quad . \tag{8.84}$$

These operators also satisfy canonical commutation relations

$$[x_\pm, p_\pm] = i\hbar \quad , \quad [x_\pm, p_\mp] = 0 \quad . \tag{8.85}$$

From (8.78) and (8.81), it follows that

$$a = \frac{1}{\sqrt{\hbar\omega_c m}} \left(p_- - \frac{im\omega_c}{2} x_+ \right) = \frac{1}{\sqrt{\hbar\omega_c m}} \left(\frac{\hbar}{i} \frac{\partial}{\partial x_-} - \frac{im\omega_c}{2} x_+ \right) \quad ,$$

and with the magnetic length

$$r_0 = \sqrt{\frac{\hbar}{m\omega_c}} \equiv \sqrt{\frac{\hbar c}{e_0 B}}$$

one finally has

$$a = \frac{1}{2r_0} \frac{1}{i} \left(x_+ + 2r_0^2 \frac{\partial}{\partial x_-} \right) \tag{8.86a}$$

and

$$a^\dagger = \frac{i}{2r_0} \left(x_- - 2r_0^2 \frac{\partial}{\partial x_+} \right) \quad . \tag{8.86b}$$

For the ground state,

$$a|0\rangle = 0 \quad \text{or} \quad \left(\frac{\partial}{\partial x_-} + \frac{x_+}{2r_0^2} \right) \psi_0 = 0 \quad ,$$

the result for the wave function is

$$\psi_0 = \exp\left\{ -\frac{x_- x_+}{2r_0^2} \right\} f(x_+) \quad . \tag{8.87}$$

We cannot take ψ to be a common eigenfunction of both X_1 and X_2, but only an eigenfunction of

$$X_1^2 + X_2^2 = 2X_+ X_- + r_0^2 \quad . \tag{8.88}$$

Here,

$$X_\pm = \frac{X_2 \pm iX_1}{\sqrt{2}} \quad , \tag{8.89}$$

and

$$[X_+, X_-] = -r_0^2 \quad . \tag{8.90}$$

The quantity X_+ corresponds to a creation operator (compare (3.6)), which raises the eigenvalue of $X_1^2 + X_2^2$ by $2r_0^2$.

The eigenfunctions belonging to the smallest eigenvalue of $X_1^2 + X_2^2$, namely r_0^2, are determined from the condition

$$X_-|n\rangle = 0 \quad . \tag{8.91}$$

Since X and $\dot{x}$ commute, this is equivalent to $X_-|0\rangle = 0$. With

$$X_\pm = \left(\frac{1}{2}x_\pm \mp r_0^2 \frac{\partial}{\partial x_\mp} \right)$$

it then follows that

$$X_-\psi_0 = 0 = \left(\frac{1}{2}x_- + r_0^2 \frac{\partial}{\partial x_+} \right) \exp\left\{ -\frac{x_-x_+}{2r_0^2} \right\} f(x_+) \quad .$$

The solution is

$$f(x_+) = \text{const} \quad ,$$

$$\psi_0 = N\, e^{-\varrho^2/4r_0^2} \quad , \tag{8.92}$$

where N is a normalization factor and $\varrho = (x_1^2 + x_2^2)^{1/2}$. Applying $a^\dagger$ gives

$$(a^\dagger)^n \psi_0 \sim \left(x_- - 2r_0^2 \frac{\partial}{\partial x_+} \right)^n \exp\left\{ -\frac{\varrho^2}{4r_0^2} \right\} \sim (2x_-)^n \exp\left\{ \frac{-\varrho^2}{4r_0^2} \right\} \quad ,$$

$$\psi_n = N\frac{1}{\sqrt{n!}}\frac{1}{r_0^n} e^{in\varphi} \varrho^n \exp\left\{ -\frac{\varrho^2}{4r_0^2} \right\} \quad , \quad \text{with } x_- = -i\varrho e^{i\varphi} \quad . \tag{8.93}$$

The remaining eigenfunctions are obtained by applying X_+:

$$X_+^k \psi_n \quad , \quad k = 1, 2, \ldots \quad . \tag{8.94}$$

These are degenerate and belong to the Landau level with the energy eigen-value

$$E_n = \hbar\omega_c(n + \tfrac{1}{2}) \quad . \tag{8.95}$$

The radial position probability in the states (8.93) is

$$\varrho|\psi_n|^2 \sim \varrho^{2n+1} e^{-\varrho^2/2r_0^2} \quad ,$$

whose maximum is determined from

$$\frac{d}{d\varrho}(\varrho|\psi_n|^2) \sim \frac{2n+1}{\varrho} - \frac{\varrho}{r_0^2} = 0$$

i.e.,

$$\varrho = r_0\sqrt{2n+1} \quad . \tag{8.96}$$

Classical circular orbits are described by an appropriate wave packet of the form

$$\int dn\, \exp\{-in(t\omega_c - \varphi)\} g(n)\varrho^n \exp\left\{ -\frac{\varrho^2}{4r_0^2} \right\} \quad ,$$

for which it follows that the velocity satisfies $v = \omega_c\varrho$.

Problems

8.1 For Schrödinger operators A, B, and C, let $[A, B] = C$. What is the commutation relation for the corresponding operators in the Heisenberg representation?

8.2 Derive the Heisenberg equations of motion for the one-dimensional harmonic oscillator

$$H = \frac{1}{2m}p^2 + \frac{1}{2}m\omega^2 x^2 \quad .$$

Compare with the classical equations of motion. Calculate the time dependence of the operators a_H, $a_H^\dagger$, p_H, and x_H. Determine $a_H(t)$ from the equation of motion and directly by use of the Baker–Hausdorff formula.

8.3 A charged particle moves in a homogeneous electric field described by the potential $\varphi(x) = -Fx$. Determine the wave function for energy E in the momentum representation. The transformation to coordinate space yields the integral representation of the Airy functions.

8.4 Calculate the matrix representation of the angular momentum operators L_x, L_y, L_z, and $\boldsymbol{L}^2$ for the values $l = 1/2$, 1, $3/2$, and 2 by using the formulae

$$\langle l', m' | \boldsymbol{L}^2 | l, m \rangle = \hbar^2 \delta_{ll'} \delta_{mm'} \, l(l+1) \quad ,$$

$$\langle l', m' | L_z | l, m \rangle = \hbar \delta_{ll'} \delta_{mm'} m \quad ,$$

$$\langle l', m' | L_- | l, m \rangle = \hbar \sqrt{(l - m + 1)(l + m)} \delta_{ll'} \delta_{m-1,m'} \quad ,$$

$$\langle l', m' | L_+ | l, m \rangle = \hbar \sqrt{(l + m + 1)(l - m)} \delta_{ll'} \delta_{m+1,m'} \quad ,$$

$$-l \le m \le l \quad .$$

8.5 Show $[H, \boldsymbol{L}] = 0$, $[H, \boldsymbol{P}] = 0$, where

$$H = \sum_{n=1}^{N} \frac{p_n^2}{2m_n} + \frac{1}{2} \sum_{n,n'} V(|\boldsymbol{x}_n - \boldsymbol{x}_{n'}|), \ \boldsymbol{L} = \sum_{n=1}^{N} \boldsymbol{x}_n \times \boldsymbol{p}_n, \ \boldsymbol{P} = \sum_{n=1}^{N} \boldsymbol{p}_n,$$

(a) by evaluating the commutators

(b) by using that $\boldsymbol{L}$ and $\boldsymbol{P}$ generate rotations and translations respectively.

9. Spin

9.1 The Experimental Discovery
of the Internal Angular Momentum

9.1.1 The "Normal" Zeeman Effect

In Sect. 7.3, we obtained for electrons in a magnetic field $\boldsymbol{B}$ the interaction term

$$H_{\text{int}} = -\frac{e}{2mc}\boldsymbol{B}\cdot\boldsymbol{L} = -\boldsymbol{\mu}\cdot\boldsymbol{B} \quad . \tag{9.1}$$

Here, the magnetic moment is

$$\boldsymbol{\mu} = \frac{e}{2mc}\boldsymbol{L} \tag{9.2}$$

and the quantity $e/2mc$ is known as the gyromagnetic ratio. This contribution to the Hamiltonian splits the $2l + 1$ angular momentum states according to

$$\mu_{\text{B}}Bm_l \quad , \tag{9.3}$$

where m_l runs over the values $-l, \ldots, l$. Experimentally, however, one finds that in atoms with odd atomic number Z, the splitting is as if m_l were half-integer. Moreover, in contrast to (9.3), the magnitude of the splitting is different for different levels.

9.1.2 The Stern–Gerlach Experiment

In the Stern–Gerlach experiment, an atomic beam traverses an inhomogeneous magnetic field (Fig. 9.1). By (9.1), the force on an atom is

$$\boldsymbol{F} = \boldsymbol{\nabla}(\boldsymbol{\mu}\cdot\boldsymbol{B}) \cong \mu_z\frac{\partial B_z}{\partial z}\boldsymbol{e}_z \quad . \tag{9.4}$$

From the preceding, one would expect a splitting into an odd number of beams (more precisely, $2l + 1$). The experiment was carried out by O. Stern and W. Gerlach in 1922 with silver atoms. Silver has a spherically symmetric charge distribution plus one $5s$-electron. Thus, the total angular momentum of silver is zero, i.e., $l = 0$; no splitting should occur. If the electron from the fifth shell were in a $5p$-state, one would then expect a splitting into three beams. The experiment gives a splitting into two beams. Consequently, the electron must possess an internal angular momentum (*spin*) with corresponding gyromagnetic ratio e/mc.

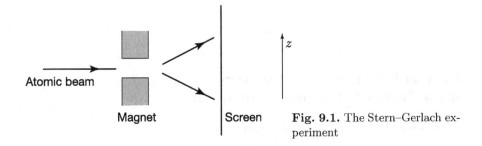

Fig. 9.1. The Stern–Gerlach experiment

A few historical remarks: Pauli: "The doublet structure of the alkali spectra, as well as the violation of the Larmor theorem, occur due to a peculiar – and not classically describable – ambiguity of the quantum theoretical properties of the valence electron."[1]

As early as 1921, Compton deduced from the properties of ferromagnetic materials that the electron must possess a magnetic moment.[2]

Uhlenbeck and Goudsmit: "The electron rotates about its own axis with angular momentum $\hbar/2$. For this value of the angular momentum, there are only two orientations of the angular momentum vector. The gyromagnetic ratio is twice as large for the rotation about its own axis as for the orbital motion."[3]

One can summarize by saying that the electron possesses an internal angular momentum (called spin) which can assume only the values $+\hbar/2$ and $-\hbar/2$ in an arbitrarily chosen direction. We will return to the magnetic moment in Sect. 9.5.

The other elementary particles also have spin. *Fermions* possess half-integral spin, *bosons* integral spin (including zero) (see Sect. 13.1.1). In the following, we will develop the theory for spin-1/2 fermions.

---

[1] W. Pauli: Z. Phys. **31**, 373 (1925)

[2] A.H. Compton: J. Franklin Inst. **192**, 144 (1921)

[3] G.E. Uhlenbeck, S. Goudsmit: Naturwiss. **13**, 953, (1925); Nature **127**, 264 (1926)

9.2 Mathematical Formulation for Spin-1/2

Let the spin operator be $\boldsymbol{S} = (S_x, S_y, S_z)$. Now, if $\boldsymbol{e}$ is a unit vector, then according to Sect. 9.1 $\boldsymbol{S} \cdot \boldsymbol{e}$ has only the two eigenvalues $\pm \hbar/2$, i.e.

$$\boldsymbol{S} \cdot \boldsymbol{e} |e, \pm\rangle = \pm \frac{\hbar}{2} |e, \pm\rangle \quad . \tag{9.5}$$

Without loss of generality, one can choose $\boldsymbol{e} = \boldsymbol{e}_3$ and introduce the following notation:

$$|e_3, \pm\rangle = \begin{cases} |\uparrow\rangle \\ |\downarrow\rangle \end{cases} \quad . \tag{9.6}$$

The eigenvalue equation then takes the form

$$S_z \begin{pmatrix} |\uparrow\rangle \\ |\downarrow\rangle \end{pmatrix} = \frac{\hbar}{2} \begin{pmatrix} +|\uparrow\rangle \\ -|\downarrow\rangle \end{pmatrix} \quad . \tag{9.7}$$

Since the spin is a physical observable, S_z is a Hermitian operator, and the states $|\uparrow\rangle$ and $|\downarrow\rangle$ belonging to distinct eigenvalues are orthogonal, that is,

$$\langle \uparrow | \downarrow \rangle = 0 \quad . \tag{9.8a}$$

We further normalize them to unity:

$$\langle \uparrow | \uparrow \rangle = \langle \downarrow | \downarrow \rangle = 1 \quad . \tag{9.8b}$$

According to the postulate stated previously, the S_i satisfy angular momentum commutation relations

$$[S_i, S_j] = \mathrm{i}\hbar\varepsilon_{ijk} S_k \quad , \quad [S_z, S_\pm] = \pm \hbar S_\pm \quad , \quad [S_+, S_-] = 2\hbar S_z \quad . \tag{9.9}$$

Here,

$$S_\pm = S_x \pm \mathrm{i} S_y \tag{9.10a}$$

with the inversion

$$S_x = \frac{1}{2}(S_+ + S_-) \quad , \quad S_y = \frac{1}{2\mathrm{i}}(S_+ - S_-) \quad . \tag{9.10b}$$

For spin $S = 1/2$, $\boldsymbol{S}^2$ has the eigenvalue $3\hbar^2/4$:

$$\boldsymbol{S}^2 |\uparrow\rangle = \tfrac{3}{4}\hbar^2 |\uparrow\rangle \quad ,$$
$$\boldsymbol{S}^2 |\downarrow\rangle = \tfrac{3}{4}\hbar^2 |\downarrow\rangle \quad . \tag{9.11}$$

From (5.15) it follows with $l \to 1/2$ and $m \to \pm 1/2$ that

$$S_+ |\uparrow\rangle = 0 \quad , \qquad S_- |\uparrow\rangle = \hbar |\downarrow\rangle \quad ,$$
$$S_+ |\downarrow\rangle = \hbar |\uparrow\rangle \quad , \quad S_- |\downarrow\rangle = 0 \quad . \tag{9.12}$$

We can now represent the spin operators in the basis of the states $|\uparrow\rangle$ and $|\downarrow\rangle$ by the spin matrices

$$S_i \rightarrow S_i = \begin{pmatrix} \langle\uparrow|\,S_i\,|\uparrow\rangle & \langle\uparrow|\,S_i\,|\downarrow\rangle \\ \langle\downarrow|\,S_i\,|\uparrow\rangle & \langle\downarrow|\,S_i\,|\downarrow\rangle \end{pmatrix} \ . \tag{9.13}$$

We immediately obtain from (9.12)

$$S_+ = \hbar \begin{pmatrix} 0 & 1 \\ 0 & 0 \end{pmatrix} \ , \quad S_- = \hbar \begin{pmatrix} 0 & 0 \\ 1 & 0 \end{pmatrix} \ , \quad S_z = \frac{\hbar}{2}\begin{pmatrix} 1 & 0 \\ 0 & -1 \end{pmatrix} \ . \tag{9.14}$$

Introducing the *Pauli spin matrices* by

$$\boldsymbol{S} = \frac{\hbar}{2}\boldsymbol{\sigma} \ , \tag{9.15}$$

we obtain for them

$$\sigma_x = \begin{pmatrix} 0 & 1 \\ 1 & 0 \end{pmatrix} \ , \quad \sigma_y = \begin{pmatrix} 0 & -i \\ i & 0 \end{pmatrix} \ , \quad \sigma_z = \begin{pmatrix} 1 & 0 \\ 0 & -1 \end{pmatrix} \ . \tag{9.16}$$

Here for the first time we have an example of a finite-dimensional state space. In the basis $|\uparrow\rangle$, $|\downarrow\rangle$, S_z is diagonal. If we had chosen as a basis the states $|e_x, \pm\rangle$ with $e_x = (1,0,0)$, then S_x would be diagonal.

9.3 Properties of the Pauli Matrices

We collect here the properties of the Pauli spin matrices:

$$\sigma_x^2 = \sigma_y^2 = \sigma_z^2 = \mathbb{1} \quad (\mathbb{1} = \text{unit matrix}) \tag{9.17a}$$

$$[\sigma_x, \sigma_y] = 2i\sigma_z \quad \text{and cyclical permutations} \ , \tag{9.17b}$$

$$\{\sigma_x, \sigma_y\} = 0 \quad \text{and cyclical permutations} \ , \tag{9.17c}$$

$$\sigma_x\sigma_y = -\sigma_y\sigma_x = i\sigma_z \quad \text{and cyclical permutations} \ , \tag{9.17d}$$

$$\sigma_x\sigma_y\sigma_z = i\cdot\mathbb{1} \ , \tag{9.17e}$$

$$\text{tr}\,\sigma_x = \text{tr}\,\sigma_y = \text{tr}\,\sigma_z = 0 \ , \tag{9.17f}$$

$$\det \sigma_x = \det \sigma_y = \det \sigma_z = -1 \ . \tag{9.17g}$$

These relations can be proven without explicit use of the representation (9.16).

It follows from (9.7) that $S_z^2\,|\uparrow\rangle = (\hbar/2)^2\,|\uparrow\rangle$ and likewise for $|\downarrow\rangle$. Thus, $\sigma_z^2 = \mathbb{1}$. Since the unit matrix remains invariant under transformation to another basis, and the z-direction is not preferred, the remaining identities in (9.17a) follow. Equation (9.17b) follows from the spin commutation relations, and (9.17c) can be seen from

$$0 = S_+^2 = (S_x + iS_y)^2 = S_x^2 - S_y^2 + i\{S_x, S_y\} = i\{S_x, S_y\} \quad .$$

Adding (9.17b) and (9.17c), one gets (9.17d). From (9.17d) and (9.17a) follows (9.17e). Since σ_z has vanishing trace and determinant -1, due to its equal (in magnitude) and opposite eigenvalues, (9.17f) and (9.17g) follow.

The relations (9.17a,b,c) can be summarized compactly in the identity

$$\sigma_i\sigma_j = \delta_{ij} + i\varepsilon_{ijk}\sigma_k \quad , \tag{9.18a}$$

from which

$$(\boldsymbol{\sigma}\cdot\boldsymbol{a})(\boldsymbol{\sigma}\cdot\boldsymbol{b}) = \mathbb{1}\,\boldsymbol{a}\cdot\boldsymbol{b} + i\boldsymbol{\sigma}\cdot(\boldsymbol{a}\times\boldsymbol{b}) \tag{9.18b}$$

immediately follows for arbitrary vectors $\boldsymbol{a}$ and $\boldsymbol{b}$ which commute with $\boldsymbol{\sigma}$.

9.4 States, Spinors

In the basis $\{|\uparrow\rangle, |\downarrow\rangle\}$ used up to now, a general spin state can be written as

$$|\rangle = \alpha_+ |\uparrow\rangle + \alpha_- |\downarrow\rangle \tag{9.19}$$

with complex coefficients $\alpha_\pm$. Normalization requires

$$|\alpha_+|^2 + |\alpha_-|^2 = 1 \quad . \tag{9.20}$$

This general state $|\rangle$ can also be represented by a two-component column vector whose components are given by projection onto the basis $\{|\uparrow\rangle, |\downarrow\rangle\}$:

$$\chi = \begin{pmatrix} \alpha_+ \\ \alpha_- \end{pmatrix} \quad , \tag{9.21a}$$

$$\alpha_+ = \langle\uparrow|\rangle \quad , \quad \alpha_- = \langle\downarrow|\rangle \quad . \tag{9.21b}$$

The vector χ is known as a *spinor*. The basis spinors corresponding to $|\uparrow\rangle$ and $|\downarrow\rangle$ are then

$$\chi_+ = \begin{pmatrix} 1 \\ 0 \end{pmatrix} \quad , \quad \chi_- = \begin{pmatrix} 0 \\ 1 \end{pmatrix} \quad . \tag{9.22}$$

The completeness relation for the basis of spin-$1/2$ space can be written either in the form

$$|\uparrow\rangle\langle\uparrow| + |\downarrow\rangle\langle\downarrow| = 1 \tag{9.23a}$$

or in matrix representation as

$$\chi_+\chi_+^\dagger + \chi_-\chi_-^\dagger = \begin{pmatrix} 1 & 0 \\ 0 & 1 \end{pmatrix} \quad . \tag{9.23b}$$

The orthonormality relations were presented in (9.8a) and (9.8b).

9.5 Magnetic Moment

As explained in Sect. 7.3, and as one knows from electrodynamics, a magnetic moment

$$\boldsymbol{\mu}_{\text{orbit}} = \frac{e}{2mc}\boldsymbol{L} \tag{9.24}$$

is associated with the orbital angular momentum $\boldsymbol{L}$ of an electron. Now, there is no reason to suppose that the magnetic moment due to spin $\boldsymbol{\mu}_{\text{spin}}$ should have the same ratio to $\boldsymbol{S}$ as that given in (9.24). We therefore write

$$\boldsymbol{\mu}_{\text{spin}} = g\frac{e}{2mc}\boldsymbol{S} \ , \tag{9.25}$$

where g is the *Landé factor* or *gyromagnetic* factor. The analysis of the Zeeman effect given in Sect. 9.1.1, which will be discussed in detail in Chap. 14, implies for electrons

$$g \approx 2 \ . \tag{9.26}$$

The total magnetic moment of the electron thus becomes

$$\boldsymbol{\mu} = \boldsymbol{\mu}_{\text{orbit}} + \boldsymbol{\mu}_{\text{spin}} = \frac{e}{2mc}(\boldsymbol{L} + 2\boldsymbol{S}) = \frac{e}{2mc}(\boldsymbol{L} + \boldsymbol{\sigma}\hbar) \ , \tag{9.27}$$

and the total interaction energy with the magnetic field is

$$H_{\text{int}} = -\boldsymbol{\mu}\cdot\boldsymbol{B} = \mu_{\text{B}}\left(\frac{\boldsymbol{L}}{\hbar} + \boldsymbol{\sigma}\right)\cdot\boldsymbol{B} \ . \tag{9.28}$$

Remarks:

(i)   The Dirac equation[4], the relativistic wave equation for spin-1/2 fermions, predicts exactly $g = 2$.

(ii)   The quantum electrodynamic correction up to $O(\alpha^3)$ gives

$$g = 2\left(1 + \frac{\alpha}{2\pi} - 0.328\,478\,445\left(\frac{\alpha}{\pi}\right)^2 + 1.183(11)\left(\frac{\alpha}{\pi}\right)^3\right)$$

$$= 2.002\,319\,304\,718(564) \tag{9.29}$$

where α is the Sommerfeld fine structure constant. The experimental value agrees with this up to the seventh decimal place, and the difference can be attributed to α^4-corrections.

The nuclear magneton is defined by

$$\mu_{\text{N}} = \frac{e_0\hbar}{2m_{\text{p}}c} \ , \tag{9.30}$$

---

[4] For references see page 225. See, e.g., F. Schwabl, QM II, Sect. 5.3.

where $m_\mathrm{p} = 1.6726 \times 10^{-24}\,\mathrm{g}$ is the mass of the proton. Due to the larger mass of the proton, $\mu_\mathrm{N} = 0.505 \times 10^{-23}\,\mathrm{erg/G} \approx 10^{-3}\mu_\mathrm{B}$ is about one-thousandth of the Bohr magneton. The quantity μ_N is the characteristic quantity for nuclear magnetism and for the magnetic moments of nucleons. The gyromagnetic ratio of the strongly interacting proton is $g_\mathrm{proton} = 5.59$. Neutral particles such as n, Λ, Σ^0 also possess a magnetic moment. It arises from the internal charge distribution, which can be understood in light of the fact that the fundamental constituents of hadrons are quarks. For the neutron,

$$\boldsymbol{\mu}_\mathrm{n} = -3.83 \frac{e_0}{2m_\mathrm{n}c} \boldsymbol{S} \quad . \tag{9.31}$$

For the deuteron, the nucleus of deuterium , $\mu_\mathrm{deut} = 0.86\mu_\mathrm{N}$, i.e., $\mu_\mathrm{deut} \approx \mu_\mathrm{prot} + \mu_\mathrm{neut}$. The spins of the proton and the neutron are parallel in the deuteron, i.e. the spin of the deuteron is 1; the antiparallel state does not give rise to binding.

9.6 Spatial Degrees of Freedom and Spin

Spin is an additional degree of freedom independent of the spatial degrees of freedom. Spin and position (or momentum) can assume precise values simultaneously and independently of one another, i.e., the corresponding operators commute:

$$[\boldsymbol{S}, \boldsymbol{x}] = 0 \quad , \quad [\boldsymbol{S}, \boldsymbol{p}] = 0 \quad , \quad [\boldsymbol{S}, \boldsymbol{L}] = 0 \quad . \tag{9.32}$$

The total state is constructed from the direct product of position and spin eigenstates. The states $|\boldsymbol{x}\rangle\,|\!\uparrow\rangle$ and $|\boldsymbol{x}\rangle\,|\!\downarrow\rangle$ can be chosen as a basis. In this basis, a general state $|\Psi\rangle$ is given by

$$|\Psi\rangle = \int d^3x\, (\psi_+(\boldsymbol{x})|\boldsymbol{x}\rangle\,|\!\uparrow\rangle + \psi_-(\boldsymbol{x})|\boldsymbol{x}\rangle\,|\!\downarrow\rangle) \quad . \tag{9.33}$$

The projections onto position and spin eigenstates are

$$\langle \boldsymbol{x}|\Psi\rangle = \psi_+(\boldsymbol{x})\,|\!\uparrow\rangle + \psi_-(\boldsymbol{x})\,|\!\downarrow\rangle \quad , \tag{9.34a}$$

$$\langle\uparrow|\,\langle \boldsymbol{x}|\Psi\rangle = \psi_+(\boldsymbol{x}) \quad , \quad \langle\downarrow|\,\langle \boldsymbol{x}|\Psi\rangle = \psi_-(\boldsymbol{x}) \quad . \tag{9.34b}$$

The quantities $|\psi_{+(-)}(\boldsymbol{x})|^2$ express the probability of finding the particle at the position $\boldsymbol{x}$ with spin in the positive (negative) z-direction. The normalization condition is

$$\langle\Psi|\Psi\rangle = \int d^3x (|\psi_+(\boldsymbol{x})|^2 + |\psi_-(\boldsymbol{x})|^2) = 1 \quad . \tag{9.35}$$

As in Sect. 9.4, we can combine the two components $\psi_\pm(\boldsymbol{x})$ into a spinor

$$\Psi(\boldsymbol{x}) = \begin{pmatrix} \psi_+(\boldsymbol{x}) \\ \psi_-(\boldsymbol{x}) \end{pmatrix} \quad . \tag{9.36}$$

These two-component spinors are also called Pauli spinors.

We now write the Hamiltonian of an electron in an external magnetic field without the spin–orbit interaction, which is to be discussed in Chap. 12:

$$H = \frac{\boldsymbol{p}^2}{2m} + V(\boldsymbol{x}) + \mu_{\rm B} \left(\frac{\boldsymbol{L}}{\hbar} + \boldsymbol{\sigma} \right) \cdot \boldsymbol{B} \quad . \tag{9.37}$$

The Schrödinger equation,

$$i\hbar \frac{\partial}{\partial t} |\Psi\rangle = H |\Psi\rangle \quad , \tag{9.38}$$

in component form is

$$i\hbar \frac{\partial}{\partial t} \begin{pmatrix} \psi_+(\boldsymbol{x},t) \\ \psi_-(\boldsymbol{x},t) \end{pmatrix}$$
$$= \left[\left(-\frac{\hbar^2}{2m} \boldsymbol{\nabla}^2 + V(\boldsymbol{x}) + \frac{\mu_{\rm B}}{\hbar} \boldsymbol{L} \cdot \boldsymbol{B} \right) \mathbb{1} + \mu_{\rm B} \boldsymbol{\sigma} \cdot \boldsymbol{B} \right] \begin{pmatrix} \psi_+(\boldsymbol{x},t) \\ \psi_-(\boldsymbol{x},t) \end{pmatrix} \quad . \tag{9.39}$$

This nonrelativistic equation is also known as the *Pauli equation*.

In a time dependent, external electromagnetic field, the Hamiltonian including the diamagnetic term is

$$H = \frac{l}{2m} \left[\boldsymbol{p} - \frac{e}{c} \boldsymbol{A}(\boldsymbol{x},t) \right]^2 + e\Phi(\boldsymbol{x},t) + \mu_{\rm B} \boldsymbol{\sigma} \cdot \boldsymbol{B} \quad , \tag{9.40}$$

and the Pauli equation reads

$$i\hbar \frac{\partial}{\partial t} \begin{pmatrix} \psi_+(\boldsymbol{x},t) \\ \psi_-(\boldsymbol{x},t) \end{pmatrix}$$
$$= \left[\left(\frac{1}{2m} \left(\frac{\hbar}{i} \boldsymbol{\nabla} - \frac{e}{c} \boldsymbol{A}(\boldsymbol{x},t) \right)^2 + e\Phi(\boldsymbol{x},t) \right) \mathbb{1} + \mu_{\rm B} \boldsymbol{\sigma} \cdot \boldsymbol{B} \right] \begin{pmatrix} \psi_+(\boldsymbol{x},t) \\ \psi_-(\boldsymbol{x},t) \end{pmatrix} \quad . \tag{9.41}$$

Problems

9.1 Show that every 2×2 matrix can be represented in terms of the unit matrix and the three Pauli matrices.

9.2 Prove the identity (9.18a) for the Pauli matrices and, using this, show (9.18b).

9.3 Show that the states

$$|e, \pm\rangle = e^{-i\alpha S_y/\hbar} \genfrac{}{}{0pt}{}{|\uparrow\rangle}{|\downarrow\rangle}$$

are eigenstates of the projection of the spin operator in the direction of

$$e = (\sin \alpha, 0, \cos \alpha) \quad .$$

Hint: Compute $S' = e^{-i\alpha S_y/\hbar} \, S e^{i\alpha S_y/\hbar}$ and show that $S'_z = S \cdot e$.

9.4 For a spatial rotation through angle φ about an axis of rotation along a unit vector n, transformations in spinor space are represented by

$$U = e^{i\varphi n \cdot S/\hbar} \quad .$$

(a) Show using (9.18b) that

$$U = \cos \varphi/2 + i (n \cdot \sigma) \sin \varphi/2 \quad .$$

(b) Using this formula for U and again exploiting the identities for the Pauli matrices, show that

$$U\sigma U^\dagger = n(n \cdot \sigma) - n \times [n \times \sigma] \cos \varphi + [n \times \sigma] \sin \varphi \quad .$$

(c) Discuss the special case $n = e_z$ and consider infinitesimal rotations.

(d) Finally, calculate for an arbitrary spinor $\binom{\alpha_+}{\alpha_-}$ the transformed spinor defined by

$$\begin{pmatrix} \alpha'_+ \\ \alpha'_- \end{pmatrix} = U \begin{pmatrix} \alpha_+ \\ \alpha_- \end{pmatrix}$$

for $n = e_z$.

9.5 Consider the precession of the spin of an electron in a homogeneous magnetic field $(0, 0, B)$.

(a) Using the Hamiltonian $H = (e_0/mc)S \cdot B$, write down the equation of motion for the spin operator in the Heisenberg picture and give the solution for the initial condition $S(t = 0) = S(0)$. The solution is

$$S_z(t) = S_z(0),$$
$$S_x(t) = \cos \omega t \, S_x(0) - \sin \omega t \, S_y(0),$$
$$S_y(t) = \sin \omega t \, S_x(0) + \cos \omega t \, S_y(0).$$

(b) Given $\Psi(0) = \binom{a}{b}$, determine the state $\Psi(t)$ at the time t.

(c) What is the probability of obtaining the value $\hbar/2$ in a measurement of S_z at the time t if at the time $t = 0$ the spin was oriented in the x-direction?

(d) The same as in problem (c) for a measurement of S_x.

(e) How can part (b) of this problem be derived from the Pauli equation?

10. Addition of Angular Momenta

10.1 Posing the Problem

In later sections it will be necessary to introduce the total angular momentum $J = L + S$ and, in the case of two electrons with their respective spins, S_1 and S_2, to consider the total spin $S = S_1 + S_2$. Generally, supposing that we have two angular momentum operators, J_1 and J_2, the problem consists in investigating the total angular momentum

$$J = J_1 + J_2 \quad . \tag{10.1}$$

Assuming that J_1 and J_2 correspond to distinct degrees of freedom, they commute with each other:

$$[J_1, J_2] = 0 \quad . \tag{10.2}$$

Together with the angular momentum commutation relations of $J_{1,2}$, this gives for the components of J

$$[J_i, J_j] = i\hbar \varepsilon_{ijk} J_k \quad . \tag{10.3}$$

Thus, all the properties of angular momenta and their eigenstates hold for the total angular momentum J.

We begin with the states $|j_1, m_1\rangle$ and $|j_2, m_2\rangle$, where the two quantum numbers j_1 and j_2 are fixed, and the m_i take the values $-j_i, \ldots, j_i$. From these states, we can construct the product states

$$|j_1 m_1 j_2 m_2\rangle = |j_1 m_1\rangle |j_2 m_2\rangle \quad , \tag{10.4}$$

which are eigenstates of the operators

$$J_1^2, \; J_{1z}, \; J_2^2, \; J_{2z} \tag{10.5a}$$

with eigenvalues

$$\hbar^2 j_1(j_1 + 1), \; \hbar m_1, \; \hbar^2 j_2(j_2 + 1), \; \hbar m_2 \quad . \tag{10.5b}$$

The product states (10.4) are of course also eigenfunctions of J_z with eigenvalue $\hbar(m_1 + m_2)$, but not eigenfunctions of J^2, since

$$[\boldsymbol{J}^2, J_{iz}] \neq 0 \quad , \quad i = 1,2 \quad . \tag{10.6}$$

One sees this either by explicit calculation of the commutators or by realizing that J_{iz} generates a rotation in the subspace i, and that $\boldsymbol{J}^2$ is only a scalar for the total rotation generated by $J_z = J_{1z} + J_{2z}$. However, for the problems mentioned above, we search for states in which $\boldsymbol{J}^2$ is also diagonal; more precisely, we seek eigenfunctions

$$|jm_jj_1j_2\rangle \tag{10.7}$$

of the four mutually commuting operators

$$\boldsymbol{J}^2 \, , \; J_z \, , \; \boldsymbol{J}_1^2 \, , \; \boldsymbol{J}_2^2 \tag{10.8}$$

with eigenvalues $\hbar^2 j(j+1), \hbar m_j, \hbar^2 j_1(j_1+1), \hbar^2 j_2(j_2+1)$.

At the same time, we have to find the values taken by j (the corresponding m_j are then $-j, \ldots, j$), and we have to represent $|jm_jj_1j_2\rangle$ as a linear combination of the product states (10.4). We postpone the general problem until Sect. 10.4 and treat for the moment the addition of two spins followed by the addition of an orbital angular momentum to a spin.

10.2 Addition of Spin-1/2 Operators

This is the simplest case, which among other things will be required for the treatment of the helium atom. Let $\boldsymbol{S}_1$ and $\boldsymbol{S}_2$ be the two spin-1/2 operators and

$$\boldsymbol{S} = \boldsymbol{S}_1 + \boldsymbol{S}_2 \tag{10.9}$$

the total spin. The four states

$$|\uparrow\uparrow\rangle = |\uparrow\rangle\,|\uparrow\rangle \quad , \quad |\downarrow\downarrow\rangle = |\downarrow\rangle\,|\downarrow\rangle \quad ,$$
$$|\uparrow\downarrow\rangle = |\uparrow\rangle\,|\downarrow\rangle \quad , \quad |\downarrow\uparrow\rangle = |\downarrow\rangle\,|\uparrow\rangle \quad , \tag{10.10}$$

in which the first (second) symbol refers to the first (second) spin, are eigenstates of $\boldsymbol{S}_1^2, \boldsymbol{S}_2^2, S_{1z}, S_{2z}$. It is reasonable to suppose that the total spin S assumes the values 1 and 0. To show this, we compute

$$S_z|\uparrow\uparrow\rangle = \hbar|\uparrow\uparrow\rangle \quad , \qquad S_z|\uparrow\downarrow\rangle = 0 \quad ,$$
$$S_z|\downarrow\downarrow\rangle = -\hbar|\downarrow\downarrow\rangle \quad , \qquad S_z|\downarrow\uparrow\rangle = 0 \quad . \tag{10.11}$$

Furthermore,

$$\begin{aligned} \boldsymbol{S}^2 &= \boldsymbol{S}_1^2 + \boldsymbol{S}_2^2 + 2\boldsymbol{S}_1 \cdot \boldsymbol{S}_2 \\ &= \tfrac{3}{2}\hbar^2 + 2S_{1z}S_{2z} + S_{1+}S_{2-} + S_{1-}S_{2+} \quad . \end{aligned} \tag{10.12}$$

Let us first consider the two maximally aligned states $|\uparrow\uparrow\rangle$ and $|\downarrow\downarrow\rangle$. For these, we find from (10.12)

$$\boldsymbol{S}^2 \left|\uparrow\uparrow\right\rangle = \left(\frac{3}{2}\hbar^2 + 2\left(\frac{\hbar}{2}\right)^2\right)\left|\uparrow\uparrow\right\rangle = 2\hbar^2 \left|\uparrow\uparrow\right\rangle \tag{10.13a}$$

and

$$\boldsymbol{S}^2 \left|\downarrow\downarrow\right\rangle = 2\hbar^2 \left|\downarrow\downarrow\right\rangle \quad . \tag{10.13b}$$

The states $\left|\uparrow\uparrow\right\rangle$ and $\left|\downarrow\downarrow\right\rangle$ therefore have total spin $S = 1$ and $S_z = \pm\,\hbar$. The as yet missing spin-1 state with $S_z = 0$ is obtained by application of S_- to $\left|\uparrow\uparrow\right\rangle$:

$$\frac{1}{\hbar\sqrt{2}}S_- \left|\uparrow\uparrow\right\rangle = \frac{1}{\hbar\sqrt{2}}(S_{1-} + S_{2-})\left|\uparrow\uparrow\right\rangle = \frac{1}{\sqrt{2}}(\left|\uparrow\downarrow\right\rangle + \left|\downarrow\uparrow\right\rangle) \quad . \tag{10.14}$$

The resulting state has been normalized to unity by inserting the factor $1/\hbar\sqrt{2}$. By (10.11), it has $S_z = 0$. Thus, we have found all three $S = 1$ states.

Using the notation $\left|S, m\right\rangle$, in which S designates the total spin and m its z-component, we have

$$\left|1, 1\right\rangle = \left|\uparrow\uparrow\right\rangle \quad , \quad \left|1, 0\right\rangle = \frac{1}{\sqrt{2}}(\left|\uparrow\downarrow\right\rangle + \left|\downarrow\uparrow\right\rangle) \quad , \quad \left|1, -1\right\rangle = \left|\downarrow\downarrow\right\rangle \quad . \tag{10.15}$$

There is an additional state, which is orthogonal to those given above:

$$\left|0, 0\right\rangle = \frac{1}{\sqrt{2}}(\left|\uparrow\downarrow\right\rangle - \left|\downarrow\uparrow\right\rangle) \quad . \tag{10.16}$$

For this state, evidently

$$S_z\left|0, 0\right\rangle = 0 \quad , \tag{10.17a}$$

and, by (10.12),

$$\boldsymbol{S}^2\left|0, 0\right\rangle = \left(\frac{3}{2}\hbar^2 - 2\left(\frac{\hbar}{2}\right)^2 - \hbar^2\right)\left|0, 0\right\rangle = 0 \quad . \tag{10.17b}$$

As implied already in (10.16), this state has spin 0. Thus, we have found all four eigenstates of $\boldsymbol{S}^2$ and S_z. The states (10.15) are also referred to as *triplet states* and (10.16) as a *singlet state*. Occasionally it turns out to be useful to introduce the projection operators

$$P_1 = \frac{3}{4} + \frac{1}{\hbar^2}\boldsymbol{S}_1 \cdot \boldsymbol{S}_2 \quad , \tag{10.18a}$$

$$P_0 = 1 - P_1 = \frac{1}{4} - \frac{1}{\hbar^2}\boldsymbol{S}_1 \cdot \boldsymbol{S}_2 \quad . \tag{10.18b}$$

P_1 projects onto the triplet space, since

$$\begin{aligned} P_1\left|1, m\right\rangle &= \left(\frac{3}{4} + \frac{1}{\hbar^2}\frac{1}{2}(\boldsymbol{S}^2 - \boldsymbol{S}_1^2 - \boldsymbol{S}_2^2)\right)\left|1, m\right\rangle \\ &= \left(\frac{3}{4} + \frac{1}{2}\left(2 - \frac{3}{4} - \frac{3}{4}\right)\right)\left|1, m\right\rangle = \left|1, m\right\rangle \end{aligned}$$

and

$$P_1|0,0\rangle = \left(\frac{3}{4} + \frac{1}{2}\left(0 - \frac{3}{4} - \frac{3}{4}\right)\right)|0,0\rangle = 0 \quad.$$

Because of $P_0 = 1 - P_1$, P_0 projects onto the singlet subspace.

10.3 Orbital Angular Momentum and Spin 1/2

Beginning with orbital angular momentum $\boldsymbol{L}$ and spin $\boldsymbol{S}$, we define the total angular momentum

$$\boldsymbol{J} = \boldsymbol{L} + \boldsymbol{S} \quad. \tag{10.19}$$

The $2(2l + 1)$ product states

$$|l, m_l\rangle |\uparrow\rangle \quad \text{and} \quad |l, m_l\rangle |\downarrow\rangle \tag{10.20}$$

with $m_l = -l, \ldots, l$ are eigenstates of $\boldsymbol{L}^2$, $\boldsymbol{S}^2$, L_z, S_z, but not of $\boldsymbol{J}^2$. We seek eigenstates of $\boldsymbol{J}^2$, $\boldsymbol{L}^2$, $\boldsymbol{S}^2$, J_z. Presumably j, the quantum number to $\boldsymbol{J}^2$, has the values

$$j = l + \tfrac{1}{2}, l - \tfrac{1}{2} \quad ;$$

this would also give the right number of states:

$$2(l + \tfrac{1}{2}) + 1 + 2(l - \tfrac{1}{2}) + 1 = 2(2l + 1) \quad.$$

For the following, recall (5.15)

$$L_\pm|l, m_l\rangle = ((l \pm m_l + 1)(l \mp m_l))^{1/2}\hbar|l, m_l \pm 1\rangle$$

and the identity

$$\boldsymbol{J}^2 = \boldsymbol{L}^2 + \boldsymbol{S}^2 + 2L_zS_z + L_+S_- + L_-S_+ \quad. \tag{10.21}$$

For the eigenstates of $\boldsymbol{J}^2$, J_z, $\boldsymbol{L}^2$, and $\boldsymbol{S}^2$ we use the notation $|j, m_j, l\rangle$, without indicating $S = 1/2$, and claim as a point of departure that

$$|l + \tfrac{1}{2}, l + \tfrac{1}{2}, l\rangle = |l, l\rangle |\uparrow\rangle \quad. \tag{10.22}$$

We prove this by means of

$$J_z|l, l\rangle |\uparrow\rangle = (L_z + S_z)|l, l\rangle |\uparrow\rangle = \hbar(l + \tfrac{1}{2})|l, l\rangle |\uparrow\rangle \tag{10.23a}$$

and, using (10.21),

$$\begin{aligned}\boldsymbol{J}^2|l, l\rangle |\uparrow\rangle &= \hbar^2(l(l + 1) + \tfrac{3}{4} + 2l\tfrac{1}{2})|l, l\rangle |\uparrow\rangle \\ &= \hbar^2(l + \tfrac{1}{2})(l + \tfrac{3}{2})|l, l\rangle |\uparrow\rangle \quad. \end{aligned} \tag{10.23b}$$

In order to obtain all the states $|l + 1/2, m_{l+1/2}, l\rangle$, one need only apply $J_- = L_- + S_-$ repeatedly to (10.22). We first find

$$J_-|l, l\rangle |\uparrow\rangle = (2l)^{1/2}\hbar|l, l - 1\rangle |\uparrow\rangle + \hbar|l, l\rangle |\downarrow\rangle \quad , \tag{10.24a}$$

and, by normalizing the right-hand side to unity,

$$|l + \tfrac{1}{2}, l - \tfrac{1}{2}, l\rangle = \sqrt{\frac{2l}{2l + 1}}\,|l, l - 1\rangle |\uparrow\rangle + \sqrt{\frac{1}{2l + 1}}\,|l, l\rangle |\downarrow\rangle \quad . \tag{10.24b}$$

Repeated application of J_- leads to the following general result, which can be easily verified by mathematical induction with respect to m_j:

$$|l + \tfrac{1}{2}, m_j, l\rangle = \sqrt{\frac{l + m_j + 1/2}{2l + 1}}\,|l, m_j - \tfrac{1}{2}\rangle |\uparrow\rangle$$
$$+ \sqrt{\frac{l - m_j + 1/2}{2l + 1}}\,|l, m_j + \tfrac{1}{2}\rangle |\downarrow\rangle \quad , \tag{10.25}$$

where m_j takes the half-integer values $l + 1/2, \ldots, -(l + 1/2)$. With (10.25), the states belonging to $j = l + 1/2$ have been found. In order to determine the remaining states, we remark that, beginning with (10.25) with $m_j = l - 1/2, \ldots, -(l - 1/2)$, one can immediately construct states which are orthogonal to all the states of (10.25), that is,

$$|l - \tfrac{1}{2}, m_j, l\rangle = -\sqrt{\frac{l - m_j + 1/2}{2l + 1}}\,|l, m_j - \tfrac{1}{2}\rangle |\uparrow\rangle$$
$$+ \sqrt{\frac{l + m_j + 1/2}{2l + 1}}\,|l, m_j + \tfrac{1}{2}\rangle |\downarrow\rangle \tag{10.26}$$

with $m_j = l - 1/2, \ldots, -(l - 1/2)$.

Since

$$J_z|l - \tfrac{1}{2}, m_j, l\rangle = \hbar m_j|l - \tfrac{1}{2}, m_j, l\rangle \tag{10.27a}$$

and

$$\boldsymbol{J}^2|l - \tfrac{1}{2}, m_j, l\rangle = \hbar^2(l - \tfrac{1}{2})(l + \tfrac{1}{2})|l - \tfrac{1}{2}, m_j, l\rangle \quad , \tag{10.27b}$$

these are the eigenstates of $\boldsymbol{J}^2$ and J_z with $j = l - \tfrac{1}{2}$ and m_j. And of course the states (10.25) and (10.26) are eigenstates of $\boldsymbol{L}^2$ and $\boldsymbol{S}^2$. Equation (10.27a) follows without calculation, and (10.27b) is left to the reader as an exercise.

For $l = 0$, there is only the state $|\tfrac{1}{2}, \pm\tfrac{1}{2}, 0\rangle$. States $|0 - \tfrac{1}{2}, m_j, 0\rangle$ of negative j do not exist.

For later use, we write the total angular momentum states as follows:

$$|l \pm \tfrac{1}{2}, m_j, l\rangle = \alpha_\pm|l, m_j - \tfrac{1}{2}\rangle |\uparrow\rangle + \beta_\pm|l, m_j + \tfrac{1}{2}\rangle |\downarrow\rangle \quad ,$$

$$\alpha_\pm = \pm\sqrt{\frac{l \pm m_j + 1/2}{2l + 1}} = \pm\beta_\mp \quad , \tag{10.28}$$

$$\beta_+^2 + \beta_-^2 = 1 \quad .$$

10.4 The General Case

After these introductory special cases, we now treat the general case of the addition of two arbitrary angular momenta $\boldsymbol{J}_1$ and $\boldsymbol{J}_2$ to form a total angular momentum $\boldsymbol{J} = \boldsymbol{J}_1 + \boldsymbol{J}_2$. In Sect. 10.1, we introduced the two complete sets of commuting angular momentum operators

$$\boldsymbol{J}_1^2,\; J_{1z},\; \boldsymbol{J}_2^2,\; J_{2z} \quad , \tag{10.5a}$$

$$\boldsymbol{J}^2,\; J_z,\; \boldsymbol{J}_1^2,\; \boldsymbol{J}_2^2 \quad . \tag{10.8}$$

The corresponding systems of eigenfunctions

$$|j_1 m_1\rangle |j_2 m_2\rangle = |j_1 m_1 j_2 m_2\rangle \quad , \tag{10.4}$$

$$|j m j_1 j_2\rangle \tag{10.7}$$

form complete orthonormal systems. Consequently, the states (10.7) can be expanded in terms of (10.4). Here, the scalar products

$$\langle j_1' m_1 j_2' m_2 | j m j_1 j_2\rangle \equiv \langle j_1' m_1 | \langle j_2' m_2 | j m j_1 j_2\rangle \tag{10.29}$$

enter as expansion coefficients. It is immediately clear that, because of

$$\langle j_1' m_1 j_2' m_2 | (\boldsymbol{J}_i^2 | j m j_1 j_2\rangle) = (\langle j_1' m_1 j_2' m_2 | \boldsymbol{J}_i^2) | j m j_1 j_2\rangle \quad ,$$

one has

$$j_i'(j_i' + 1)\langle j_1' m_1 j_2' m_2 | j m j_1 j_2\rangle = j_i(j_i + 1)\langle j_1' m_1 j_2' m_2 | j m j_1 j_2\rangle \quad .$$

Therefore, only the coefficients with $j_1' = j_1$ and $j_2' = j_2$ are nonvanishing. If one further considers the matrix element of $J_z = J_{1z} + J_{2z}$,

$$\begin{aligned}\langle j_1 m_1 j_2 m_2 | J_z | j m j_1 j_2\rangle &= (m_1 + m_2)\langle j_1 m_1 j_2 m_2 | j m j_1 j_2\rangle \\ &= m\langle j_1 m_1 j_2 m_2 | j m j_1 j_2\rangle \quad ,\end{aligned}$$

one finds that the coefficients (10.29) are nonvanishing only for $m = m_1 + m_2$. Thus, the expansion of $|j m j_1 j_2\rangle$ in the basis $|j_1 m_1 j_2 m_2\rangle$ can be reduced to

$$|j m j_1 j_2\rangle = \sum_{\substack{m_1 \\ m_2 = m - m_1}} |j_1 m_1 j_2 m_2\rangle \langle j_1 m_1 j_2 m_2 | j m j_1 j_2\rangle \quad . \tag{10.30}$$

The $\langle j_1 m_1 j_2 m_2 | j m j_1 j_2\rangle$ are called *Clebsch–Gordan coefficients*. We now wish to determine the values of j for given j_1 and j_2. The values of m_1 and m_2 corresponding to j_1 and j_2 are

$$m_1 = j_1, j_1 - 1, \ldots, -j_1 \quad \text{and} \quad m_2 = j_2, j_2 - 1, \ldots, -j_2 \quad .$$

Since $m = m_1 + m_2$, the values of m and all the corresponding m_1 and m_2 can immediately be given. This is presented in Table 10.1, where we assume without loss of generality $j_1 \geq j_2$.

Table 10.1. Values of m

m	(m_1, m_2)		Degeneracy of m
$j_1 + j_2$	(j_1, j_2)		1
$j_1 + j_2 - 1$	$(j_1 - 1, j_2), (j_1, j_2 - 1)$		2
$j_1 + j_2 - 2$	$(j_1 - 2, j_2), (j_1 - 1, j_2 - 1), (j_1, j_2 - 2)$		3
$\vdots$	$\vdots$	$\ddots$	$\vdots$
$j_1 - j_2$	$(j_1 - 2j_2, j_2), \ldots,$	$(j_1, j_2 - 2j_2)$	$2j_2 + 1$
$j_1 - j_2 - 1$	$(j_1 - 2j_2 - 1, j_2), \ldots,$	$(j_1 - 1, -j_2)$	$2j_2 + 1$
$\vdots$	$\vdots$	$\vdots$	$\vdots$
$-(j_1 - j_2)$	$(-j_1, j_2), \ldots,$	$(-j_1 + 2j_2, -j_2)$	$2j_2 + 1$
$-(j_1 - j_2) - 1$	$(-j_1, j_2 - 1), \ldots,$	$(-j_1 + 2j_2 - 1, -j_2)$	$2j_2$
$\vdots$		$\vdots$	$\vdots$
$-(j_1 + j_2)$		$(-j_1, -j_2)$	1

Using Table 10.1, one can easily convince oneself that the degeneracy is given according to Table 10.2.

Table 10.2. Degeneracy of m

m	Degree of degeneracy
$m \geq |j_1 - j_2|$	$j_1 + j_2 - m + 1$
$-|j_1 - j_2| < m < |j_1 - j_2|$	$j_1 + j_2 - |j_1 - j_2| + 1$
$m \leq -|j_1 - j_2|$	$j_1 + j_2 - |m| + 1$

The possible j-values can now be determined by beginning with the maximally aligned states. Since $j_1 + j_2$ is the maximal m-value, $j_1 + j_2$ must also be the largest value of the total angular momentum j. By the general properties of angular momentum operators, the corresponding m-values are then $j_1 + j_2, j_1 + j_2 - 1, \ldots, -j_1 - j_2$. These values of m are thus already used up. Since the value $m = j_1 + j_2 - 1$ is doubly degenerate, it now is the highest value of m, so that the next highest value of the total angular momentum is $j = j_1 + j_2 - 1$. The corresponding m-values are $j_1 + j_2 - 1, \ldots, -j_1 - j_2 + 1$. Since the value $m = j_1 + j_2 - 2$ has three-fold degeneracy, this m-value is now the highest, so that one can take the next multiplet to be $j = j_1 + j_2 - 2$. The argument can be continued in this manner, until one finally obtains the multiplet structure shown in Table 10.3.

Table 10.3. Multiplets of j

Values of j	Corresponding m-Values
$j_1 + j_2$	$j_1 + j_2, \ldots, -(j_1 + j_2)$
$j_1 + j_2 - 1$	$j_1 + j_2 - 1, \ldots, -(j_1 + j_2 - 1)$
$\vdots$	$\vdots$
$|j_1 - j_2|$	$|j_1 - j_2|, \ldots, -|j_1 - j_2|$

The possible values of j are thus given by the triangular relation

$$|j_1 - j_2| \leq j \leq j_1 + j_2 \quad , \tag{10.31}$$

where the values of j proceed in integer steps. Evidently, one obtains a half-integral angular momentum upon adding an integral and a half-integral angular momentum. The two limiting values in (10.31) correspond to angular momenta aligned parallel and antiparallel.

Remark: As a check, we calculate the number of states $|jmj_1j_2\rangle$ for fixed values $j_1 \geq j_2$:

$$
\begin{aligned}
\sum_{j=|j_1-j_2|}^{j_1+j_2} (2j + 1) &= \sum_{k=0}^{2j_2} 2(j_1 - j_2 + k) + 1 \\
&= (2(j_1 + j_2) + 1)(2j_2 + 1) - 2j_2(2j_2 + 1) \\
&= (2j_1 + 1)(2j_2 + 1) \quad ,
\end{aligned}
$$

which clearly agrees with the number of states $|j_1m_1\rangle|j_2m_2\rangle$.

Instead of the Clebsch–Gordan coefficients, the *Wigner 3j-symbol* is also used:

$$\begin{pmatrix} j_1 & j_2 & j_3 \\ m_1 & m_2 & m_3 \end{pmatrix} = (-1)^{j_1-j_2-m_3} \frac{\langle j_1m_1j_2m_2|j_3 - m_3j_1j_2\rangle}{\sqrt{2j_3+1}} \quad . \tag{10.32}$$

For the addition of spin and orbital angular momentum treated earlier $j_1 = l$, $j_2 = s = 1/2$, $j = l + 1/2$, $l - 1/2$, and for this case we collect the Clebsch–Gordan coefficients $\langle jml\frac{1}{2}|lm_1\frac{1}{2}m_2\rangle$ for $l \geq 1$ in Table 10.4.

Table 10.4. Clebsch–Gordan coefficients $\langle jml\frac{1}{2}|lm_1, \frac{1}{2}m_2\rangle$ for $l \geq 1$

m_2 j	$\frac{1}{2}$	$-\frac{1}{2}$
$l + \frac{1}{2}$	$\left(\frac{l+m+1/2}{2l+1}\right)^{1/2}$	$\left(\frac{l-m+1/2}{2l+1}\right)^{1/2}$
$l - \frac{1}{2}$	$-\left(\frac{l-m+1/2}{2l+1}\right)^{1/2}$	$\left(\frac{l+m+1/2}{2l+1}\right)^{1/2}$

Problems

10.1 Let the Hamiltonian of a two-spin system be given by

$$H = A + B\boldsymbol{S}_1 \cdot \boldsymbol{S}_2 + C\left(S_{1,z} + S_{2,z}\right) \quad .$$

Determine the eigenvalues and eigenfunctions if the particles are identical and have spin $1/2$.

10.2 Suppose that a system consists of two distinguishable particles each with spin $S = 1/2$ and is described by the Hamiltonian

$$H = -\frac{a+b}{2}\left(S_{1,z} + S_{2,z}\right)B - \frac{a-b}{2}\left(S_{1,z} - S_{2,z}\right)B + J\boldsymbol{S}_1 \cdot \boldsymbol{S}_2 \quad ,$$

where a, b, and J are constants, and B represents the external magnetic field. Determine the energy eigenvalues.

10.3 Derive (10.25).

10.4 Prove the validity of (10.27b).

11. Approximation Methods for Stationary States

Although we have succeeded in solving some important and interesting quantum mechanical problems, an exact solution is not possible in complicated situations, and we must then resort to approximation methods. For the calculation of stationary states and energy eigenvalues, these include *perturbation theory, the variational method*, and *the WKB approximation*. Perturbation theory is applicable if the problem differs from an exactly solvable problem by a small amount. The variational method is appropriate for the calculation of the ground state energy if one has a qualitative idea of the form of the wave function, and finally the WKB method is applicable in the nearly classical limit.

11.1 Time Independent Perturbation Theory (Rayleigh–Schrödinger)

Let the Hamiltonian H consist of two parts,

$$H = H_0 + \lambda H_1 \quad . \tag{11.1}$$

Let the eigenvalues E_n^0 and eigenfunctions $|n^0\rangle$ of the operator H_0 be known exactly,

$$H_0|n^0\rangle = E_n^0|n^0\rangle \quad , \tag{11.2}$$

and the "perturbation" λH_1 be, in comparison to H_0, a small additional term. One seeks the discrete stationary states $|n\rangle$ and eigenvalues E_n of H,

$$H|n\rangle = E_n|n\rangle \quad . \tag{11.3}$$

We now assume that the eigenvalues and eigenfunctions can be expanded in a power series in the parameter λ

$$E_n = E_n^0 + \lambda E_n^1 + \lambda^2 E_n^2 + \ldots \quad ,$$

$$|n\rangle = |n^0\rangle + \lambda|n^1\rangle + \lambda^2|n^2\rangle + \ldots \quad , \tag{11.4}$$

where in both expansions the first term is the "unperturbed" one.

Remarks:

(i)   This series is frequently not convergent. However, in many cases it is an asymptotic expansion[1], i.e., the first few terms nevertheless give reliable results. The crucial point is whether in the limit $\lambda \to 0$, $E_n \to E_n^0$ and $|n\rangle \to |n^0\rangle$ are valid. Perturbation theory works whenever the state with finite λ does not differ qualitatively from the state with $\lambda = 0$.

(ii)   In some cases, E_n and $|n\rangle$ are not expandable in λ. For example, the binding energy of Cooper pairs is $\Delta \approx \omega_D \exp\{-1/VN(0)\}$. The bound states of a potential cannot be obtained from the continuum states by means of perturbation theory.

(iii) The smallness of the perturbation is by no means always evident from a small coupling parameter λ, but can be hidden in the structure of H_1.

11.1.1 Nondegenerate Perturbation Theory

We first develop perturbation theory for nondegenerate unperturbed states $|n^0\rangle$ of the discrete part of the spectrum. From the time independent Schrödinger equation (11.3) and from (11.4)

$$(H_0 + \lambda H_1)(|n^0\rangle + \lambda|n^1\rangle + \lambda^2|n^2\rangle + \ldots)$$
$$= (E_n^0 + \lambda E_n^1 + \lambda^2 E_n^2 + \ldots)(|n^0\rangle + \lambda|n^1\rangle + \lambda^2|n^2\rangle + \ldots) \quad ,$$

one obtains for λ by comparison of the coefficients of $\lambda^0, \lambda^1, \lambda^2 \ldots$

$$H_0|n^0\rangle = E_n^0|n^0\rangle \quad , \tag{11.5a}$$
$$H_0|n^1\rangle + H_1|n^0\rangle = E_n^0|n^1\rangle + E_n^1|n^0\rangle \quad , \tag{11.5b}$$
$$H_0|n^2\rangle + H_1|n^1\rangle = E_n^0|n^2\rangle + E_n^1|n^1\rangle + E_n^2|n^0\rangle \quad . \tag{11.5c}$$

$$\vdots$$

It is most convenient to fix the normalization of $|n\rangle$ by

$$\langle n^0|n\rangle = 1 \quad , \tag{11.6a}$$

i.e.,

$$\lambda\langle n^0|n^1\rangle + \lambda^2\langle n^0|n^2\rangle + \ldots = 0 \quad ,$$

whence follows

$$\langle n^0|n^1\rangle = \langle n^0|n^2\rangle = \ldots = 0 \quad . \tag{11.6b}$$

---

[1] An asymptotic expansion for a function $f(\lambda)$, $f(\lambda) = \sum_{k=0}^m a_k\lambda^k + R_m(\lambda)$, is characterized by the following behavior of the remainder: $\lim_{\lambda \to 0}(R_m(\lambda)/\lambda^m) = 0$, $\lim_{m \to \infty} R_m(\lambda) = \infty$.

We would now like to determine the expansion coefficients. To this end, we multiply (11.5b) by $\langle n^0|$ and use (11.5a), with the result

$$E_n^1 = \langle n^0|H_1|n^0\rangle \quad . \tag{11.7a}$$

Since the unperturbed states $|m^0\rangle$ form a complete orthonormal set, by (11.6b) we have the expansion

$$|n^1\rangle = \sum_{m\neq n} c_m|m^0\rangle \tag{11.8a}$$

with

$$c_m = \langle m^0|n^1\rangle \quad . \tag{11.8b}$$

Multiplying (11.5b) by $\langle m^0|$ (different from $\langle n^0|$), we find

$$c_m(E_n^0 - E_m^0) = \langle m^0|H_1|n^0\rangle$$

and thus the first correction to the state $|n^0\rangle$:

$$|n^1\rangle = \sum_{m\neq n} \frac{\langle m^0|H_1|n^0\rangle}{E_n^0 - E_m^0}|m^0\rangle \quad . \tag{11.8c}$$

The energy in second order is obtained by multiplying equation (11.5c) by $\langle n^0|$ and using (11.6b) and (11.8c):

$$E_n^2 = \langle n^0|H_1|n^1\rangle = \sum_{m\neq n} \frac{|\langle m^0|H_1|n^0\rangle|^2}{E_n^0 - E_m^0} \quad . \tag{11.7b}$$

Remarks:

(i)   For the ground state, the second-order shift E_0^2 is negative.
(ii) If the matrix elements of H_1 are of comparable magnitude, neighboring levels make a larger contribution in second-order perturbation theory than distant levels.
(iii) If an important (large matrix element, small distance) level E_m^0 lies above E_n^0, then E_n is pushed down and E_m is pushed up; the levels spread apart as if they repelled each other.
(iv) In (11.7b), in the continuous part of the spectrum, the sum over m should be replaced by an integral.

11.1.2 Perturbation Theory for Degenerate States

Let $|n_a^0\rangle, |n_b^0\rangle, \ldots, |n_k^0\rangle$ be degenerate, i.e.,

$$H_0|n_i^0\rangle = \varepsilon|n_i^0\rangle \quad .$$

(11.9)

Since perturbation theory is an expansion in $\lambda\langle m^0|H_1|n^0\rangle/(E_n^0 - E_m^0)$, in place of this basis a basis $|n_\alpha^0\rangle$ must be used in which

$$\langle n_\alpha^0|H_1|n_\beta^0\rangle = H_1^{(\alpha)}\delta_{\alpha\beta} \quad ,$$

(11.10)

thus vanishing for $\alpha \neq \beta$, in order that no zero energy denominators (which would lead to divergences) occur.

The matrix elements

$$H_{1ij} = \langle n_i^0|H_1|n_j^0\rangle$$

(11.11)

form a Hermitian matrix. The new states

$$|n_\alpha^0\rangle = \sum_i c_{i\alpha}|n_i^0\rangle$$

(11.12)

appropriate for perturbation theory give matrix elements

$$H_{1\alpha\beta} = \langle n_\alpha^0|H_1|n_\beta^0\rangle = \sum_{i,j} c_{i\alpha}^* H_{1ij} c_{j\beta} \quad .$$

(11.13)

Since any Hermitian matrix can be diagonalized by a unitary transformation, it is always possible to choose the $c_{i\alpha}$ in such a way that (11.10) is satisfied. In general, one must solve an eigenvalue problem, but often the correct states can be guessed. From (11.10) and (11.13), one obtains

$$\sum_{i,j} c_{i\alpha}^* H_{1ij} c_{j\beta} = H_1^{(\alpha)}\delta_{\alpha\beta} \quad ,$$

(11.14a)

and, multiplying by $c_{i\alpha}$ and using unitarity ($\sum_i c_{i\alpha}^* c_{i\beta} = \delta_{\alpha\beta}, \sum_\alpha c_{i\alpha} c_{j\alpha}^* = \delta_{ij}$), one obtains the eigenvalue equation

$$\sum_j H_{1ij} c_{j\beta} = H_1^{(\beta)} c_{i\beta} \quad .$$

(11.14b)

The solvability condition implies the vanishing of the secular determinant

$$\det\left(H_{1ij} - H_1^{(\beta)}\delta_{ij}\right) = 0 \quad .$$

(11.14c)

From (11.14c) and (11.14b), one obtains the $H_1^{(\beta)}$ and $c_{i\beta}$ in the usual way and thus the appropriate unperturbed states (11.12) for the application of the perturbation theory developed in Sect. 11.1.1.

Here, an additional remark may be of use. In any nontrivial problem, H_0 and H_1 do not commute. These two operators cannot therefore be diagonalized simultaneously. However, in the subspace of a group of degenerate states, the reduced matrix corresponding to H_0 is a unit matrix, and hence it is possible to find a new basis in this subspace in which H_0 and H_1 are both diagonal. Of course, even after this transformation has been carried out, there are still nonvanishing nondiagonal matrix elements of H_1 for states of differing E_n^0.

11.2 The Variational Principle

For a Hamiltonian H with basis $|n\rangle$, one has for arbitrary $|\psi\rangle$

$$\langle\psi|H|\psi\rangle = \sum_n \langle\psi|n\rangle\langle n|H|\psi\rangle = \sum_n E_n\langle\psi|n\rangle\langle n|\psi\rangle$$
$$\geq E_0 \sum_n |\langle\psi|n\rangle|^2 = E_0\langle\psi|\psi\rangle$$

and thus

$$E_0 \leq \frac{\langle\psi|H|\psi\rangle}{\langle\psi|\psi\rangle} \quad . \tag{11.15}$$

The Ritz variational principle consists in specifying $|\psi(\mu)\rangle$ to be a function of one or more parameters μ and seeking the minimum of the expression

$$E(\mu) = \frac{\langle\psi(\mu)|H|\psi(\mu)\rangle}{\langle\psi(\mu)|\psi(\mu)\rangle} \quad . \tag{11.16}$$

The minimum of $E(\mu)$ is then an upper bound for the ground state energy.

In this procedure, an error in the wave function manifests itself at quadratic order in the energy; i.e., let

$$|\psi\rangle = |n\rangle + |\varepsilon\rangle \tag{11.17}$$

with $\langle n|\varepsilon\rangle = 0$; then

$$\frac{\langle\psi|H|\psi\rangle}{\langle\psi|\psi\rangle} = \frac{E_n + \langle\varepsilon|H|\varepsilon\rangle}{\langle n|n\rangle + \langle\varepsilon|\varepsilon\rangle} = E_n + O(\varepsilon^2) \quad . \tag{11.18}$$

The energy is thus determined more accurately than the wave function under application of a variational principle.

Aside from its function as an approximation method, the variational principle is also an important tool in mathematical physics in the proof of exact inequalities.

11.3 The WKB (Wentzel–Kramers–Brillouin) Method

We now wish to consider the stationary states of a potential for energies sufficiently high such that the typical wavelength of the state is small compared to the characteristic distance over which the potential varies significantly. In this "quasi-classical" limit, we expect that the wave function can be characterized by a position dependent wave number.

To analyze this behavior systematically, we represent the wave function by an amplitude A and a phase S,

$$\psi(\boldsymbol{x}) = A(\boldsymbol{x})e^{iS(\boldsymbol{x})/\hbar} \quad . \tag{11.19}$$

Substituting this into the time independent Schrödinger equation

$$\frac{-\hbar^2}{2m}\boldsymbol{\nabla}^2\psi = (E - V(\boldsymbol{x}))\psi \quad ,$$

one finds

$$A(\boldsymbol{\nabla}S)^2 - i\hbar A\boldsymbol{\nabla}^2 S - 2i\hbar(\boldsymbol{\nabla}A)(\boldsymbol{\nabla}S) - \hbar^2\boldsymbol{\nabla}^2 A = 2m(E - V)A \quad . \tag{11.20}$$

Comparing the first two terms, we expect that the quasi-classical region is given by

$$(\boldsymbol{\nabla}S)^2 \gg \hbar\boldsymbol{\nabla}^2 S \quad . \tag{11.21}$$

We take the real and imaginary parts of (11.20):

$$(\boldsymbol{\nabla}S)^2 = 2m(E - V) + \hbar^2(\boldsymbol{\nabla}^2 A)/A \quad , \tag{11.22a}$$

$$-\boldsymbol{\nabla}^2 S = 2\boldsymbol{\nabla}S \cdot \boldsymbol{\nabla}\log A \quad . \tag{11.22b}$$

Below, we consider only one-dimensional problems, which of course also include radial motion in central potentials. One can then express (11.22b) in the form

$$\frac{d}{dx}\left(\frac{1}{2}\log\frac{dS}{dx} + \log A\right) = 0 \quad ,$$

and one finds

$$A = \frac{C}{\sqrt{S'}} \quad . \tag{11.23}$$

In (11.22a), we neglect the term $\hbar^2(d^2A/dx^2)/A$ compared to $(dS/dx)^2$. The resulting equation

$$\left(\frac{dS}{dx}\right)^2 = 2m(E - V(x)) \tag{11.24}$$

can then easily be integrated:

$$S(x) = \pm \int^x dx' \sqrt{2m(E - V(x'))} \quad . \tag{11.25}$$

Substituting (11.25) and (11.23) into (11.19), one finds

$$\psi(x) = \sum_{\pm} \frac{C_{\pm}}{\sqrt{p(x)}} \exp\{\pm i \int dx\, p(x)/\hbar\} \tag{11.26}$$

with momentum

$$p(x) = \sqrt{2m(E - V(x))} \quad . \tag{11.27}$$

For $E < V$, the situation in tunneling problems, the solutions (11.26) become exponentially growing or decreasing, i.e.:

$$\psi(x) = \sum_{\pm} \frac{C'_{\pm}}{\sqrt{\kappa(x)}} \exp\{\mp \int dx\, \kappa(x)/\hbar\} \tag{11.26'}$$

with

$$\kappa(x) = \sqrt{2m(V(x) - E)} \quad . \tag{11.27'}$$

We now wish to determine the *bound states* in the potential $V(x)$. For the energy E, let the classical turning points be b and a (Fig. 11.1, $V(a) = V(b) = E$). By an appropriate choice of the coordinate system ($V - E \to V$, $x - b \to x$) the potential near the turning point b (expansion to first order in x) is

$$V = V'x \quad \text{with} \quad V' < 0 \quad . \tag{11.28}$$

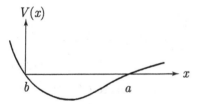

Fig. 11.1. Turning points b and a

The Schrödinger equation near b

$$\frac{d^2}{dx^2}\psi = -c^2 x\psi \tag{11.29}$$

with

$$c = (-2mV')^{1/2}/\hbar \tag{11.30}$$

is solved by the *Airy functions*, which are linear combinations of

$$\psi(x) = x^{1/2} J_{\pm 1/3}\left(\frac{2c}{3}x^{3/2}\right) \quad . \tag{11.31}$$

Here, $J_n(z)$ is the nth order Bessel function, which frequently arises in cylindrically symmetrical problems. It satisfies *Bessel's differential equation*

$$\left[\frac{d^2}{dz^2} + \frac{1}{z}\frac{d}{dz} + \left(1 - \frac{n^2}{z^2}\right)\right] J_n(z) = 0 \quad . \tag{11.32}$$

In the neighborhood of the turning point, the relevant characteristic length is

$$l_0 = (3/2c)^{2/3} \approx (\hbar^2/m|V'|)^{1/3} \quad . \tag{11.33}$$

Far from the turning point, that is for $x \gg l_0$, the asymptotic behavior is

$$x^{1/2} J_{\pm 1/3}\left(\frac{2c}{3}x^{3/2}\right) \propto x^{-1/4}\cos\left(\frac{2c}{3}x^{3/2} \mp \frac{\pi}{6} - \frac{\pi}{4}\right) \quad . \tag{11.34}$$

For $x \to -\infty$, the quantity $J_{\pm 1/3}(\mathrm{i}|x|^{3/2} \dots)$ contains an exponentially growing and an exponentially decreasing part. The two solutions must be combined in such a way that only the decreasing part remains. Since

$$\int dx\, p(x)/\hbar = \int dx \sqrt{2m(-V')x}/\hbar = \tfrac{2}{3}cx^{3/2} \quad , \tag{11.35}$$

one can then determine the coefficients in (11.26). The result is

$$\psi(x) = \frac{C}{\sqrt{p(x)}}\cos\left(\frac{1}{\hbar}\int_b^x dx'\, p(x') - \frac{\pi}{4}\right) \quad . \tag{11.36}$$

At the other turning point, reflection of (11.36) implies

$$\begin{aligned}
\psi(x) &= \frac{C'}{\sqrt{p(x)}}\cos\left(\frac{1}{\hbar}\int_x^a dx'\, p(x') - \frac{\pi}{4}\right) \\
&= \frac{C'}{\sqrt{p(x)}}\cos\left(\frac{1}{\hbar}\int_b^x dx'\, p(x') - \left(\frac{1}{\hbar}\int_b^a dx'\, p(x') - \frac{\pi}{4}\right)\right) \quad .
\end{aligned} \tag{11.37}$$

The condition that (11.37) should agree with (11.36), where $C' = \pm C$, implies for the loop integral of the momentum

$$\frac{1}{2\pi\hbar}\oint dx\, p(x) = n + \frac{1}{2} \quad . \tag{11.38}$$

This is the *Bohr–Sommerfeld quantization condition,* from which one can determine the allowed energy levels of the bound states. According to our original supposition, (11.38) is applicable for large number of nodes n.

We now discuss briefly the domain of validity of the WKB approximation. In the transition from (11.22a) to (11.24), it was assumed that

$$\hbar^2(d^2 A/dx^2)/A \ll (dS/dx)^2 \quad . \tag{11.39}$$

The supposition (11.21) can also be written in the form

$$\left|\frac{dp}{dx}\right| \ll \frac{1}{\hbar}p^2 = 2\pi\frac{p}{\lambda} \tag{11.40}$$

with $\lambda = 2\pi\hbar/p$. Using (11.23), one easily sees that (11.40) also implies the validity of (11.39).

Remark: Using (11.27), one can also write the condition (11.40) in the form $m\hbar|V'|/p^3 \ll 1$. To the right of the turning point b, one has $p = \sqrt{2m|V'|}x^{1/2}$ and

$$\frac{\hbar}{(m|V'|)^{1/2}} \ll x^{3/2} \quad .$$

This is identical to the domain of validity $x \gg l_0$ of the asymptotic expansion of the Airy function. Thus, the two conditions are compatible. However, in this x-region, $V''x \ll |V'|$ and in particular $V''l_0 \ll |V'|$ must also hold. For atoms, $V = Ze^2/r$, $V' \approx -Ze^2/a^2$, $V'' \approx Ze^2/a^3$ and one obtains $l_0 \ll a$. Since $l_0 \propto Z^{-1/3}$, the quasiclassical approximation is reliable for large Z.

11.4 Brillouin–Wigner Perturbation Theory

We now discuss the Brillouin–Wigner perturbation theory, in which the perturbation series for the stationary states contains the exact energy eigenvalues. We begin with the time independent Schrödinger equation

$$(E_n - H_0)|n\rangle = \lambda H_1|n\rangle \quad , \tag{11.41a}$$

where we again fix the normalization of $|n\rangle$ by

$$\langle n^0|n\rangle = 1 \tag{11.41b}$$

and multiply (11.41a) by $\langle n^0|$. This yields the following relation for the energy eigenvalue E_n:

$$E_n = E_n^0 + \lambda\langle n^0|H_1|n\rangle \quad . \tag{11.42}$$

In order to determine the state $|n\rangle$, we now multiply (11.41a) by $\langle m^0|$,

$$\left(E_n - E_m^0\right)\langle m^0|n\rangle = \lambda\langle m^0|H_1|n\rangle \quad . \tag{11.43}$$

Substituting this into the expansion of $|n\rangle$ in unperturbed states

$$|n\rangle = |n^0\rangle + \sum_m{}'|m^0\rangle\langle m^0|n\rangle \quad , \tag{11.44}$$

we obtain

$$|n\rangle = |n^0\rangle + {\sum_m}' |m^0\rangle \frac{1}{E_n - E_m^0} \lambda \langle m^0|H_1|n\rangle \quad . \tag{11.44'}$$

The prime indicates that in the sum over m the term $m = n$ is excluded. We can solve this equation for the state $|n\rangle$, which appears on both the left and right hand sides, by iteration (i.e., by repeatedly replacing $|n\rangle$ on the right side by the entire right side):

$$|n\rangle = |n^0\rangle + \lambda {\sum_m}' |m^0\rangle \frac{1}{E_n - E_m^0} \langle m^0|H_1|n^0\rangle$$
$$+ \lambda^2 {\sum_{j,m}}' |j^0\rangle \frac{1}{E_n - E_j^0} \langle j^0|H_1|m^0\rangle \frac{1}{E_n - E_m^0} \langle m^0|H_1|n^0\rangle$$
$$+ \dots \quad . \tag{11.45}$$

Here, the exact energy E_n arises. If E_n is known – e.g., from a variational calculation – the Brillouin–Wigner series for $|n\rangle$ provides a more rapidly converging series than the Rayleigh–Schrödinger series for $|n\rangle$. If one substitutes for E_n the Rayleigh–Schrödinger perturbation series, (11.45) again yields the Rayleigh–Schrödinger series for $|n\rangle$.

Problems

11.1 For the anharmonic, one-dimensional oscillator described by the potential energy

$$V(x) = \frac{m}{2}\omega^2 x^2 + bx^3 + cx^4 \quad ,$$

compute the energy eigenvalues in first-order perturbation theory. Include second-order perturbation theory with respect to the contribution ax^3. Hint: The matrix elements occurring here can be computed most simply by transforming the position coordinate to a and $a^\dagger$.

11.2 Two coupled, identical rotators are described by the Hamiltonian

$$H = A\left(p_{\Theta_1}^2 + p_{\Theta_2}^2\right) - B\cos\left(\Theta_1 - \Theta_2\right) \quad ,$$

where A and $B > 0$, $p_{\Theta_i} = (\hbar/i)\partial/\partial\Theta_i$, and $\Theta_i + 2\pi$ is equivalent to Θ_i. For the case $B \ll A\hbar^2$, calculate the energy eigenvalues using perturbation theory. In the opposite case $B \gg A\hbar^2$, assume that only small oscillations occur, and construct the solution in analogy to the harmonic oscillator. Hint: Introduce new variables $x = \Theta_1 + \Theta_2$ and $y = \Theta_1 - \Theta_2$. In terms of these variables, the Hamiltonian becomes

$$H = -2A\hbar^2\left(\frac{\partial^2}{\partial x^2} + \frac{\partial^2}{\partial y^2}\right) - B\cos y \quad .$$

11.3 In dimensionless variables $\left(\hbar = 1, m = 1, \text{ and thus } p_j = \frac{1}{i}\frac{\partial}{\partial x_j}\right)$ the Hamiltonian of a two-dimensional oscillator takes the form

$$H = \frac{1}{2}\left(p_x^2 + p_y^2\right) + \frac{1}{2}(1 + \delta xy)(x^2 + y^2) \quad ,$$

where we suppose that $\delta \ll 1$. Determine the wave functions for the three lowest lying energy levels in the case $\delta = 0$. Calculate the shift of these levels for $\delta \neq 0$ in first-order perturbation theory. Note the degeneracy which occurs.

11.4 A one-dimensional harmonic oscillator carrying charge e is located in an external electric field of strength E pointing in the positive x-direction:

$$H = -\frac{\hbar^2}{2m}\frac{d^2}{dx^2} + \frac{m\omega^2}{2}x^2 - eEx \quad .$$

Calculate the energy levels in second-order and the wave function in first-order perturbation theory and compare with the exact result.

11.5 Estimate, using the variational principle, the ground state energy of the one-dimensional harmonic oscillator. Use as a test function $\psi(\mu) = N e^{-\mu x^2}$, where the parameter $\mu > 0$.

11.6 (a) Using the Bohr–Sommerfeld quantization rule, determine the energy levels of a charged, spinless particle which is under the influence of a homogeneous magnetic field.

(b) Show that the magnetic flux enclosed in a semi classical Bohr-Sommerfeld orbit is a multiple of $\hbar c/e$.

11.7 Consider the potential from Problem 3.13.

(a) Find the ground state energy by the variational ansatz

$$\psi_0(x) = x\, e^{-\kappa_0 x} \quad .$$

(b) Find the energy of the first excited state by the variational ansatz

$$\psi_1(x) = x(x - n)\, e^{-\kappa_1 x} \quad ,$$

where the coefficients are to be chosen such that ψ_1 is orthogonal to ψ_0.

(c) Compare with the exact result.

(d) What is the result of the variational ansatz

$$\psi_0 = x\, e^{-\kappa_0 x^2} \quad ?$$

11.8 Motion of a particle in a gravitational field and reflecting floor: The potential takes the form

$$V(z) = \begin{cases} gz & z > 0 \\ \infty & z < 0 \end{cases} \quad .$$

(a) Solve the corresponding Schrödinger equation exactly using the Airy functions. (Remark: the same problem occurs in the motion of electrons in a semiconducting half-space under the action of an electric field.)

(b) Carry out a variational ansatz as in Problem 11.7 for the ground state.

11.9 The energy levels following from the secular equation (11.14c) are non degenerate in most cases. Consider a Hamiltonian H_0 with two degenerate eigenstates $|a\rangle$ and $|b\rangle$.

(a) Show that these remain degenerate iff $\langle a|H_1|a\rangle = \langle b|H_1|b\rangle$ and $\langle a|H_1|b\rangle = 0$.

(b) To lift the degeneracy then one has to turn to second order. Derive from (11.5c) a secular equation of the type (11.14b) with the replacement

$$H_{1ij} \rightarrow H_{1ij} + \sum_{n \neq a,b} \frac{\langle i|H_1|n\rangle\langle n|H_1|j\rangle}{E_i^0 - E_n^0} \quad,$$

where $E_i^0 \equiv E_a^0 = E_b^0$.

12. Relativistic Corrections

The relativistic corrections – also known as *fine-structure corrections* – consist of the following contributions:

(i)   relativistic kinetic energy
(ii) spin–orbit coupling
(iii) the Darwin term

These follow from the Dirac equation in an expansion in $(v/c)^2$. According to our estimate in Sect. 6.3, $v/c \approx Z\alpha$, these corrections are smaller than the Coulomb energy by a factor of $(Z\alpha)^2$. We will discuss each of the terms in turn by explaining their physical origin heuristically and by taking their precise form from the theory of the Dirac equation.[1]

12.1 Relativistic Kinetic Energy

On the basis of the relativistic energy–momentum relation

$$E = \sqrt{p^2c^2 + m^2c^4} = mc^2 + \frac{p^2}{2m} - \frac{1}{8}\frac{(p^2)^2}{m^3c^2} + \cdots \quad ,$$

in addition to the Hamiltonian of the hydrogen atom of Chap. 6

$$H_0 = \frac{p^2}{2m} - \frac{Ze^2}{r} \quad , \tag{12.1}$$

a perturbation

$$H_1 = -\frac{1}{8}\frac{(p^2)^2}{m^3c^2} \tag{12.2}$$

arises.

In comparison to $p^2/2m$, H_1 is smaller by a factor of $p^2/m^2c^2 = v^2/c^2 \approx (Z\alpha)^2$, i.e., H_1 is only a small perturbation for small atomic number. With the identity

$$H_1 = -\frac{1}{2mc^2}\left(H_0 + \frac{Ze^2}{r}\right)^2 \tag{12.3}$$

---

[1] For references see p. 225. See, e.g., F. Schwabl, QM II, Sect. 9.2.

one obtains for the energy shift of the state $|nlm\rangle$ in first-order perturbation theory

$$\Delta E_{nlm} = \langle nlm|H_1|nlm\rangle$$

$$= -\frac{1}{2mc^2}\left(E_n^2 + 2E_n Ze^2 \left\langle\frac{1}{r}\right\rangle_{nl} + (Ze^2)^2\left\langle\frac{1}{r^2}\right\rangle_{nl}\right) \quad . \tag{12.4}$$

With the expectation values

$$\left\langle\frac{1}{r}\right\rangle_{nl} = \frac{Z}{an^2} \quad , \tag{12.4a}$$

$$\left\langle\frac{1}{r^2}\right\rangle_{nl} = \frac{Z^2}{a^2n^3(l+1/2)} \tag{12.4b}$$

determined at the end of Sect. 12.3 and (6.24′), (12.4) becomes

$$\Delta E_{nlm} = -\frac{mc^2(Z\alpha)^2}{2n^2}\frac{(Z\alpha)^2}{n^2}\left(\frac{n}{l+1/2} - \frac{3}{4}\right) \quad . \tag{12.5}$$

Since H_1 is negative definite, $\langle H_1\rangle_{nl} < 0$ holds for all n and l, as one easily confirms in (12.5).

Remarks:

(i)   The correction is of the order of magnitude $Ry\alpha^2$, where α is the fine-structure constant,

$$\alpha = \frac{e_0^2}{\hbar c} = \frac{1}{137.036} \quad .$$

(ii)   Perturbation theory with the states $|nlm\rangle$ is permissible since

$$\langle nlm|H_1|nl'm'\rangle = 0 \quad \text{for} \quad l \neq l' \quad \text{or} \quad m \neq m' \quad . \tag{12.6}$$

Proof:

One has $[H_1, \boldsymbol{L}^2] = 0$, whence

$$\hbar^2[l'(l'+1) - l(l+1)]\langle nlm|H_1|nl'm'\rangle = 0 \tag{12.7a}$$

follows. Similarly, from $[H_1, L_z] = 0$ the equation

$$\hbar(m'-m)\langle nlm|H_1|nl'm'\rangle = 0 \tag{12.7b}$$

follows. For the states $|njj_z ls\rangle$, to be introduced later, the nondiagonal matrix elements of H_1 vanish as well.

12.2 Spin–Orbit Coupling

From the relativistic Dirac equation, the spin–orbit coupling

$$H_2 = \frac{1}{2m^2c^2} \boldsymbol{S} \cdot \boldsymbol{L} \frac{1}{r} \frac{d}{dr} V(r) \tag{12.8}$$

follows, where $V(r) = e\Phi(r)$ is the potential energy in the electrostatic potential $\Phi(r)$ of the nucleus. In order to understand (12.8) heuristically, we note that the electric field of the nucleus is $\boldsymbol{E} = -\boldsymbol{\nabla}\Phi = -(\boldsymbol{x}/r)d\Phi/dr$. In the rest frame of the electron, the nucleus orbits about the electron, which feels a magnetic field $\boldsymbol{B} = -\boldsymbol{v} \times \boldsymbol{E}/c$.

The energy of the magnetic moment of the electron is then

$$
\begin{aligned}
-\frac{e}{mc}\boldsymbol{S} \cdot \boldsymbol{B} &= -\frac{e}{mc^2}\boldsymbol{S} \cdot (\boldsymbol{v} \times \boldsymbol{x})\frac{1}{r}\frac{d}{dr}\Phi(r) \\
&= \frac{1}{m^2c^2}\boldsymbol{S} \cdot \boldsymbol{L}\frac{1}{r}\frac{d}{dr}V(r) \quad .
\end{aligned}
\tag{12.9}
$$

However, this result is too large by a factor of 2 compared to (12.8). The discrepancy occurs because the rest frame of the electron is not an inertial frame. If one takes into account the effect of the Thomas precession[2], an additional kinematic term arises which together with (12.9) gives the correct result (12.8).

For the hydrogen atom, the spin–orbit interaction is

$$H_2 = \frac{1}{2m^2c^2}\boldsymbol{S} \cdot \boldsymbol{L}\frac{Ze^2}{r^3} \quad . \tag{12.10}$$

It now turns out to be useful to use the total angular momentum studied in Sect. 10.3

$$\boldsymbol{J} = \boldsymbol{L} + \boldsymbol{S} \tag{12.11}$$

and to substitute

$$\boldsymbol{S} \cdot \boldsymbol{L} = \tfrac{1}{2}(\boldsymbol{J}^2 - \boldsymbol{L}^2 - \boldsymbol{S}^2) \tag{12.12}$$

into (12.10). The states $|l \pm 1/2, m_j, l\rangle$ diagonalize the operator $\boldsymbol{S} \cdot \boldsymbol{L}$:

$$
\begin{aligned}
\boldsymbol{S} \cdot \boldsymbol{L}|l \pm \tfrac{1}{2}, m_j, l\rangle &= \frac{\hbar^2}{2}\left(\begin{array}{c} l^2 + 2l + 3/4 - 3/4 - l^2 - l \\ l^2 - 1/4 - 3/4 - l^2 - l \end{array}\right)|l \pm \tfrac{1}{2}, m_j, l\rangle \\
&= \frac{\hbar^2}{2}\left(\begin{array}{c} l \\ -l - 1 \end{array}\right)|l \pm \tfrac{1}{2}, m_j, l\rangle \quad .
\end{aligned}
\tag{12.13}
$$

---

[2] See for example, J.D. Jackson: *Classical Electrodynamics*, 2nd edn. (Wiley, New York 1975).

Because of the spin–orbit term H_2 in the total Hamiltonian

$$H = H_0 + H_1 + H_2 + H_3 \quad,$$

L_z does not commute with H. Thus $H, \boldsymbol{L}^2, L_z, \boldsymbol{S}^2, S_z$ do not form a complete set of operators. Instead, the operators $H, \boldsymbol{J}^2, J_z, \boldsymbol{L}^2, \boldsymbol{S}^2$ now form a complete set, and the eigenfunctions can be characterized by their eigenvalues. H_3 is defined in (12.20).

The correct unperturbed states for perturbation theory are therefore

$$\langle \boldsymbol{x} | n, j = l \pm \tfrac{1}{2}, m_j, l \rangle$$
$$= R_{nl}(r) \left(\alpha_\pm Y_{l m_j - 1/2}(\vartheta, \varphi) | \uparrow \rangle + \beta_\pm Y_{l m_j + 1/2}(\vartheta, \varphi) | \downarrow \rangle \right) \quad.$$
$$(12.14)$$

The coefficients $\alpha_\pm$ and $\beta_\pm$ are defined in (10.28). The states (12.14) are eigenstates of H_0 with energy eigenvalues E_n, which are $2n^2$-fold degenerate; however, all the off-diagonal matrix elements of the perturbed part of the Hamiltonian $H_1 + H_2 + H_3$ vanish.

The correction from first-order perturbation theory yields

$$\langle H_2 \rangle_{n, j = l \pm 1/2, l, m_j} = \frac{1}{2m^2c^2} \frac{\hbar^2}{2} \begin{pmatrix} l \\ -l - 1 \end{pmatrix} Z e_0^2 \left\langle \frac{1}{r^3} \right\rangle_{nl} \quad. \qquad (12.15)$$

Substituting

$$\left\langle \frac{1}{r^3} \right\rangle_{nl} = \frac{Z^3}{a^3 n^3 l (l + 1/2)(l + 1)} = \frac{m^3 c^3 \alpha^3 Z^3}{\hbar^3 n^3 l (l + 1/2)(l + 1)} \quad, \qquad (12.16)$$

we obtain

$$\langle H_2 \rangle_{n, j = l \pm 1/2, l} = \frac{mc^2 (Z\alpha)^4}{4 n^3 l (l + 1/2)(l + 1)} \begin{pmatrix} l \\ -l - 1 \end{pmatrix} \quad. \qquad (12.17)$$

Together with the term that resulted from the relativistic kinetic energy, this gives

$$\langle H_1 + H_2 \rangle_{n, j = l \pm 1/2, l} = \frac{mc^2 (Z\alpha)^2}{2 n^2} \frac{(Z\alpha)^2}{n^2} \left\{ \frac{3}{4} - \frac{n}{j + 1/2} \right\} \quad. \qquad (12.18)$$

Remark:

The states with $l = 0$ require further discussion; $\langle 1/r^3 \rangle \sim 1/l$ diverges for $l = 0$. Because of the factor l in the matrix element of $\boldsymbol{S} \cdot \boldsymbol{L}$, this divergence was not manifest. If one assumes in place of the Coulomb potential an extended nucleus, then $\langle (1/r) dV/dr \rangle$ is no longer singular for $l = 0$, i.e.,

$$\langle H_2 \rangle_{n, j = 1/2, l = 0} = 0 \quad. \qquad (12.19)$$

Therefore, (12.18) only holds for $l \geq 1$.

12.3 The Darwin Term

A further term results from the Dirac equation, the so-called Darwin term

$$H_3 = \frac{\hbar^2}{8m^2c^2}\boldsymbol{\nabla}^2 V = -\frac{\hbar^2}{8m^2c^2}(4\pi e Q_{\text{nuclear}}(\boldsymbol{x}))$$

$$= \frac{\pi\hbar^2 Z e_0^2}{2m^2c^2}\delta^{(3)}(\boldsymbol{x}) \tag{12.20}$$

with nuclear charge density $Q_{\text{nuclear}}(\boldsymbol{x})$. This term can be understood from the "Zitterbewegung" of the electron. According to the relativistic theory, the position of a localized electron fluctuates with

$$\delta r = \frac{\hbar}{mc} = \lambda_{\text{c}} \quad .$$

The electron therefore feels on the average a potential

$$\langle V(\boldsymbol{x}+\delta\boldsymbol{x})\rangle = V(\boldsymbol{x}) + \langle\delta\boldsymbol{x}\boldsymbol{\nabla}V\rangle + \tfrac{1}{2}\langle(\delta\boldsymbol{x}\boldsymbol{\nabla})(\delta\boldsymbol{x}\boldsymbol{\nabla})V(\boldsymbol{x})\rangle$$

$$= V(\boldsymbol{x}) + \tfrac{1}{6}(\delta r)^2\boldsymbol{\nabla}^2 V(\boldsymbol{x}) \quad . \tag{12.21}$$

The correction thus calculated is in qualitative agreement in form, sign, and magnitude with the Darwin term (12.20).

Because of the δ-function, a contribution

$$\langle H_3\rangle_{n,j,l} = \frac{\pi\hbar^2 Z e_0^2}{2m^2c^2}|\psi_{nl}(0)|^2 = \frac{mc^2(Z\alpha)^4}{2n^3}\delta_{l,0} \tag{12.22}$$

arises for s-wave states only; $\langle H_3\rangle$ is formally identical to $\langle H_2\rangle_{n,j=1/2,l=0}$, from (12.17).

We thus have for the energy shift due to all three relativistic corrections the result

$$\Delta E_{n,j=l\pm 1/2,l} = \frac{\text{Ry}\,Z^2}{n^2}\frac{(Z\alpha)^2}{n^2}\left\{\frac{3}{4}-\frac{n}{j+1/2}\right\} \tag{12.23}$$

in first-order perturbation theory. Formally identical formulae were given by Sommerfeld on the basis of the old quantum theory with only the p^4-term. However, in fact the correction is due to three effects.

Below, for a level with spin s, orbital angular momentum L, and total angular momentum j, we use spectroscopic notation $^{2s+1}L_j$. In the hydrogen atom, $2s+1=2$. Figure 12.1 shows the shift (kinetic relativistic energy less than zero) and splitting of the levels with $l \geq 1$ in $j = l \pm 1/2$ (spin–orbit coupling greater than zero for $j = l+1/2$, less than zero for $j = l-1/2$).

The levels $^2S_{1/2}$ and $^2P_{1/2}$ are still degenerate. This remains valid in Dirac theory to all orders in α. The fine-structure splitting between $2^2P_{1/2}$ and $2^2S_{1/2}$ on the one hand and $2^2P_{3/2}$ on the other is 0.45×10^{-4} eV or equivalently 1.09×10^4 MHz.

$n=2,\ l=0,1$ fine structure + Lamb shift + hyperfine structure

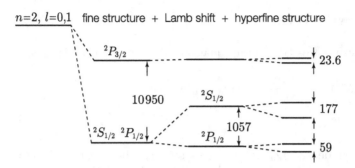

Fig. 12.1. Splitting of the energy levels of the hydrogen atom in MHz due to relativistic corrections, the Lamb shift, and the hyperfine structure

We now give the computation of the expectation values (12.4a,b) and (12.16). These could be calculated directly using the Laguerre polynomials, but the calculation is tedious. Here, we instead make use of algebraic methods[3].

To calculate $\langle 1/r \rangle$, we use the *virial theorem* for the Hamiltonian $H = p^2/2m + V(\boldsymbol{x})$, which we now prove. We start from

$$[H, \boldsymbol{x} \cdot \boldsymbol{p}] = -\mathrm{i}\hbar \left(\frac{\boldsymbol{p}^2}{m} - \boldsymbol{x} \cdot \boldsymbol{\nabla} V(\boldsymbol{x}) \right) \quad . \tag{12.24}$$

Since for eigenstates $|\psi\rangle$ of the Hamiltonian one has $\langle \psi | [H, \boldsymbol{x} \cdot \boldsymbol{p}] | \psi \rangle = 0$, the *virial theorem*

$$\left\langle \psi \left| \frac{\boldsymbol{p}^2}{m} \right| \psi \right\rangle - \langle \psi | \boldsymbol{x} \cdot \boldsymbol{\nabla} V(\boldsymbol{x}) | \psi \rangle = 0 \tag{12.25}$$

results. Specializing to the Coulomb potential, one obtains from (12.25)

$$2\langle \psi | H | \psi \rangle + \left\langle \psi \left| \frac{Ze^2}{r} \right| \psi \right\rangle = 0 \quad , \tag{12.26}$$

or $2E_n = -Ze^2 \langle \psi | 1/r | \psi \rangle$, whence

$$\left\langle \frac{1}{r} \right\rangle_{nl} = \frac{Z}{an^2} \quad . \tag{12.4a}$$

For the remaining expectation values, we write

$$\left\langle \frac{1}{r^k} \right\rangle_{nl} = \int_0^\infty dr\, r^2 \frac{1}{r^k} R_{nl}^2 = \left(u_{nl}, \frac{1}{r^k} u_{nl} \right) \tag{12.27}$$

---

[3] R. Becker, F. Sauter: *Electromagnetic Fields and Interactions II. Quantum Theory of Atoms and Radiation* (Blackie, Glasgow 1964)

and rewrite the radial Schrödinger equation (6.11) in terms of the dimension-less variable $y = r/a$, obtaining

$$Hu(y) = \varepsilon u(y) \tag{12.28}$$

with

$$\varepsilon = -Z^2/(N + l + 1)^2$$

and

$$H = -\frac{d^2}{dy^2} + \frac{l(l+1)}{y^2} - \frac{2Z}{y} \quad . \tag{12.29}$$

We now differentiate the Schrödinger equation (12.28) with respect to l,

$$\frac{\partial H}{\partial l} u + H \frac{\partial u}{\partial l} = \frac{\partial \varepsilon}{\partial l} u + \varepsilon \frac{\partial u}{\partial l} \quad ,$$

and construct the scalar product with u,

$$\left(u, \frac{\partial H}{\partial l} u \right) + \left(u, H \frac{\partial u}{\partial l} \right) = \frac{\partial \varepsilon}{\partial l} + \varepsilon \left(u, \frac{\partial u}{\partial l} \right) \quad . \tag{12.30}$$

Substituting

$$\left(u, H \frac{\partial u}{\partial l} \right) = \left(Hu, \frac{\partial u}{\partial l} \right) = \varepsilon \left(u, \frac{\partial u}{\partial l} \right) \quad , \tag{12.31}$$

we obtain from (12.30)

$$\left(u, \frac{\partial H}{\partial l} u \right) = \frac{\partial \varepsilon}{\partial l} \quad . \tag{12.32}$$

With

$$\frac{\partial H}{\partial l} = \frac{2l + 1}{y^2} \quad \text{and} \quad \frac{\partial \varepsilon}{\partial l} = \frac{2Z^2}{n^3}$$

(12.32) yields

$$\left\langle \frac{1}{r^2} \right\rangle_{nl} = \frac{2Z^2}{(2l + 1)n^3 a^2} \quad . \tag{12.4b}$$

We finally differentiate the Schrödinger equation with respect to y, obtaining

$$\frac{\partial H}{\partial y} u + H \frac{\partial u}{\partial y} = \varepsilon \frac{\partial u}{\partial y} \quad .$$

The scalar product with u is

$$\left(u, \frac{\partial H}{\partial y} u \right) + \left(u, H \frac{\partial u}{\partial y} \right) = \varepsilon \left(u, \frac{\partial u}{\partial y} \right)$$

and gives, using (12.28),

$$\left(u, \frac{\partial H}{\partial y} u \right) = 0 \quad . \tag{12.33}$$

Now, substituting

$$\frac{\partial H}{\partial y} = -\frac{2l(l+1)}{y^3} + \frac{2Z}{y^2} \quad ,$$

into (12.33) and combining with (12.4b), we finally obtain

$$\left\langle \frac{1}{r^3} \right\rangle_{nlm} = \frac{2Z^3}{l(l+1)(2l+1)n^3a^3} \quad . \tag{12.16}$$

12.4 Further Corrections

12.4.1 The Lamb Shift

Due to the zero-point fluctuations of the quantized electromagnetic field, a shift in the position of the electron occurs, causing a perturbation similar in structure to the Darwin term, where here

$$\langle (\delta \boldsymbol{x})^2 \rangle \approx \frac{2\alpha}{\pi} \left(\frac{\hbar}{mc} \right)^2 \log \frac{1}{\alpha Z} \quad , \tag{12.34}$$

$$\Delta E_{\text{Lamb}} \approx \frac{4}{3\pi} \frac{mc^2 Z^4 \alpha^5}{n^3} \log \frac{1}{\alpha Z} \delta_{l,0} \quad . \tag{12.35}$$

This rough computation gives for hydrogen a shift of $\Delta \approx 660$ MHz from $2S_{1/2}$ towards $2P_{1/2}$. The observed shift is 1057.862 ± 0.020 MHz[4]. The complete quantum electrodynamic theory of radiative corrections gives 1057.864 ± 0.014 MHz[5]. Compared to the Darwin term, the radiative corrections are smaller by a factor of $\alpha \log 1/\alpha$. The radiative corrections also include $\alpha(Z\alpha)^4$-terms, which are numerically somewhat smaller, and shifts of the levels $l \neq 0$.

12.4.2 Hyperfine Structure

The nuclear spin $\boldsymbol{I}$ leads to a nuclear magnetic moment

$$\boldsymbol{M} = \frac{Ze_0 g_N}{2M_N c} \boldsymbol{I} \quad , \tag{12.36}$$

which generates a vector potential

$$\boldsymbol{A} = -\boldsymbol{M} \times \boldsymbol{\nabla}(1/r) = \boldsymbol{M} \times \boldsymbol{x} \frac{1}{r^3} \tag{12.37a}$$

and a magnetic field

[4] W.E. Lamb, Jr., R.C. Retherford: Phys. Rev. **72**, 241 (1947)
[5] See for example, C. Itzykson, J.-B. Zuber: *Quantum Field Theory* (McGraw-Hill, New York 1980) pp. 359, 80.

$$B = \nabla \times A = -\left\{ M\nabla^2 \frac{1}{r} - \nabla(M \cdot \nabla)\frac{1}{r} \right\} \quad . \tag{12.37b}$$

We consider first just the s-electrons. The interaction of the magnetic moment of the electrons with the field B of the nucleus gives rise to the hyperfine interaction

$$H_{\text{hyp}} = \frac{e_0}{mc} S \cdot B = \frac{Ze_0^2 g_N}{2mM_N c^2} S\left[-I\nabla^2 \frac{1}{r} + \nabla(I \cdot \nabla)\frac{1}{r} \right] \quad . \tag{12.38}$$

Since $\nabla^2(1/r) = -4\pi\delta^{(3)}(x)$ and

$$\int d^3x \left[\nabla(I \cdot \nabla)\frac{1}{r} \right] (\psi_{n,0}(r))^2 = \frac{I}{3} \int d^3x \left(\nabla^2 \frac{1}{r} \right) (\psi_{n,0}(r))^2$$

for s (radially symmetric) wave functions, as a first step in first-order perturbation theory the spatial expectation value in the state $|n, j = 1/2, l = 0\rangle$ becomes

$$\left\langle H_{\text{hyp}} \right\rangle_{n,1/2,0} = \frac{4}{3} g_N \frac{m}{M_N} (Z\alpha)^4 mc^2 \frac{1}{n^3} \frac{S \cdot I}{\hbar^2} \quad . \tag{12.39}$$

Comparison with (12.22) shows that $\langle H_{\text{hyp}} \rangle$ is smaller than the fine structure by a factor of m/M_N. It remains to evaluate the expectation value with respect to the spin degrees of freedom.

Analogously to the total angular momentum J of the electron, which we introduced in our discussion of the spin–orbit coupling in order to diagonalize $S \cdot L$, we now introduce the total spin F

$$F = S + I \quad .$$

We then have

$$\frac{1}{\hbar^2} S \cdot I = \frac{1}{2\hbar^2}(F^2 - S^2 - I^2) = \tfrac{1}{2}[F(F+1) - \tfrac{3}{4} - I(I+1)]$$

$$= \begin{cases} \tfrac{1}{2}I & \text{for } F = I + \tfrac{1}{2} \\ \tfrac{1}{2}(-I-1) & \text{for } F = I - \tfrac{1}{2} \end{cases} \quad .$$

For the hydrogen atom, $g_N = g_P = 5.56$ and $I = 1/2$. The s-wave states in the hydrogen atom are therefore either in a singlet state ($F = 0$, ground state) or in a triplet state ($F = 1$, excited state).

The splitting in the nth level for s-electrons is therefore

$$\Delta E_{n,1/2,0}^{\text{hyp}} = \frac{4}{3} g_N \frac{m}{M_N} (Z\alpha)^4 \frac{mc^2}{n^3} \frac{(2I+1)}{2} \quad ,$$

$$\Delta E_{n,1/2,0}^{\text{hyp}} \text{ (H-atom, s-electron)} = \frac{4}{3} 5.56 \frac{1}{1840} \frac{1}{(137)^4} \frac{mc^2}{n^3} \quad . \tag{12.40}$$

For the lowest levels, one has

$$1S_{1/2}: \quad \nu = 1420\,\text{MHz} \quad , \quad \lambda = 21.4\,\text{cm} \quad ;$$
$$2S_{1/2}: \quad \nu = \tfrac{1}{8} \times 1420\,\text{MHz} = 177\,\text{MHz} \quad ;$$
$$2P_{1/2}: \quad \nu = \tfrac{1}{24} \times 1420\,\text{MHz} = 59\,\text{MHz} \quad .$$

The 21-cm radiation is very important in astronomy. From its intensity, Doppler broadening, and Doppler shift, one obtains information concerning the density, temperature, and motion of interstellar and intergalactic hydrogen clouds. Due to the hyperfine interaction, all of the levels in hydrogen are split into doublets.

Hyperfine interaction including the orbital part

We begin once again with equations (12.36) and (12.37a,b). The interaction of the orbital moment of the electron with the magnetic moment of the nucleus is

$$H_{\text{orbit}} = \frac{i\hbar e}{mc} \boldsymbol{A} \cdot \boldsymbol{\nabla} = -\frac{e}{mc}\frac{\boldsymbol{M}}{r^3} \cdot \boldsymbol{x} \times \boldsymbol{p} = -\frac{e}{mc}\frac{1}{r^3}\boldsymbol{M} \cdot \boldsymbol{L} \quad .$$

Using

$$\boldsymbol{B} = -\left\{ \boldsymbol{M}\delta^{(3)}(\boldsymbol{x})\left(\frac{-8\pi}{3}\right) + \frac{\boldsymbol{M}}{r^3} - \frac{3(\boldsymbol{M}\cdot\boldsymbol{x})\boldsymbol{x}}{r^5} \right\} \quad , \tag{12.41}$$

one finds for the total hyperfine interaction[6]

$$H_{\text{hyp}} = H_{\text{orbit}} + \frac{e_0}{mc}\boldsymbol{S} \cdot \boldsymbol{B}$$

and finally

$$H_{\text{hyp}} = \frac{Ze_0^2 g_N}{2M_N mc^2}\left[\frac{1}{r^3}\boldsymbol{I}\cdot\boldsymbol{L} + \frac{8\pi}{3}\delta^{(3)}(\boldsymbol{x})\boldsymbol{I}\cdot\boldsymbol{S} - \frac{\boldsymbol{I}\cdot\boldsymbol{S}}{r^3} + \frac{3(\boldsymbol{I}\cdot\boldsymbol{x})(\boldsymbol{S}\cdot\boldsymbol{x})}{r^5} \right] \quad .$$

$$\tag{12.42}$$

Remark:
The magnetic field of a dipole is singular at the origin. Its effect can be decomposed into a part within a small infinitesimal sphere with radius ϵ and a part outside this sphere. The former is represented by the $\delta^{(3)}(\boldsymbol{x})$-function in (12.41)[7] and leads to the so called Fermi-contact interaction in (12.42). The latter is represented by the last two terms of Eq. (12.42), and enters in all matrix elements with a factor $\Theta(|\boldsymbol{x}| - \epsilon)$ excluding the infinitesimal sphere. The Hamiltonian (12.42) is the basis

---

[6] H. A. Bethe, E. E. Salpeter: *Quantum Mechanics of One- and Two-Electron Atoms* (Springer, Berlin, Heidelberg 1957). Measurement of the hyperfine splitting permits the determination of I and g_N.

[7] J. D. Jackson, *Classical Electrodynamics*, Second Edition, John Wiley & Sons, New York, 1975, p. 184

for important experimental methods in condensed matter physics such as nuclear spin resonance, muon spin resonance etc.

Further reading:

The relativistic Dirac equation, which is outside the scope of this book, is treated in:

J. D. Bjorken, S. D. Drell: *Relativistic Quantum Mechanics* (McGraw-Hill, New York 1964)

C. Itzykson, J.-B. Zuber: *Quantum Field Theory* (McGraw-Hill, New York 1980)

F. Schwabl, *Advanced Quantum Mechanics* (Springer, Berlin, Heidelberg, New York 3rd ed., 2005)

Problems

12.1 Relativistic corrections in positronium (bound state of electron and positron): How do the (relative) orders of magnitude change (in comparison to the hydrogen atom) for the corrections due to the relativistic mass, the spin–orbit interaction, and the hyperfine splitting in positronium?

12.2 Study the influence of the spin–orbit interaction

$$H_2 = \frac{1}{2m^2c^2} S \cdot L \frac{1}{r} \frac{dV(r)}{dr}$$

on the energy spectrum $\left(E_{n,l} = \hbar\omega \left(l + 2n + \frac{3}{2} \right), \text{compare problem 17.1} \right)$
of a three-dimensional isotropic harmonic oscillator. Discuss the degeneracy of the energy levels without and with the spin–orbit interaction. Note: $\psi_{nlm}(\boldsymbol{x}) = R_{nl}(r) Y_{lm}(\vartheta, \varphi)$.

13. Several-Electron Atoms

13.1 Identical Particles

13.1.1 Bosons and Fermions

We consider N identical particles (e.g., electrons, π-mesons). The Hamiltonian

$$H = H(1, 2, \ldots, N) \tag{13.1}$$

is symmetric in the variables $1, 2, \ldots$. Here, $1 \equiv \boldsymbol{x}_1, \chi_1$ includes position and spin degrees of freedom. Likewise, we write a wave function in the form

$$\psi = \psi(1, 2, \ldots, N) \quad .$$

The permutation operator P_{ij} interchanges $i \leftrightarrow j$; its action on an arbitrary N-particle wave function is

$$P_{ij}\psi(\ldots i, \ldots, j \ldots) = \psi(\ldots j, \ldots, i \ldots) \quad . \tag{13.2}$$

Since $P_{ij}^2 = 1$, P_{ij} has the eigenvalues ± 1. By the symmetry of the Hamiltonian, for every element P of the permutation group, the relation

$$PH = HP \tag{13.3}$$

holds.

Remark: Every element P can be represented as the product of transpositions P_{ij}. An element P is called even (odd) if there are an even (odd) number of the P_{ij}.

Let $\psi(1, \ldots, N)$ be an eigenfunction of H with eigenvalue E; this then holds for $P\psi(1, \ldots, N)$ as well.

Proof: $H\psi = E\psi \rightarrow HP\psi = PH\psi = EP\psi$.

For any symmetrical operator $S(1, \ldots, N)$, one has by (13.3)

$$[P, S] = 0$$

and[1]

$$\langle P\psi|S|P\psi\rangle = \langle \psi|P^\dagger SP|\psi\rangle = \langle \psi|P^\dagger PS|\psi\rangle = \langle \psi|S|\psi\rangle \quad .$$

The expectation value and, more generally, the matrix elements in the states ψ and $P\psi$ of a symmetrical operator S are equal.[2] Since identical particles are influenced equivalently by any physical process, all physical operators are symmetric and thus the states

$$\psi(1, \ldots, N) \quad \text{and} \quad P\psi(1, \ldots, N)$$

cannot be distinguished from one another. The question then arises whether all these states are realized in nature. On aesthetic grounds, one might suspect that the *completely symmetric* states and the *completely antisymmetric* states occupy a privileged position. They are defined by

$$P_{ij}\psi(\ldots i, \ldots, j \ldots) = \pm \psi(\ldots i, \ldots, j \ldots) \tag{13.4}$$

for all P_{ij}, e.g., for two particles,

$$\psi_{\mathrm{s}}(1,2) = \psi(1,2) + \psi(2,1) \; , \quad \psi_{\mathrm{a}}(1,2) = \psi(1,2) - \psi(2,1) \quad . \tag{13.5}$$

A completely (anti)symmetric state remains so for all time. (Perturbation V : $\psi \to \psi + V\psi + \ldots$, $PV\psi = VP\psi = \pm V\psi$. Quite generally Eq. (16.9′) $\psi(t) = T \exp\{\mathrm{i} \int_0^t dt' H(t')/\hbar\}\psi(0)$ implies $P\psi(t) = T \exp\{\mathrm{i} \int_0^t dt' H(t')/\hbar\}P\psi(0)$.)

Experimental findings imply that there exist two types of particles, *bosons* and *fermions*, whose states are completely symmetric and completely antisymmetric, respectively. Fermions have half-integral spin, whereas bosons have integral spin. This connection between spin and symmetry follows from the spin-statistics theorem of quantum field theory. Significant consequences of these two symmetries of many-particle physics are Fermi–Dirac statistics and Bose–Einstein statistics.

fermions	*bosons*
leptons ν_{e}, ν_μ, ν_τ	mesons π, K, ϱ, ω,…
e, μ, τ	photon γ
baryons p, n	
Λ, Σ, Ξ, Ω,…	

---

[1] For any permutation P and arbitrary ψ and φ one has $\langle \psi|\varphi\rangle = \langle P\psi|P\varphi\rangle$ implying $\langle \psi|\varphi\rangle = \langle \psi|P^\dagger P\varphi\rangle$ and the unitarity $P^\dagger = P^{-1}$.

[2] The opposite is also true. The requirement that a permutation of identical particles must not have any observable consequences implies that observables O are symmetric (permutation invariant). Proof: $\langle \psi|O|\psi\rangle = \langle P\psi|O|P\psi\rangle = \langle \psi|P^\dagger OP|\psi\rangle$ holds for arbitrary ψ and thus $P^\dagger OP = O$. Consequently $PO = OP$.

Table 13.1. The fundamental fermions

Charge	Lepton doublets				Antilepton doublets	Charge
0	electron-neutrino	ν_e		$\overline{\nu}_e$	electron-antineutrino	0
−1	electron	e^-		e^+	positron	+1
0	muon-neutrino	ν_μ		$\overline{\nu}_\mu$	muon-antineutrino	0
−1	muon	μ^-		μ^+	anti-muon	+1
0	tau-neutrino	ν_τ		$\overline{\nu}_\tau$	tau-antineutrino	0
−1	tau	τ^-		τ^+	anti-tau	+1
Charge	Quark doublets				Antiquark doublets	Charge
+2/3	up-quark	u		$\overline{u}$	up-antiquark	−2/3
−1/3	down-quark	d		$\overline{d}$	down-antiquark	+1/3
+2/3	charm-quark	c		$\overline{c}$	charm-antiquark	−2/3
−1/3	strange-quark	s		$\overline{s}$	strange-antiquark	+1/3
+2/3	top-quark (truth)	t		$\overline{t}$	top-antiquark	−2/3
−1/3	bottom-quark (beauty)	b		$\overline{b}$	bottom-antiquark	+1/3

The hadrons, i.e., baryons and mesons, are composed of quarks. The fundamental particles at this level are (see also Table 13.1)

leptons	Yang-Mills gauge bosons[3]
quarks	Higgs bosons

From $P_{ij}\psi(\dots i, \dots, j \dots) = \psi(\dots j, \dots, i \dots) = -\psi(\dots i, \dots, j \dots)$, it follows for fermions that

$$\psi(\dots, \boldsymbol{x}\sigma, \dots, \boldsymbol{x}\sigma, \dots) = 0 \quad . \tag{13.6}$$

Thus, two fermions in the same spin state cannot occupy the same position. This is the *Pauli exclusion principle*[4].

Remark: The completely symmetric and the completely antisymmetric N-particle states form the basis of two one-dimensional representations of the permutation group $\mathcal{S}_N$. This is seen from (13.4) and from the fact that every permutation can be represented by products of transpositions. Since the P_{ij} do not all commute with each other for more than two particles, there are also wave functions for which not all the P_{ij} are diagonal. These are basis functions of higher-dimensional representations of the permutation group. These states do not occur in nature.[5]

---

[3] The Yang-Mills gauge bosons of the electroweak interaction are W^+, W^-, Z^0 and the photon γ, and those of the strong interaction are the gluons.

[4] W. Pauli: Z. Phys. **31**, 765 (1925)

[5] A.M.L. Messiah, O.W. Greenberg: Phys. Rev. **136**, B 248 (1964); **138**, B 1155 (1965)

13.1.2 Noninteracting Particles

For N identical, noninteracting particles, the Hamiltonian

$$H = \sum_{i=1}^{N} H(i) \tag{13.7}$$

is the sum of N identical one-particle Hamiltonians $H(i)$. From the solutions of the one-particle Schrödinger equation

$$H(i)\varphi_{\alpha_i}(i) = E_{\alpha_i}\varphi_{\alpha_i}(i) \quad, \tag{13.8}$$

where α_i numbers the one-particle states, we first form the product states

$$\varphi_{\alpha_1}(1)\varphi_{\alpha_2}(2) \ldots \varphi_{\alpha_N}(N) \quad, \tag{13.9}$$

which are eigenstates of H with energy eigenvalue

$$E = E_{\alpha_1} + \ldots + E_{\alpha_N} \quad. \tag{13.10}$$

The states (13.9) are in general neither antisymmetric nor symmetric. We now construct states of these two types.

For fermions

For two particles, the antisymmetric state is

$$\psi_a(1,2) = \frac{1}{\sqrt{2}}(\varphi_{\alpha_1}(1)\varphi_{\alpha_2}(2) - \varphi_{\alpha_2}(1)\varphi_{\alpha_1}(2)) \quad, \tag{13.11}$$

and generally for N particles

$$\psi_a(1,\ldots,N) = \frac{1}{\sqrt{N!}}\sum_P (-1)^P P\varphi_{\alpha_1}(1) \ldots \varphi_{\alpha_N}(N)$$

$$= \frac{1}{\sqrt{N!}} \begin{vmatrix} \varphi_{\alpha_1}(1) & \cdots & \varphi_{\alpha_1}(N) \\ \vdots & & \vdots \\ \varphi_{\alpha_N}(1) & \cdots & \varphi_{\alpha_N}(N) \end{vmatrix} . \tag{13.12}$$

These determinants of one-particle states are called *Slater determinants*. The antisymmetry of (13.12) is immediately apparent, since interchange of any two columns introduces a factor of –1. The normalization factor is $1/\sqrt{N!}$, because (13.12) consists of $N!$ mutually orthogonal terms. For even (odd) permutations, $(-1)^P = \pm 1$. One has $\psi_a(1,\ldots N) = 0$ for $\varphi_{\alpha_i} = \varphi_{\alpha_j}$. No state may be multiply occupied (Pauli exclusion principle).

For bosons

$$\psi_s(1,\ldots,N) = \sqrt{\frac{N_1!N_2! \cdots}{N!}}\sum_{P'} P'\varphi_{\alpha_1}(1) \ldots \varphi_{\alpha_N}(N) \quad, \tag{13.13}$$

N_1 is the multiplicity of α_1, etc. The summation $\sum_{P'}$ is only over permutations leading to distinct terms and includes $N!/N_1!N_2! \ldots$ different terms.

Free particles in a box

Let N interaction-free particles be enclosed in a volume $V = L^3$. The one-particle wave functions for free particles are plane waves

$$\psi_p \sim e^{i\boldsymbol{p}\cdot\boldsymbol{x}/\hbar} \quad,$$

$$E_p = \frac{\boldsymbol{p}^2}{2m} \quad, \quad E = \sum_1^N \frac{\boldsymbol{p}_i^2}{2m} \quad. \tag{13.14}$$

Choosing periodic boundary conditions, one obtains the discrete momentum values

$$\boldsymbol{p} = \hbar\frac{2\pi}{L}(n_1, n_2, n_3) \quad; \quad n_i \text{ integer} \quad.$$

The ground state of N *bosons* is given by

$$\boldsymbol{p}_i = 0 \tag{13.15}$$

for all N particles, and the total energy is $E = 0$.

Fermions are subject to the Pauli exclusion principle, according to which each momentum state can only be doubly occupied ($s_z = \pm\hbar/2$). The ground state is obtained by putting the N fermions one after another into the lowest available states. By (13.14), the occupied states lie within a sphere in momentum space inside of

$$|\boldsymbol{p}_i| \leq p_F \quad, \tag{13.16}$$

the Fermi sphere (Fig. 13.1), whose radius is denoted by p_F.

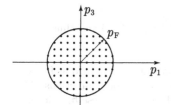

Fig. 13.1. The Fermi sphere – the discrete momentum values have separation $2\pi\hbar/L$

The number of states within the Fermi sphere is given by

$$2\frac{4\pi(p_F/\hbar)^3/3}{(2\pi/L)^3} = \frac{(p_F/\hbar)^3}{3\pi^2}L^3 = N \quad.$$

Hence, the particle number density $n = N/L^3$ is

$$n = \frac{(p_F/\hbar)^3}{3\pi^2} \quad. \tag{13.17}$$

The ground state energy is obtained by summing up the one-particle energies

$$E = 2 \sum_p \frac{p^2}{2m} = 2 \left(\frac{L}{2\pi\hbar} \right)^3 \int_0^{p_F} d^3p \frac{p^2}{2m} = N \frac{3}{5} \varepsilon_F \quad . \tag{13.18}$$

The quantities $\varepsilon_F = p_F^2/2m$ and p_F are known respectively as the Fermi energy and the Fermi momentum.

The crucial difference between bosons and fermions, which for free (= non-interacting) particles is manifest in (13.15) and (13.16), leads to characteristic differences in the low-temperature behavior of such particles (Bose–Einstein condensation, linear specific heat, etc.)

Remarks: (i) First of all, concerning the boundary conditions: While counting states, we used periodic boundary conditions $\psi_p(x + L) = \psi_p(x)$. If instead we assume infinitely high barriers, the wave functions are

$$\psi_{p'} = \sin \frac{p'}{\hbar} x \quad , \quad \frac{p'}{\hbar} = \frac{\pi}{L} k \quad , \quad k = 1, 2, 3, \ldots \quad .$$

These momentum values are equivalent to

$$\frac{p'}{\hbar} = \begin{cases} \dfrac{2\pi}{L} n & \text{for positive } n \\[2mm] \dfrac{2\pi}{L} n - \dfrac{\pi}{L} & \text{for negative } n \end{cases} \quad .$$

The densities of the momentum values for these two boundary conditions are thus equal.

(ii) We now consider one-dimensional bosons in a ring of circumference L, so that $0 \leq \varphi < L$. Is the state $\psi = 1/\sqrt{L}$ compatible with the uncertainty relation at all?

For this state

$$\langle \psi | p_\varphi | \psi \rangle = \langle \psi | p_\varphi^2 | \psi \rangle = 0 \quad ,$$

where $p_\varphi = (\hbar/i)\partial/\partial\varphi$, or $\Delta p_\varphi = 0$, while $\langle \varphi \rangle = L/2$, $\langle \varphi^2 \rangle = L^2/3$ and thus

$$\Delta\varphi\Delta p_\varphi = 0 \quad .$$

The apparent contradiction to the uncertainty relation can be explained as follows: $(\hbar/i)\,\partial/\partial\varphi$ is a Hermitian operator in the space of the periodic functions $e^{i\varphi k}$, $k = 2\pi/L$. However, the operator φ takes us out of the space of these functions,

$$\varphi e^{i\varphi k}|_0 \neq \varphi e^{i\varphi k}|_L \quad ,$$

and for such functions $(\hbar/i)\,\partial/\partial\varphi$ is not Hermitian. Therefore, the proof of the uncertainty relation does not hold for φ and $(\hbar/i)\,\partial/\partial\varphi$.

Composite particles

Example – the H-atom: An H-atom consists of two fermions: a proton p and an electron e. In order to see whether the H-atom is a fermion or a boson, we investigate what happens to the wave function $\psi(p_1, e_1; p_2, e_2)$ when two H-atoms are interchanged. Since p and e are fermions,

$$\psi(p_1, e_1; p_2, e_2) = -\psi(p_2, e_1; p_1, e_2) = \psi(p_2, e_2; p_1, e_1) \quad .$$

Hence, the H-atom is a boson.

In general, if the number of fermions in a composite particle is odd, then it is a fermion, otherwise a boson; e.g., baryons consist of three quarks and are thus fermions, while mesons consist of two quarks and are therefore bosons. ^{3}He is a fermion, ^{4}He a boson.

13.2 Helium

In this section, the simplest multielectron atom is treated, consisting of two electrons in the field of a nucleus with nuclear charge Z. For helium, $Z = 2$, and for H$^-$, $Z = 1$. Neglecting for the moment the spin–orbit interaction and the motion of the nucleus ($m_n \gg m_e$), one can write the Hamiltonian in the form

$$H = \frac{1}{2m}p_1^2 + \frac{1}{2m}p_2^2 - \frac{Ze^2}{r_1} - \frac{Ze^2}{r_2} + \frac{e^2}{|\boldsymbol{x}_1 - \boldsymbol{x}_2|} \quad . \tag{13.19}$$

We introduce the abbreviations

$$H = H(1) + H(2) + V \quad , \tag{13.19a}$$

$$H(i) = \frac{p_i^2}{2m} - \frac{Ze^2}{r_i} \quad , \quad i = 1, 2 \quad , \tag{13.19b}$$

$$V = \frac{e^2}{|\boldsymbol{x}_1 - \boldsymbol{x}_2|} \quad . \tag{13.19c}$$

Here, $H(i)$ is the hydrogenic Hamiltonian, i.e., that of a nucleus with nuclear charge Z and only one electron, while V gives the electrostatic repulsion of the two electrons.

13.2.1 Without the Electron–Electron Interaction

Neglecting for now the mutual electrostatic repulsion of the electrons, one can write H as the sum of two one-particle Hamiltonians. The product states

$$|\psi\rangle = |n_1 l_1 m_1\rangle |n_2 l_2 m_2\rangle$$

are then eigenstates of H, where the states $|n_i l_i m_i\rangle$ are eigenstates of $H(i)$, i.e., hydrogenic states with $Z = 2$ (for He). This implies

$$[H(1) + H(2)]|\psi\rangle = (E_{n_1} + E_{n_2})|\psi\rangle$$

with

$$E_n = -Z^2 \operatorname{Ry} \frac{1}{n^2} \tag{13.20}$$

Table 13.2. Energy values of the He atom for various n_1 and n_2 (13.21)

n_1	n_2	$E^0(\mathrm{Ry})$	$E^0(\mathrm{eV})$
1	1	-8	-108.8
1	2	-5	-68.0
1	3	$-40/9$	-60.4
$\vdots$	$\vdots$	$\vdots$	$\vdots$
1	∞	-4	-54.5
2	2	-2	-27.2

and with energy

$$E^0_{n_1,n_2} \equiv E_{n_1} + E_{n_2} \ . \tag{13.21}$$

In Table 13.2, several values of the total energy are given for various principal quantum numbers n_1 and n_2 (for $Z = 2$).

The ionization energy of a helium atom in the ground state is

$$E^0_{\mathrm{Ion}} = (E_1 + E_\infty) - 2E_1 = 4\,\mathrm{Ry} \ .$$

The $(n_1, n_2) = (2, 2)$-state has a higher energy than the singly ionized $(1, \infty)$-state and is not a bound state, as is true of all the further ones. In the bound, excited states, one of the electrons has the principal quantum number 1. One sees the $(2,2)$-state however as a resonance in the He^{+}– e scattering cross section. The energy spectrum of Fig. 13.2 thus results.

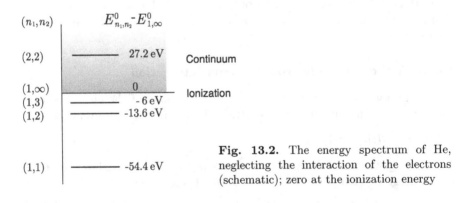

Fig. 13.2. The energy spectrum of He, neglecting the interaction of the electrons (schematic); zero at the ionization energy

We must now take into account the *Pauli principle*, that is, the antisymmetry of the total wave function, which is the product of the coordinate function and the spin state $|S, m_s\rangle$. The spin states of the two electrons are the three symmetric triplet states and the antisymmetric singlet state. The spatial part of the state must be respectively an antisymmetric or a symmetric combination of the spatial eigenfunctions $|n_1 l_1 m_1\rangle |n_2 l_2 m_2\rangle$.

For historical reasons, the following designations for these two classes of states are customary:

Parahelium: Spatial wave function symmetric, spin singlet (antisymmetric)

$$|0\rangle = |100\rangle |100\rangle |0, 0\rangle \quad ,$$

$$\frac{1}{\sqrt{2}} (|100\rangle |2lm\rangle + |2lm\rangle |100\rangle) |0, 0\rangle \quad . \tag{13.22a}$$

$$\vdots$$

Orthohelium: Spatial part of the wave function antisymmetric, spin triplet (symmetric). The antisymmetrization of $|100\rangle |100\rangle$ gives zero; therefore, the lowest spin triplet state is

$$\frac{1}{\sqrt{2}} (|100\rangle |2lm\rangle - |2lm\rangle |100\rangle) |1, m_s\rangle \quad . \tag{13.22b}$$

Remark: The two Slater determinants

$$\frac{1}{\sqrt{2}} (|100\rangle |2lm\rangle |\uparrow\downarrow\rangle - |2lm\rangle |100\rangle |\downarrow\uparrow\rangle) \text{ and } \frac{1}{\sqrt{2}} (|100\rangle |2lm\rangle |\downarrow\uparrow\rangle - |2lm\rangle |100\rangle |\uparrow\downarrow\rangle)$$

satisfy the Pauli principle as well, but they are not eigenfunctions of the total spin. In both of these degenerate states, the Coulomb interaction V is not diagonal. In contrast, the nondiagonal matrix elements of V vanish for the degenerate eigenstates of the total spin (13.22a) and (13.22b). Therefore, (13.22a,b) are appropriate unperturbed states for perturbation theory. See Sect. 11.1.2.

13.2.2 Energy Shift
Due to the Repulsive Electron–Electron Interaction

13.2.2.1 The Ground State

We now take into account the Coulomb repulsion of the electrons, (13.19c), perturbatively. The energy shift ΔE of the ground state becomes in first-order perturbation theory

$$\Delta E = \langle 0|V|0\rangle = \langle 100|\langle 100|V|100\rangle |100\rangle$$

$$= e^2 \int d^3 x_1 \, d^3 x_2 \frac{|\psi_{100}(\boldsymbol{x}_1)|^2 |\psi_{100}(\boldsymbol{x}_2)|^2}{|\boldsymbol{x}_1 - \boldsymbol{x}_2|} \quad . \tag{13.23}$$

The wave function ψ_{100} is given according to (6.43) by

$$\psi_{100}(\boldsymbol{x}) = \frac{1}{\sqrt{\pi}} \left(\frac{Z}{a}\right)^{3/2} e^{-Zr/a} \quad .$$

One thus obtains

$$\Delta E = \left[\frac{(Z/a)^3}{\pi}\right]^2 e^2 \int_0^\infty dr_1 r_1^2 e^{-2Zr_1/a} \int_0^\infty dr_2 r_2^2 e^{-2Zr_2/a}$$
$$\times \int d\Omega_1\, d\Omega_2 \frac{1}{|\boldsymbol{x}_1 - \boldsymbol{x}_2|} \quad . \tag{13.24}$$

Using the formula, problem 13.1(a)

$$\int d\Omega_1\, d\Omega_2 \frac{1}{|\boldsymbol{x}_1 - \boldsymbol{x}_2|} = (4\pi)^2 \frac{1}{\max(r_1, r_2)} \quad ,$$

one obtains the final result

$$\Delta E = \frac{5}{4}\frac{Ze^2}{2a} = \frac{5}{4}Z\frac{mc^2\alpha^2}{2} \quad , \tag{13.25}$$

$$\Delta E = 2.5\,\text{Ry} = 34\,\text{eV} \quad \text{for} \quad Z = 2 \quad . \tag{13.26}$$

The total ground state energy $E_{1,1} = E_{1,1}^0 + \Delta E$ becomes

$$E_{1,1} = -74.8\,\text{eV} = -5.5\,\text{Ry} \quad , \tag{13.27a}$$

while the experimental value is

$$(E_{1,1})_{\text{exp}} = -78.975\,\text{eV} = -5.807\,\text{Ry} \quad . \tag{13.27b}$$

13.2.2.2 Excited States

The energy shift $\Delta E_{nl}^{\text{s,t}}$ is given in first-order perturbation theory for singlet and triplet states by

$$\Delta E_{nl}^{\text{s,t}} = \frac{1}{2}\int d^3x_1 d^3x_2 |\psi_{100}(\boldsymbol{x}_1)\psi_{nl0}(\boldsymbol{x}_2) \pm \psi_{nl0}(\boldsymbol{x}_1)\psi_{100}(\boldsymbol{x}_2)|^2 \frac{e^2}{|\boldsymbol{x}_1 - \boldsymbol{x}_2|}$$
$$= e^2\left[\int d^3x_1 d^3x_2 \frac{|\psi_{100}(\boldsymbol{x}_1)|^2|\psi_{nl0}(\boldsymbol{x}_2)|^2}{|\boldsymbol{x}_1 - \boldsymbol{x}_2|}\right.$$
$$\left.\pm \int d^3x_1 d^3x_2 \frac{\psi_{100}^*(\boldsymbol{x}_1)\psi_{nl0}^*(\boldsymbol{x}_2)\psi_{100}(\boldsymbol{x}_2)\psi_{nl0}(\boldsymbol{x}_1)}{|\boldsymbol{x}_1 - \boldsymbol{x}_2|}\right]$$
$$\equiv J_{nl} \pm K_{nl} \quad . \tag{13.28}$$

It suffices to compute $\Delta E_{nl}^{\text{s,t}}$ for $m = 0$, since the result is independent of m due to the vanishing commutator $[\boldsymbol{L}, 1/|\boldsymbol{x}_1 - \boldsymbol{x}_2|] = 0$, where $\boldsymbol{L} = \boldsymbol{L}_1 + \boldsymbol{L}_2$ is the total angular momentum. The two terms in the energy shift J_{nl} and K_{nl} can be interpreted as follows: J_{nl} is the electrostatic interaction of the two charge distributions $|\psi_{100}(\boldsymbol{x}_1)|^2$ and $|\psi_{nl0}(\boldsymbol{x}_2)|^2$, and it is of course positive. K_{nl} is the exchange term, which comes from the antisymmetrization of the wave function.

Using

$$2S_1 \cdot S_2/\hbar^2 = S(S+1) - \tfrac{3}{2} = \begin{cases} -\tfrac{3}{2} & \text{singlet} \\ \tfrac{1}{2} & \text{triplet} \end{cases} \quad ,$$

one can also write the energy shift (13.28) in the form

$$\Delta E_{nl}^{s,t} = J_{nl} - \tfrac{1}{2}(1 + \boldsymbol{\sigma}_1 \cdot \boldsymbol{\sigma}_2)K_{nl} \quad . \tag{13.29}$$

Because of the Pauli principle, the energy shift depends on the spin state. However, this effective spin–spin interaction is of purely electrostatic origin. The quantity K_{nl} is also positive, since the antisymmetric spatial wave function must have lower energy due to the smaller electron–electron interaction. This must of course be checked formally (see remark).

Remark: Substituting

$$|\boldsymbol{x}_1 - \boldsymbol{x}_2|^{-1} = \sum_{l'=0}^{\infty} P_{l'}(\cos\vartheta)\frac{\min(r_1,r_2)^{l'}}{\max(r_1,r_2)^{l'+1}}$$

and

$$P_{l'}(\cos\vartheta) = \frac{4\pi}{2l'+1}\sum_{m=-l'}^{l'} Y_{l'm}(\Omega_1)Y_{l'm}^*(\Omega_2)$$

into (13.28), one finds using the orthogonality relation for the spherical harmonics

$$K_{nl} = \frac{e^2}{2l+1}\int_0^\infty dr_1 r_1^2 \int_0^\infty dr_2 r_2^2 R_{10}(r_1)R_{nl}(r_2)R_{10}(r_2)R_{nl}(r_1)\frac{\min(r_1,r_2)^l}{\max(r_1,r_2)^{l+1}}.$$

Since $R_{n,n-1}$ does not have a node, it immediately follows that $K_{n,n-1} > 0$; by explicit computation, one arrives at the same result for other values of l.

13.2.2.3 Comparison of the 1s2s- and 1s2p-States and the Influence of the Spin–Orbit Interaction

The main effect of the Coulomb repulsion of the electrons may be interpreted in the following manner. The $2s$ and $2p$ electrons feel a Coulomb potential screened by the $1s$ electron. Since the $2s$ electron has a nonzero probability of being at the nucleus, it is effected by the screening less than the $2p$ electron, $E_{1s2p} > E_{1s2s}$. The additional splitting of the 1s2s- and 1s2p-levels is caused by the exchange interaction (13.29). See Fig. 13.3. The triplet levels are below the corresponding singlets, due to the antisymmetric spatial wave functions and consequent diminuation of the Coulomb interaction.

Due to the spin–orbit coupling $\boldsymbol{S}\cdot\boldsymbol{L} = (\boldsymbol{J}^2 - \boldsymbol{L}^2 - \boldsymbol{S}^2)/2$, the parahelium levels are not split ($2S+1=1$), whereas the orthohelium levels with $L \geq 1$ are split threefold ($2S+1=3$), e.g., $L = 1$, $J = 2,1,0$. In Fig. 13.3, we

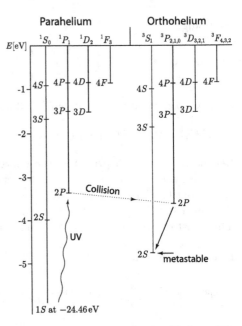

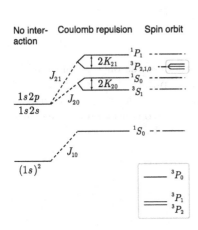

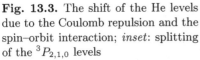

Fig. 13.3. The shift of the He levels due to the Coulomb repulsion and the spin–orbit interaction; *inset*: splitting of the $^3P_{2,1,0}$ levels

Fig. 13.4. Energy levels of helium, including the Coulomb repulsion of the electrons. The triplet levels ^3P, ^3D, ^3F are further split into 3 levels

illustrate the level shift and in Fig. 13.4, the level diagram up to the $4F$ states. We use spectroscopic notation $^{2S+1}L_J$, where L is the total orbital angular momentum, S the total spin, and J the total angular momentum. The splitting by the exchange interaction is of $O(10^{-1}\,\text{eV})$ and by the spin–orbit coupling $O(10^{-4}\,\text{eV})$.

The energy levels of helium, and in particular the absence of a triplet $1S$-state, could not be understood at all in the context of Bohr's theory; it was not until the advent of quantum theory together with the Pauli principle that an explanation became possible. In the early days of spectroscopy and atomic theory, it was suspected that helium was a mixture of two kinds of helium atoms. This came from the distinct splittings (singlet and triplet) and from the fact that no radiative transitions were found between para- and orthohelium. The triplet splitting of the radiative transitions within the orthostates was easier to reconcile with the Bohr quantum theory; thus the designation ortho and para. In Chap. 15 we will discuss the coupling to the radiation field $\boldsymbol{p} \cdot \boldsymbol{A}(\boldsymbol{x}, t)$. This does not contain the spin, and hence one has for electric dipole and quadrupole, as well as magnetic transitions the selection rule

$$\Delta S = 0 \quad,$$

where $\Delta S = S_f - S_i$ is the difference between the spin quantum numbers after (S_f) and before (S_i) the transition. Transitions from ortho- to parahelium are thus impossible, and in particular the $^3S_1$ state is (meta-) stable. The magnetic coupling of the spin to the radiation field

$$-\mu_B \sum S_i \cdot B(x_i, t) \quad ,$$

which would allow a change in S, does not give a contribution in first order either, since para- and orthohelium states have different spatial symmetry and B is symmetric. It has already been mentioned that the $(1s2s)$ $^3S_1$ state is metastable. We would like to indicate which further processes and higher transitions of low transition probability are possible. Orthohelium can form from parahelium due to bombardment with electrons. In the atmosphere it is important that irradiation of parahelium with UV light takes the state $(1s1s)$ $^1S_0$ into $(1s2p)$ $^1P_1$ (see Fig. 13.4). Collisions of $(1s2p)$ $^1P_1$ atoms can give the orthohelium state $(1s2p)$ ^3P. This state can make an electric dipole transition to the $(1s2s)$ $^3S_1$ level, because the electric dipole transitions are restricted only by the selection rules

$$\Delta l = l_f - l_i = \pm 1 \quad ,$$

$$\Delta m = m_f - m_i = \begin{cases} 1 & \text{right circularly polarized} \\ 0 & \text{linearly polarized} \\ -1 & \text{left circularly polarized} \quad . \end{cases}$$

The lowest ortholevel $(1s2s)$ $^3S_1$ is metastable with lifetime $\tau = 10^4$ s. The transition into the ground state occurs through a spin dependent, relativistic magnetic dipole transition.[6] The $(1s2s)$ $^1S_0$ level is also metastable with the lifetime $\tau = 19.7$ ms, since no electric dipole transition to $(1s1s)$ $^1S_0$ is possible. The most important decay mechanism is a two-photon electric dipole transition.[7]

Remark illustrating the exchange interaction:

$$H = J - \tfrac{1}{2}K(1 + \sigma_1 \cdot \sigma_2) = (J + K)|s\rangle\langle s| + (J - K)|t\rangle\langle t|$$

Ignoring the different spatial wave functions, the stationary states are

$$|s\rangle = (|\uparrow\downarrow\rangle - |\downarrow\uparrow\rangle)\frac{e^{-i\omega_s t}}{\sqrt{2}} \quad , \quad \omega_s = (J + K)/\hbar \quad ,$$

$$|t\rangle = (|\uparrow\downarrow\rangle + |\downarrow\uparrow\rangle)\frac{e^{-i\omega_t t}}{\sqrt{2}} \quad , \quad \omega_t = (J - K)/\hbar \quad .$$

[6] G. Feinberg, J. Sucher: Phys. Rev. Lett. **26**, 681 (1971); G.W.F. Drake: Phys. Rev. **A3**, 908 (1971)

[7] R. S. Van Dyck, Jr., C. E. Johnson, H. A. Shugart: Phys. Rev. A4, 1327 (1971)

A general state is constructed by superposition:

$$|\psi\rangle = a_s|s\rangle + a_t|t\rangle \quad .$$

Taking $a_s = a_t = 1/\sqrt{2}$, we find

$$|\psi\rangle = \frac{1}{\sqrt{2}}\left\{ \left(e^{-i\omega_s t} + e^{-i\omega_t t}\right) \frac{1}{\sqrt{2}}|\uparrow\downarrow\rangle + \left(-e^{-i\omega_s t} + e^{-i\omega_t t}\right) \frac{1}{\sqrt{2}}|\downarrow\uparrow\rangle \right\}$$

$$= e^{-i(\omega_s + \omega_t)t/2}\left(|\uparrow\downarrow\rangle \cos\left(\frac{\omega_t - \omega_s}{2}t\right) + i|\downarrow\uparrow\rangle \sin\left(\frac{\omega_t - \omega_s}{2}t\right)\right) \quad .$$

In the course of time, an exchange of the two particles takes place between the two spin states $|\uparrow\downarrow\rangle$ and $|\downarrow\uparrow\rangle$ with frequency $K/\hbar$.

13.2.3 The Variational Method

The ground state energy can be calculated much more precisely by means of the Ritz variational method of Sect. 11.2 than by means of perturbation theory. In guessing an appropriate form for the wave function, we wish to take into account the fact that each electron sees an effectively lower charge number Z^* due to screening by the other electron. We therefore make the variational ansatz

$$|\psi\rangle = |100\rangle|100\rangle|0,0\rangle \quad ,$$

$$\psi_{100}(x) = \frac{1}{\pi^{1/2}}\left(\frac{Z^*}{a}\right)^{3/2} e^{-Z^* r/a} \quad , \tag{13.30}$$

with variational parameter Z^*. It is to be determined in such a way that $\langle\psi|H|\psi\rangle$ is minimized. From (13.19), it follows that

$$\langle\psi|H|\psi\rangle = 2E_0(Z^*) - 2\langle\psi|\frac{e^2(Z - Z^*)}{|x_1|}|\psi\rangle + \langle\psi|\frac{e^2}{|x_1 - x_2|}|\psi\rangle \quad . \tag{13.31}$$

Substituting

$$E_0(Z^*) = -\text{Ry}\, Z^{*2} \quad , \tag{13.32a}$$

$$\langle 100|\frac{e^2 Z^*}{|x|}\frac{(Z - Z^*)}{Z^*}|100\rangle = 2\,\text{Ry}\, Z^{*2}\frac{(Z - Z^*)}{Z^*} \quad , \tag{13.32b}$$

$$\langle\psi|\frac{e^2}{|x_1 - x_2|}|\psi\rangle = \text{Ry}\,\frac{5}{4}Z^* \tag{13.32c}$$

(see (13.25)) into (13.31), we find

$$\langle\psi|H|\psi\rangle = -2\,\text{Ry}\left(-Z^{*2} + 2Z^* Z - \tfrac{5}{8}Z^*\right) \quad . \tag{13.33}$$

The minimum of (13.33) is at

$$Z^* = Z - \tfrac{5}{16} \quad ,$$

which gives for the ground state energy upon substitution into (13.33)

$$E_0 = -2 \left(Z - \tfrac{5}{16} \right)^2 \mathrm{Ry} = \left[-2Z^2 + \tfrac{5}{4}Z - 2 \left(\tfrac{5}{16} \right)^2 \right] \mathrm{Ry} \quad . \tag{13.34}$$

The first two terms coincide with first-order perturbation theory, while the third lowers the energy relative to first-order perturbation theory.

For $Z = 2$ (He), $E_0 = -5.7\,\mathrm{Ry} = -77.48\,\mathrm{eV}$, is a significant improvement compared to first-order perturbation theory (13.27a,b). However, for H^-, the result is still qualitatively wrong, because $-0.945\,\mathrm{Ry} > -1\mathrm{Ry}$, i.e., H^- would be unstable to the decay $\mathrm{H}^- \to \mathrm{H} + \mathrm{e}^-$, whereas in fact H^- is just barely stable. Incidentally, in the framework of nonrelativistic quantum mechanics, it can be proven exactly that H^{--} is unstable and H^- is stable.

In order to improve the variational computation, one should take into account the dependence of the wave function on the distance between the particles. Such computations have been carried out using a large number of variational parameters (~ 200) with fantastic precision[8]. It is then also necessary to take into account the motion of the nucleus. After transforming to the center-of-mass frame, the electron mass is replaced by the reduced mass ($\mu = Mm/(M + m)$), and a term $(1/M)\boldsymbol{P}_1 \cdot \boldsymbol{P}_2$ enters, in which M is the nuclear mass. However, going beyond helium, the number of variational parameters gets larger and larger with increasing electron number, and instead it turns out to be more efficient to utilize methods in which the influence of the remaining electrons on a particular electron is represented by an average field (potential).

13.3 The Hartree and Hartree–Fock Approximations (Self-consistent Fields)

We now treat the most important aspects of the theory of (perhaps ionized) atoms with N electrons and the nuclear charge number Z. Assuming a fixed nucleus, we have the Hamiltonian

$$H = \sum_{i=1}^{N} \left(\frac{\boldsymbol{p}_i^2}{2m} - \frac{Ze^2}{r_i} \right) + \sum_{i > j} \frac{e^2}{|\boldsymbol{x}_i - \boldsymbol{x}_j|} \quad , \tag{13.35}$$

and the corresponding time independent Schrödinger equation for the N-electron wave function $\psi(1, \dots, N)$,

$$H\psi(1, \dots, N) = E\psi(1, \dots, N) \quad . \tag{13.36}$$

---

[8] H.A. Bethe, R. Jackiw: *Intermediate Quantum Mechanics*, Lecture Notes in Physics (Benjamin, New York 1958)

Considering our experience with the helium atom, it is hopeless to look for an exact solution of the Schrödinger equation for more than two electrons. The situation is simplest in atoms which are similar to hydrogen, i.e., excitations of Li, Na, ... or for highly excited states (Rydberg states). Such an electron, because of screening due to the other electrons, moves effectively in a Coulomb field with charge number 1. The corrections due to the extension of the screening electron cloud lead to the replacement in (6.24′) for the hydrogen levels of n with $n + \Delta_l$, where Δ_l depends only on l but not on n. We will not go into this special case, which was important in the initial phase of atomic physics, but now discuss the method of the self-consistent field, that is, the Hartree and Hartree–Fock approximations.

One starts from the physical picture that an arbitrary electron of the atom effectively feels, in addition to the nuclear potential, a potential due to the rest of the electrons, so that each electron can be described by a one-particle Schrödinger equation. The potential in these Schrödinger equations depends on the wave functions of the other electrons and must be determined self-consistently. In the Hartree approximation, the wave function is assumed to take the form of a product of one-particle wave functions. The wave function is not antisymmetric. The Pauli exclusion principle is taken into account to the extent that all factors are distinct from one another. In the Hartree–Fock approximation, the wave function is a Slater determinant.

13.3.1 The Hartree Approximation

Here one writes the wave function in the form of a product,

$$\psi(1, \dots, N) = \varphi_1(1)\varphi_2(2) \dots \varphi_N(N) \quad , \tag{13.37a}$$

where the one-particle wave functions

$$\varphi_i(i) = \varphi_i(\boldsymbol{x}_i)\chi_i(m_{s_i}) \tag{13.37b}$$

are products of spatial and spin states. The state (13.37a) is not antisymmetric. In order to take the Pauli exclusion principle at least partly into account, the one-particle states must be distinct and orthogonal. We now determine the wave function by means of the Ritz variational principle, where the normalization

$$\int d^3x |\varphi_i(\boldsymbol{x})|^2 = 1 \tag{13.38}$$

constitutes a constraint, which is taken into account by a Lagrange multiplier ε_i. We must then minimize

$$\langle \tilde{H} \rangle = \langle H \rangle - \sum_i \varepsilon_i \left(\int d^3x |\varphi_i(\boldsymbol{x})|^2 - 1 \right) \quad , \tag{13.39}$$

where the expectation value is formed using the Hartree state (13.37a). We seek here the state which is stationary, not only under variation of parameters, but also under variation of the one-particle wave functions. We must thus construct the functional derivative $\delta/\delta\varphi_i(x)$ of (13.39).

We recall the definition of the functional derivative

$$\frac{\delta G[\varphi_i(x')]}{\delta\varphi_j(x)} = \lim_{\varepsilon \to 0} \frac{G[\varphi_i(x') + \varepsilon\delta_{ij}\delta(x' - x)] - G[\varphi_i(x')]}{\varepsilon} \tag{13.40}$$

with the important special case

$$\frac{\delta\varphi_i(x')}{\delta\varphi_j(x)} = \delta(x' - x)\delta_{ij} \quad . \tag{13.41}$$

In (13.40), $G[\varphi_i(x')]$ designates a functional of the functions φ_i, $i = 1, \ldots, N$; for example, $\int dx' F(\varphi_1(x'), \ldots, \varphi_N(x'))$.

Substituting the Hamiltonian (13.35) and the wave function of (13.37a) into (13.39), we find

$$\langle \tilde{H} \rangle = \sum_i \left\{ \int d^3x \left[\varphi_i^*(\boldsymbol{x}) \left(-\frac{\hbar^2}{2m}\boldsymbol{\nabla}^2 - \frac{Ze^2}{|\boldsymbol{x}|} - \varepsilon_i \right) \varphi_i(\boldsymbol{x}) \right] + \varepsilon_i \right\}$$

$$+ \sum_{i>j} \int d^3x \int d^3y\, \varphi_i^*(\boldsymbol{x})\varphi_j^*(\boldsymbol{y}) \frac{e^2}{|\boldsymbol{x} - \boldsymbol{y}|} \varphi_i(\boldsymbol{x})\varphi_j(\boldsymbol{y}) \quad . \tag{13.42}$$

We can now easily form the functional derivative with respect to $\varphi_i^*(\boldsymbol{x})$. This gives the *Hartree equations* for the wave function φ_i:

$$\left(-\frac{\hbar^2}{2m}\boldsymbol{\nabla}_i^2 - \frac{Ze^2}{r_i} + V_i(\boldsymbol{x}_i) \right) \varphi_i(\boldsymbol{x}_i) = \varepsilon_i\varphi_i(\boldsymbol{x}_i) \quad , \tag{13.43}$$

where

$$V_i(\boldsymbol{x}_i) = \sum_{j\neq i} \int d^3x_j \frac{e^2}{|\boldsymbol{x}_i - \boldsymbol{x}_j|} |\varphi_j(\boldsymbol{x}_j)|^2 \quad . \tag{13.44}$$

Equation (13.43) is easy to interpret and could have been guessed even without the variational calculation. The first two terms are the kinetic energy and the nuclear potential, and the third term is the electrostatic potential due to the charge distribution of the other electrons. The Hartree equation for φ_i contains in the potential all the other wave functions. These nonlinear equations can be solved self-consistently for atoms only by numerical methods, i.e., one assumes functions $\varphi_1 \ldots \varphi_N$, determines the V_i using (13.44), and then solves (13.43). With the new wave functions, one again computes the V_i and continues the iteration until no further change occurs. (The Hartree approximation is trivial only for translationally invariant problems.) A simplification of the solution of the Hartree equations results if $V_i(\boldsymbol{x})$ is approximated by the spherically symmetric potential

$$\frac{1}{4\pi} \int d\Omega_i V_i(\boldsymbol{x}_i)$$

with the element of solid angle $d\Omega_i$.

The constants ε_i entered (13.39) as Lagrange multipliers; in the Hartree equations, they have the appearance of one-particle energy eigenvalues. What is their physical meaning?

Multiplying (13.43) by φ_i^* and integrating over x, we find

$$\varepsilon_i = \int d^3x \left(\frac{\hbar^2}{2m} |\boldsymbol{\nabla}\varphi_i(\boldsymbol{x})|^2 - \frac{Ze^2}{r} |\varphi_i(\boldsymbol{x})|^2 \right)$$
$$+ \sum_{j \neq i} \int d^3x \int d^3y |\varphi_i(\boldsymbol{x})|^2 |\varphi_j(\boldsymbol{y})|^2 \frac{e^2}{|\boldsymbol{x} - \boldsymbol{y}|} \quad . \tag{13.45}$$

Since ε_i consists of just those terms of $\langle H \rangle$ (see (13.42)) containing φ_i, the quantity $-\varepsilon_i$ represents the ionization energy under the approximative assumption that the other states do not change when the electron in the one particle state φ_i is removed. Using (13.42), (13.43), and (13.44), one finds for the total energy

$$E \equiv \langle H \rangle = \sum_{i=1}^{N} \varepsilon_i - \sum_{i<j} \int d^3x \, d^3y \, \frac{e^2}{|\boldsymbol{x} - \boldsymbol{y}|} |\varphi_i(\boldsymbol{x})|^2 |\varphi_j(\boldsymbol{y})|^2 \quad . \tag{13.46}$$

13.3.2 The Hartree–Fock Approximation

In the Hartree–Fock approximation, the wave function including the spin is assumed to take the form of a Slater determinant of one-particle wave functions (or orbitals)

$$\psi(1, 2, \ldots, N) = \frac{1}{\sqrt{N!}} \begin{vmatrix} \varphi_1(1) & \cdots & \varphi_1(N) \\ \vdots & & \vdots \\ \varphi_N(1) & \cdots & \varphi_N(N) \end{vmatrix} , \tag{13.47}$$

where the φ_i are again of the form (13.37b) and are normalized. The expectation value of the Hamiltonian (13.35) in the state (13.47) is

$$\langle H \rangle = \sum_i \int d^3x \left[\frac{\hbar^2}{2m} |\boldsymbol{\nabla}\varphi_i(\boldsymbol{x})|^2 - \frac{Ze^2}{r} |\varphi_i(\boldsymbol{x})|^2 \right]$$
$$+ \frac{1}{2} \sum_{ij} \int d^3x \, d^3x' \frac{e^2}{|\boldsymbol{x} - \boldsymbol{x}'|} |\varphi_i(\boldsymbol{x})|^2 |\varphi_j(\boldsymbol{x}')|^2$$
$$- \frac{1}{2} \sum_{ij} \delta_{m_{s_i} m_{s_j}} \int d^3x \, d^3x' \frac{e^2}{|\boldsymbol{x} - \boldsymbol{x}'|} \varphi_i^*(\boldsymbol{x})\varphi_i(\boldsymbol{x}')\varphi_j^*(\boldsymbol{x}')\varphi_j(\boldsymbol{x}) \quad .$$

$$\tag{13.48}$$

Remark: To show (13.48), we first consider the expectation value of the kinetic energy

$$\left(\psi, \left(-\sum_{k=1}^{N} \nabla_k^2\right)\psi\right) = -\sum_{i=1}^{N}(\varphi_i, \nabla^2 \varphi_i) = \sum_{i=1}^{N}\int d^3x |\nabla \varphi_i(x)|^2 \quad , \qquad (13.49)$$

where we use the fact that (13.47) contains $N!$ mutually orthogonal terms, and that therefore the contribution with the wave function φ_i occurs $N!$ times in (13.49). Similarly we see that

$$\left(\psi, \sum_{k=1}^{N}\frac{1}{|x_k|}\psi\right) = \sum_{i=1}^{N}\int d^3x \frac{1}{|x|}|\varphi_i(x)|^2 \quad . \qquad (13.50)$$

After the expectation values of the one-particle operators, we calculate the expectation value of the Coulomb interaction, i.e., of a two-particle operator. For example, consider a contribution $1/|x_1 - x_2|$ and a term in (13.47) depending on φ_1 and φ_2 with argument 1 or 2,

$$\frac{1}{\sqrt{N!}}(\varphi_1(1)\varphi_2(2) - \varphi_2(1)\varphi_1(2)) \ldots \quad . \qquad (13.51)$$

The points refer to the remaining factors. The contribution of (13.51) to the expectation value of $1/|x_1 - x_2|$ is

$$\frac{1}{N!}\left(\varphi_1(1)\varphi_2(2) - \varphi_2(1)\varphi_1(2), \frac{1}{|x_1 - x_2|}(\varphi_1(1)\varphi_2(2) - \varphi_2(1)\varphi_1(2))\right). \qquad (13.52)$$

Now, each wave function pair such as φ_1 and φ_2 occurs exactly $N!/2$ times in (13.47), so that the total expectation value is given by

$$\left(\psi, \sum_{k>l}\frac{e^2}{|x_k - x_l|}\psi\right) = \frac{1}{2}\sum_{i,j}\int d^3x\, d^3x'\, \frac{e^2}{|x - x'|}$$
$$\times \left(|\varphi_i(x)|^2|\varphi_j(x')|^2 - \varphi_i^*(x)\varphi_i(x')\varphi_j^*(x')\varphi_j(x)\delta_{m_{s_i} m_{s_j}}\right) \quad . \qquad (13.53)$$

The Kronecker delta comes from the spin scalar product of the mixed terms in (13.52). Thus, (13.48) has been shown.

Using the Ritz variational procedure with the constraint (13.38), we find by differentiating (13.48) with respect to φ_i^* the stationarity condition

$$\left(-\frac{\hbar^2}{2m}\nabla^2 - \frac{Ze^2}{r}\right)\varphi_i(x) + \int d^3x' \frac{e^2}{|x - x'|}$$
$$\times \sum_j \varphi_j^*(x')\left[\varphi_j(x')\varphi_i(x) - \varphi_j(x)\varphi_i(x')\delta_{m_{s_i} m_{s_j}}\right] = \varepsilon_i\,\varphi_i(x) \quad .$$
$$\qquad (13.54)$$

These are the *Hartree–Fock equations*. The Hartree–Fock equation for φ_i differs from the Hartree equation by the term

$$\int d^3x' \frac{e^2}{|\boldsymbol{x} - \boldsymbol{x}'|} \left(|\varphi_i(\boldsymbol{x}')|^2 \varphi_i(\boldsymbol{x}) - \sum_j \varphi_j^*(\boldsymbol{x}')\varphi_j(\boldsymbol{x})\varphi_i(\boldsymbol{x}')\delta_{m_{s_i} m_{s_j}} \right)$$

$$= -\sum_{j \neq i} \int d^3x' \frac{e^2}{|\boldsymbol{x} - \boldsymbol{x}'|} \varphi_j^*(\boldsymbol{x}')\varphi_j(\boldsymbol{x})\varphi_i(\boldsymbol{x}')\delta_{m_{s_i} m_{s_j}} \quad . \tag{13.55}$$

The first three terms in (13.54) excluding $j = i$ can be interpreted as in the Hartree equation as the Hamiltonian of an electron in the Coulomb potential screened by the charge density of the other electrons. The fourth term is the *exchange term*. This is a nonlocal term, since here φ_i occurs with argument $\boldsymbol{x}' \neq \boldsymbol{x}$. The exchange term is nonzero only for $m_{s_i} = m_{s_j}$. The expression in square brackets in (13.54) is then the probability amplitude for finding i and j at the positions $\boldsymbol{x}$ and $\boldsymbol{x}'$.

Remark: One easily sees that the Hartree–Fock states are orthogonal: The state $\varphi_i(\boldsymbol{x}_i)\chi_i(m_{s_i})$ is orthogonal to all states with $m_{s_j} \neq m_{s_i}$. The spatial wave functions φ_j with $m_{s_j} = m_{s_i}$ all satisfy a Schrödinger equation with exactly the same potential,

$$\left(-\frac{\hbar^2}{2m}\nabla^2 + u(x) \right) \varphi_i(x) + \int dx' u(x, x')\varphi_i(x') = \varepsilon_i\varphi_i(x) \quad ,$$

where

$$u(x, x') = \sum_j \varphi_j^*(x') \frac{-e^2}{|x - x'|} \varphi_j(x) = u(x', x)^* \quad .$$

Forming $\int dx\, \varphi_j^*(x)$, one obtains

$$\left(\varphi_j, \left(-\frac{\hbar^2}{2m}\nabla^2 + u(x) \right) \varphi_i \right) + \int dx \int dx' u(x, x')\varphi_j(x)\varphi_i(x') = \varepsilon_i(\varphi_j, \varphi_i) \quad .$$

Subtracting from this the equation in which $i \leftrightarrow j$ are interchanged, and using the Hermiticity, one finds $(\varphi_i, \varphi_j) = 0$ for $\varepsilon_i \neq \varepsilon_j$. For degenerate eigenvalues, one can orthogonalize, so that one finally obtains

$$(\varphi_i, \varphi_j) = \delta_{ij} \quad . \tag{13.56}$$

We now show that $-\varepsilon_i$ is also the ionization energy in Hartree–Fock theory. For this, we form the scalar product of φ_i with the right and left side of (13.54):

$$\varepsilon_i = \int d^3x \left(\frac{\hbar^2}{2m}|\nabla\varphi_i|^2 - \frac{Ze^2}{r}|\varphi_i|^2 \right) + \int d^3x\, d^3x' \frac{e^2}{|\boldsymbol{x} - \boldsymbol{x}'|}\varphi_i^*(\boldsymbol{x})$$

$$\times \sum_j \varphi_j^*(\boldsymbol{x}') \left[\varphi_j(\boldsymbol{x}')\varphi_i(\boldsymbol{x}) - \varphi_j(\boldsymbol{x})\varphi_i(\boldsymbol{x}')\delta_{m_{s_i} m_{s_j}} \right] \quad . \tag{13.57}$$

These are just the terms in $\langle H \rangle$ from (13.48) which contain the wave function φ_i. Thus, $-\varepsilon_i$ is the energy needed to remove a particle in the state $\varphi_i{}^9$, under

---

[9] *Koopmans-Theorem*, T.H. Koopmans: Physica **1**, 104 (1933)

the assumption that the other wave functions do not change. The larger the particle number, the better this is satisfied. A corresponding result was found for ε_i in the Hartree theory. The Hartree–Fock equations give in comparison to the Hartree equations an improvement of 10–20 %. The exchange term lowers the energy.

13.4 The Thomas–Fermi Method

For atoms with many electrons, computation of the Hartree or Hartree–Fock wave functions is very tedious. On the other hand, physical considerations give rise to a simplification. Because of the large number of electrons, each one feels just the same effective potential, constant in time, arising due to the other electrons and the nucleus. Furthermore, most of the electrons are in states of high energy, i.e., large principal quantum number, so that the wavelength is small and the change of the potential relative to its value is slight within one wavelength. We can then utilize a semiclassical approximation and assume that there are many electrons within volume elements in which the potential is nearly constant, that their states are locally plane waves, and that their distribution can be determined on the basis of the results (13.17) for free fermions.

We start from the Hartree potential $V_i(\boldsymbol{x})$, Eqn. (13.44), and, because of the large number of electrons, we may include in the sum also the state $\varphi_i(\boldsymbol{x})$. Then one and the same potential acts on all electrons

$$V(\boldsymbol{x}) = \int d^3x' \frac{e^2}{|\boldsymbol{x} - \boldsymbol{x}'|} \sum_j \varrho_j(\boldsymbol{x}') - \frac{Ze^2}{r} \quad , \quad \text{i.e.} \tag{13.58a}$$

$$V(\boldsymbol{x}) = \int d^3x' \frac{e^2}{|\boldsymbol{x} - \boldsymbol{x}'|} n(\boldsymbol{x}') - \frac{Ze^2}{r} \quad , \tag{13.58b}$$

where $\varrho_j(\boldsymbol{x}) = |\varphi_j(\boldsymbol{x})|^2$ is the probability density of the jth electron and $n(\boldsymbol{x}) = \sum_j \varrho_j(\boldsymbol{x})$ is the particle number density at the position $\boldsymbol{x}$.

Locally, the states are semiclassical plane waves $e^{i\boldsymbol{p}(\boldsymbol{x})\cdot\boldsymbol{x}/\hbar}$, where $\hbar/|\boldsymbol{p}(\boldsymbol{x})|$ is small compared to the spatial variation of $V(\boldsymbol{x})$, except in the vicinity of the classical turning points. The connection between energy and local momentum is

$$\varepsilon = \frac{\boldsymbol{p}^2(\boldsymbol{x})}{2m} + V(\boldsymbol{x}) \quad . \tag{13.59}$$

At each point $\boldsymbol{x}$, the states with local momentum between 0 and a maximal value $|\boldsymbol{p}_{\mathrm{F}}(\boldsymbol{x})|$ are occupied. Denoting the energy of the highest occupied state by ε_{F}, we find for the Fermi momentum

$$|p_{\mathrm{F}}(\boldsymbol{x})| = (2m(\varepsilon_{\mathrm{F}} - V(\boldsymbol{x}))^{1/2} \quad . \tag{13.60}$$

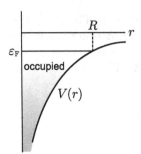

Fig. 13.5. The effective potential, Fermi energy, and occupied states

The particle number density at the point $\boldsymbol{x}$ then becomes, according to (13.17),

$$n(\boldsymbol{x}) = \frac{p_F^3(\boldsymbol{x})}{3\pi^2\hbar^3} \quad . \tag{13.61}$$

We next determine ε_F. The atomic radius R is obtained from $V(R) = \varepsilon_F$. Now, in a neutral atom, the electrons screen off the nuclear charge, and $V(r) = 0$ must hold for $r > R$ (Fig. 13.5), which implies $\varepsilon_F = 0$. (In the case of an ion, $V(r) = -(Z-N)e^2/r$ for $r > R$.)

Below, we consider only neutral atoms. Substituting $\varepsilon_F = 0$ into (13.60) and (13.61), we find

$$n(\boldsymbol{x}) = \frac{1}{3\pi^2\hbar^3}(-2mV(\boldsymbol{x}))^{3/2} \quad . \tag{13.62}$$

The density or the potential must now be determined self-consistently. For this purpose, one could solve the integral equation (13.58b) in conjunction with (13.62). Equivalently, applying the operator $\boldsymbol{\nabla}$ to (13.58b) one derives the Poisson equation

$$\Delta V(\boldsymbol{x}) = -4\pi e^2 n(\boldsymbol{x}) + 4\pi Z e^2 \delta^{(3)}(\boldsymbol{x}) \,, \tag{13.63}$$

which gives together with (13.62) $n(\boldsymbol{x})$ or $V(\boldsymbol{x})$, respectively. By radial symmetry, one has for finite r

$$\frac{1}{r^2}\frac{\partial}{\partial r}r^2\frac{\partial}{\partial r}(-V(r)) = \frac{4e^2}{3\pi\hbar^3}[-2mV(r)]^{3/2} \quad . \tag{13.63'}$$

Since $V(r)$ must go over to the potential of the nucleus for small distances, the boundary condition

$$\lim_{r \to 0} V(r) = -\frac{Ze^2}{r} \tag{13.64}$$

results. We therefore introduce the substitution

$$V(r) = -\frac{Ze^2}{r}\chi(x) \tag{13.65}$$

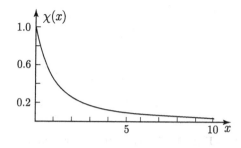

Fig. 13.6. Solution of the Thomas–Fermi equation for neutral atoms

with

$$r = \frac{bx}{Z^{1/3}} \quad , \quad b = \frac{1}{2}\left(\frac{3\pi}{4}\right)^{2/3} a = 0.885a \tag{13.66}$$

and $a = \hbar^2/me^2$, the Bohr radius. We obtain the differential equation

$$\frac{d^2\chi}{dx^2} = x^{-1/2}\chi^{3/2}(x) \tag{13.67}$$

with boundary conditions $\chi(0) = 1$, $\chi(\infty) = 0$. Numerical solution of (13.67) gives (Fig. 13.6)

$$\chi(x) = \begin{cases} 1 - 1.59\,x & x \to 0 \\ \dfrac{144}{x^3} & x \to \infty \end{cases} . \tag{13.68}$$

The density $n(x)$ resulting from (13.62) and (13.66) is

$$n(x) = \frac{(2mZe^2)^{3/2}}{3\pi^2\hbar^3}\left(\frac{1}{r}\chi\left(\frac{r}{Z^{-1/3}b}\right)\right)^{3/2} . \tag{13.69}$$

For small distances, the potential is

$$V(r) = -\frac{Ze^2}{r} + 1.8\,Z^{4/3}\frac{e^2}{a} . \tag{13.70}$$

The extension of the atom is infinite, an obviously unphysical feature. The form of $V(r)$ and $n(r)$ is the same for all atoms, because χ is a universal function. The typical length scale is $\sim Z^{-1/3}$. Numerical solution shows that half of the electrons are within the radius $1.33\,aZ^{-1/3}$.

We can now estimate the validity of the Thomas–Fermi approximation. The radius is $\sim Z^{-1/3}$ and decreases with growing Z, the potential at a fixed distance is $\sim Z^{4/3}$, and therefore, from (13.60), the typical wavelength is $\sim Z^{-2/3}$. The characteristic length over which the potential changes significantly is $\sim Z^{-1/3}$. Hence, the ratio of the latter two quantities is $Z^{-1/3}$. The larger Z, the smaller this ratio becomes. In this limit, the statistical

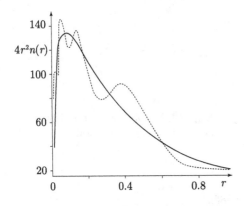

Fig. 13.7. The radial distribution function $4r^2n(r)$ for mercury Hg; (——): Thomas–Fermi, (---): Hartree

treatment as a free electron gas is also better justified. Thomas–Fermi theory becomes exact in the limit $Z \to \infty$[10].

The radial distribution function for the Thomas–Fermi approximation $D(r) = 4r^2n(r)$ is shown in Fig. 13.7 and compared to the Hartree approximation.

The total energy of the Thomas–Fermi atom is given by $E = E_{\text{kin}} + E_{\text{pot}} = 1/2E_{\text{pot}}$, where the virial theorem has been used. The potential energy is composed of two parts:

$$E_{\text{pot}} = -Ze^2 \int d^3x \frac{n(\boldsymbol{x})}{r} + \frac{1}{2}e^2 \iint d^3x\, d^3x' \frac{n(\boldsymbol{x})n(\boldsymbol{x}')}{|\boldsymbol{x} - \boldsymbol{x}'|} \quad ;$$

by $\int d^3x n(\boldsymbol{x}) = Z$, one finds the proportionality $E \sim E_{\text{pot}} \sim -Z^{7/3}$. Numerical computation gives $E = -20.8\, Z^{7/3}$ eV.

Thomas–Fermi theory has the following shortcomings: It is inaccurate for small r, since the variation of the potential is too strong there. It is also inaccurate for large distances, since there the wavelength is no longer small, because the difference between the energy and the potential goes to zero. Moreover, the density gets small, and a statistical treatment becomes unjustified. This is the origin of the unphysical infinitely large atomic radius. However, most of the electrons lie in the region $a/Z < r < a$, and in this region Thomas–Fermi theory is reliable.

One can also build exchange effects into Thomas–Fermi theory; the result is then the *Thomas–Fermi–Dirac* equation[11]:

$$\chi'' = x\left(\sqrt{\frac{\chi}{x}} + \beta\right)^3 \quad , \tag{13.71a}$$

[10] E. Lieb, B. Simon: Adv. Math. **23**, 22 (1977)

[11] H.A. Bethe, R. Jackiw: *Intermediate Quantum Mechanics*, Lecture Notes in Physics (Benjamin, New York 1958)

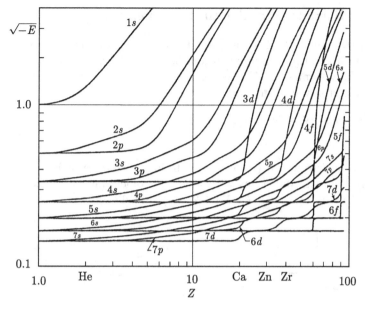

Fig. 13.8. One-electron energy levels in a modified Thomas–Fermi–Dirac potential[12]

with

$$\beta = \sqrt{\frac{b}{aZ^{4/3}}} \frac{1}{\pi\sqrt{2}} = 0.2118\,Z^{-2/3} \quad . \tag{13.71b}$$

For $Z \to \infty$, it is identical to the Thomas–Fermi equation. The solutions of (13.71) are no longer universal, but rather depend on Z. We also mention a computation of the one-electron states as a function of Z for the Thomas–Fermi(–Dirac) potential[12]. The energy of the orbitals for the Thomas–Fermi–Dirac potential is shown in Fig. 13.8 as a function of Z. This theory shows the filling of the energy levels with increasing Z in the order $1s$, $2s$, $2p$, $3s$, $3p$, $4s$, $3d$. One also sees the approximate degeneracy of $4s$ and $3d$ in the region $20 < Z < 30$ (iron group). The filling of the levels continues between $Z = 31$ (Ga) and $Z = 36$ (Kr) with the $4p$-shell. The $5s$-shell then follows, and then the $4d$-shell. For the Thomas–Fermi potential, instead the $5p$-shell would be filled first after $Z = 39$.

[12] R. Latter: Phys. Rev. **99**, 510 (1955)

13.5 Atomic Structure and Hund's Rules

The effective potential of the Hartree and Hartree–Fock equations is not a $1/r$-potential. In a good first approximation, however, the potential can be assumed spherically symmetric. The energy eigenvalues ε_{nl} now depend on l, and the one-electron wave functions are

$$\varphi_i = R_{nl}(r)Y_{lm_l}(\vartheta, \varphi)\chi(m_s) \quad ,$$

$$\chi(m_s) = \chi_\pm \quad , \tag{13.72}$$

where R_{nl} is to be determined (respectively) from the Hartree or Hartree–Fock equations. The eigenfunctions with angular momentum l are labeled with increasing energy by n $(n = l + 1, l + 2, \ldots)$ (Fig. 13.9).

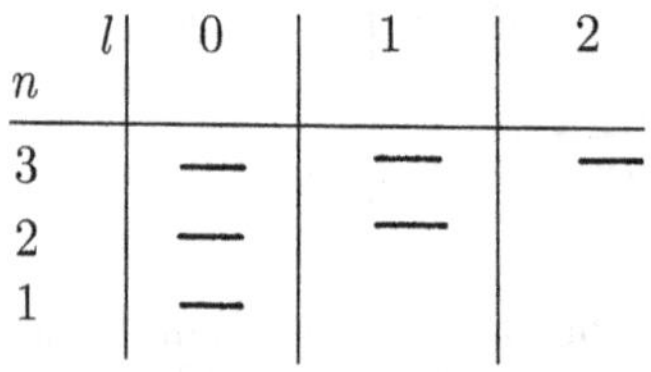

l	0	1	2
n			
3	—	—	—
2	—	—	
1	—		

Fig. 13.9. Numbering of energy levels

For each fixed pair of n and l, there are $(2S + 1)(2l + 1) = 2(2l + 1)$ different states (orbitals) (Table 13.3). The set of these $2(2l+1)$ one-electron states is called a shell. (Notation: (n, l), where, as in the hydrogen atom, for $l = 0, 1, 2, \ldots$ one uses the symbols $s, p, d, \ldots$). For example, the 1s-shell has two states, 2p-shell six states, etc.

By filling the orbitals one after the other in successive shells, one gets all the elements of the *periodic table*. See page 413. The electronic state of an atom is characterized by specifying the occupied orbitals, or, as it is often called, the *configuration*. For example, in carbon C $(1s)^2(2s)^2(2p)^2$ two electrons are in the 1s-shell, two in the 2s-shell and two in the 2p-shell. The configurations of the lightest elements are

H	$1s$
He	$(1s)^2$
Li	$(1s)^2 2s$
Be	$(1s)^2(2s)^2$
B	$(1s)^2(2s)^2 2p.$

Table 13.3. Degree of degeneracy of the first shells

l	$s, 0$	$p, 1$	$d, 2$	$f, 3$
Degeneracy	2	6	10	14

For given n, orbitals $s, p, \ldots$ are successively filled. The smaller l is, the larger the probability density at the nucleus, and the less the electron in question is affected by the screening due to the other electrons. Below, we no longer give the orbitals of the inner filled shells. Continuing the configurations of the periodic table begun above, we skip the elements carbon C to fluorine F, in which the $2p$-shell is successively filled. With neon, the $2p$-shell is closed: Ne $(2p)^6$.

From sodium Na to argon Ar the $3s$ and $3p$-shells are filled, in analogy to $2s$ and $2p$. However, the further progression does not begin with the $3d$-shell, but rather at potassium K a $4s$-electron follows, and the periodic table proceeds further in the following manner ($E_{4s} \approx E_{3d}$):

K $4s$

Ca $(4s)^2$

Sc $(4s)^2 3d.$

In the transition metals of row IV, from scandium Sc to zinc Zn, the $3d$-shell is filled. Now, after all the orbitals with $n = 3$ have been used up, the occupation continues with the $4p$-shell. The corresponding elements are chemically similar to the $3p$-elements above them in the periodic table.

Row V of the periodic table is filled analogously, i.e., the $4f$-states remain empty at first.

Row VI begins with Cs $6s$, Ba $6s^2$, La $6s^2 5d$; however, it is not continued with the completion of the $5d$-shell but rather the $4f$-shell is successively built up. The first element after lanthanum is Ce $6s^2 5d\,4f$. After this filling process has been completed with Lu $6s^2 5d\,4f^{14}$, the transition metals of the $5d$-shell are formed in the sequence from Hf $6s^2 5d^2 4f^{14}$ onwards.

The structure of row VII corresponds to that of row VI. After actinium Ac $7s^2 6d$, from protactinium Pa $6d\,5f^2$ onwards the $5f$-shell is filled in competition with $6d$.

The elements in which d or f-shells are being filled are called *transition elements*. The elements in which the d-shells are being filled are called *transition metals* ($3d$-iron group, $4d$-paladium group and $5d$-platinum group).

The elements following lanthanum are the lanthanides or rare earths, and those following actinium are the actinides. Note that in the rare earth elements Gd and Tb a $5d$-orbital is again occupied. The important point is that the chemical behavior of the transition elements is determined by the most distant s-electrons. A consequence of this is that the rare earths are closely related chemically and are easily substituted in crystals. In general, chemical behavior is determined by the outer electrons and is therefore similar in a given column of the periodic table.

We now discuss the *ionization potential*, which is the binding energy of the most weakly bound electron.

Atoms with an s-electron outside of closed shells have the lowest binding energy (H, Li, Na, $\ldots$). The second s-electron is more strongly bound (He,

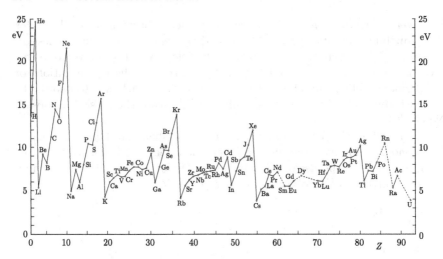

Fig. 13.10. The ionization potential as function of Z

Be, Mg, ...) because of the higher nuclear charge. The next electron occupies
the otherwise empty p-shell, which is associated with a drop in the binding
energy (B, Al). As the p-shell is built up, the binding energy grows at first,
until with the fourth electron, which must be added to the half-filled shell,
another drop occurs. This circumstance can be explained by the fact that up
to the third electron all the spins are aligned in parallel and thus the three
different spatial wave functions can become completely antisymmetric. The
fourth electron has oppositely aligned spin, and the spatial wave function is
no longer totally antisymmetric. From Ga on, a similar picture again occurs
(Fig. 13.10).

Now that we know the electron configurations of the elements, we must
determine the quantum number L of the total orbital angular momentum

$$\boldsymbol{L} = \sum_{i=1}^{N} \boldsymbol{L}_i$$

and the quantum number S of the total spin

$$\boldsymbol{S} = \sum_{i=1}^{N} \boldsymbol{S}_i \quad .$$

If a shell is not completely filled, several values for S and L are possible.

For the moment, we ignore the spin–orbit coupling, so that the Hamiltonian is given by (13.35); then the angular momentum commutes with the
Hamiltonian, $[H, \boldsymbol{L}] = 0$, and L is thus a good quantum number. The eigenstates of H are simultaneously eigenstates of $\boldsymbol{L}^2$ and L_z; by rotational invariance, the energy does not depend on L_z.

Although $[H, \boldsymbol{S}_i] = 0$ for every single $\boldsymbol{S}_i$, after antisymmetrization the eigenfunctions are only eigenfunctions of $\boldsymbol{S}^2$ and S_z. The energy also depends on S, because the larger S, the more antisymmetric the spatial wave function and the smaller the mutual Coulomb repulsion among the electrons.

Thus far, only the Coulomb interaction has been taken into account. With *spin–orbit coupling*, the quantum numbers are L, S, J, where $\boldsymbol{J} = \boldsymbol{L} + \boldsymbol{S}$ is the total angular momentum. The values of J are $L+S, \ldots |L-S|$. These are $2S+1$ values for $L > S$ and $2L+1$ for $L < S$. For $L \geq S$, the multiplicity of the energy terms with orbital angular momentum L is $2S+1$. Since the energy depends on J, the levels belonging to an L and S split into a $2S+1$-multiplet.

Which of these states has the lowest energy? The answer is important for an understanding of the magnetic properties of atoms and ions. By means of the *Hund's* semiempirical rules, which are based on simple physical arguments, one can determine the *ground state*.

Hund's Rules

Taking into account the Pauli exclusion principle, the states are occupied such that

1. S is maximal,
2. L is maximal,
3. for shells, not more than half-filled $^{2S+1}L_{J=|L-S|}$, while for more than half-filled shells $^{2S+1}L_{J=L+S}$.

We indicate the reasons for Hund's rules.

1. The larger the spin, the more antisymmetric the spatial part of the wave function and the smaller the Coulomb interaction, since the spatial wave function vanishes for vanishing particle separation, which is where the Coulomb interaction would have been strongest.

2. The larger the total orbital angular momentum, L, the farther the electrons are separated from the center and therefore from each other as well. Since the Coulomb interaction decreases with $1/r$, the Coulomb repulsion becomes smaller, though not as markedly as in 1.

3. According to (12.8), the spin–orbit interaction is of the form

$$V_{\mathrm{SO}} = \sum_i \alpha_i \boldsymbol{L}_i \cdot \boldsymbol{S}_i \quad \text{with} \quad \alpha_i = \frac{\hbar^2}{2m^2c^2r_i}\frac{dV(r_i)}{dr_i} \quad ,$$

where $V(r_i)$ is the potential energy of the self-consistent field.

Calculating first-order perturbation theory, one finds for a state with quantum numbers L and S

$$\langle \boldsymbol{L}_i \rangle \propto \langle \boldsymbol{L} \rangle \quad , \quad \langle \boldsymbol{S}_i \rangle \propto \langle \boldsymbol{S} \rangle \quad .$$

The Hamiltonian is thus equivalent to $V_{SO} = A\,\boldsymbol{S}\cdot\boldsymbol{L}$ with coefficient A. From

$$\boldsymbol{S}\cdot\boldsymbol{L} = \tfrac{1}{2}[J(J+1) - L(L+1) - S(S+1)]$$

it follows that for $A > 0$ the most favorable value of J is $|L - S|$, whereas for $A < 0$ the most favorable value of J is $L + S$. In computing A, we note that the α_i are positive and independent of m_l. If only one shell is incompletely filled, and if this one is at most half full, then according to Hund's first rule the spins are parallel, i.e., $\boldsymbol{S}_i = \boldsymbol{S}/n$, where n is the number of electrons. Hence, the spin–orbit Hamiltonian becomes

$$V_{SO} = \sum_i \alpha_i \boldsymbol{L}_i \cdot \boldsymbol{S}/n \sim \frac{1}{n}\alpha \boldsymbol{L}\cdot\boldsymbol{S}\ \ .$$

Therefore

$$A = \frac{\alpha}{n} = \frac{\alpha}{2S} > 0\ \ .$$

For a shell which is more than half full, we imagine the empty configurations to be first completed and then removed again. For the filled shell, one would have $V_{SO} = 0$. There remains $V_{SO} = -\sum_i \alpha_i \boldsymbol{L}_i \boldsymbol{S}_i$, where one sums over the holes. The total spin and the total angular momentum are $\boldsymbol{S} = -\sum_i \boldsymbol{S}_i$ and $\boldsymbol{L} = -\sum_i \boldsymbol{L}_i$. By the same argument as previously, it follows that $A = -\alpha/2S$. For an at most half-filled shell, $A > 0$ and $J = |L - S|$, whereas for a more than half-filled shell, $A < 0$ and $J = L + S$. For a half-filled shell of orbitals with angular momentum l, one has according to the first two of Hund's rules $S = (2l + 1)/2$ and $L = 0$, and therefore $L + S = |L - S| = S$.

Let us apply Hund's rules to a few examples:

- He $(1s)^2$: In the ground state, $S = 0$, because of the antisymmetrization, rather than $S = 1$, i.e., $^1S_0$.
- B $(1s)^2(2s)^2 2p$: The closed $1s$ and $2s$ shells have $J = L = S = 0$. Thus, $S = 1/2$ and $L = 1$; it remains to compare $^2P_{1/2}$ and $^2P_{3/2}$. Since the $2p$-shell is less than half full, one finds according to Hund's third rule $^2P_{1/2}$.
- C $(1s)^2(2s)^2(2p)^2$: Hund's first rule demands $S = 1$. Now, by the Pauli exclusion principle, L cannot be 2, since the two orbital angular momentum quantum numbers m_l must differ from one another. With $m_l = 0$ and 1 as the maximal L, the value $L = 1$ results. Since the shell is less than half full, Hund's third rule selects from $J = 0, 1, 2$ the smallest value, i.e., $^3P_0$.
- N $(1s)^2(2s)^2(2p)^3$: The maximal value of the spin is $3/2$. The possible values of L are 3, 2, 1, 0. Since the spin function is completely symmetric, the spatial wave function must be completely antisymmetric, for which one requires all three wave functions with $m_l = 1, 0, -1$. Thus $L = 0$ and $J = 3/2$, i.e., $^4S_{3/2}$.

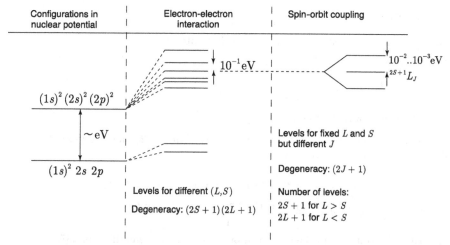

Fig. 13.11. Splitting of the energy levels (schematic); the energy scale increases to the right

The splitting of the levels within a configuration comes from the Coulomb interaction of the electrons and the spin–orbit interaction. The splitting is illustrated schematically in Fig. 13.11.

In the above discussion, we have assumed that the Coulomb interaction of the electrons is much larger than the spin–orbit coupling. In that case, L and S are good quantum numbers and one speaks of $L - S$ coupling or *Russel–Saunders coupling*. This situation obtains in light atoms. If the spin–orbit coupling dominates compared to L and S dependent parts of the Coulomb energy, then each electron is characterized by its total angular momentum j. The j are combined to form a J of the atom. This is called *j–j coupling*. The *j–j* coupling is never realized in its pure form, even in heavy atoms.

Further Literature:

B. H. Bransden and C. J. Joachain: *Physics of Atoms and Molecules* (Longman, New York 1983)

H. A. Bethe and E. E. Salpeter: *Quantum Mechanics of One- and Two-Electron Atoms* (Plenum, New York 1957)

H. Friedrich: *Theoretical Atomic Physics*, 3rd. ed. (Springer, Berlin, Heidelberg 2005)

L. D. Landau and E. M. Lifshitz: *Quantum Mechanics: Nonrelativistic Theory* (Addison-Wesley, Reading, Mass. 1958)

Problems

13.1 On the perturbation theory of the helium atom:

(a) Show that

$$\int \frac{d\Omega}{|\boldsymbol{x}_1 - \boldsymbol{x}_2|} = \frac{4\pi}{\max\{r_1, r_2\}}$$

for the integral over solid angle ($r_i \equiv |\boldsymbol{x}_i|$).

(b) Starting from (13.24), compute the energy shift (13.25) for the ground state due to the electron–electron interaction.

13.2 Show that for $l = n - 1$ the exchange term defined in (13.28),

$$K_{nl} = \int d^3x_1 \int d^3x_2 \frac{\psi_{100}^*(\boldsymbol{x}_1)\,\psi_{nl0}^*(\boldsymbol{x}_2)\,\psi_{100}(\boldsymbol{x}_2)\,\psi_{nl0}(\boldsymbol{x}_1)}{|\boldsymbol{x}_1 - \boldsymbol{x}_2|}$$

is positive by completing the computational steps in the remark on p. 237.

13.3 With the aid of Hund's rules, determine the values of the total spin S, the total orbital angular momentum L, and the total angular momentum J for the $3d$-shells of the transition metals Cu, Ni, Co, Fe, Mn, Cr, V, Ti, and Sc.

13.4 Explain on the basis of Hund's rules why the ground state of carbon is $^3P_0$ and that of oxygen is $^3P_2$.

13.5 Let the one-particle states $\varphi_{\alpha_i}(i)$ form a complete orthonormal set. Show that the states $\psi_{\rm a}(1, \ldots, N)$ and $\psi_{\rm s}(1, \ldots, N)$ of (13.12) and (13.13) form complete orthonormal sets in the state space of purely antisymmetric and purely symmetric states.

13.6 Consider an "atom" described by the Hamiltonian

$$H = T + V \quad,$$

$$T = \sum_i \frac{\boldsymbol{p}_i^2}{2m_i} \quad, \quad V = -\sum_i \frac{e_i e_{\rm N}}{|\boldsymbol{x}_i|} + \frac{1}{2} \sum_{i \neq j} \frac{e_i e_j}{|\boldsymbol{x}_i - \boldsymbol{x}_j|} \quad.$$

Let $|\psi\rangle$ be a stationary state, i.e., $H|\psi\rangle = E|\psi\rangle$. Prove the *virial theorem*

$$\langle \psi|T|\psi \rangle = -\frac{1}{2}\langle \psi|V|\psi \rangle \quad,$$

$$\langle \psi|H|\psi \rangle = \frac{1}{2}\langle \psi|V|\psi \rangle \quad.$$

Hint: Apply the dilatation operator

$$U(\beta)^{-1}\, x\, U(\beta) = x e^{\beta} \quad,$$

$$U(\beta)^{-1}\, p\, U(\beta) = p e^{-\beta}$$

to the Hamiltonian. Consider the derivative of the transformed Hamiltonian with respect to β at $\beta = 0$.

13.7 Find an explicit representation for the dilatation operator of Problem 13.6. Hint: Employ the Baker–Hausdorff relation.

14. The Zeeman Effect and the Stark Effect

We now return to our treatment of the theory of atomic levels in a magnetic field. In Sect. 7.3, we took only the orbital magnetic moment into account. The lack of agreement of the so-called normal Zeeman effect with experiment led to the discovery of spin. We now consider the complete theory of the Zeeman effect, first for the hydrogen atom, and then for multi-electron atoms. In addition, the effect of an electric field on the energy levels of atoms will be investigated in this chapter.

14.1 The Hydrogen Atom in a Magnetic Field

The Hamiltonian

$$H = H_{\text{Coul}} + H_{\text{rel}} + H_Z \tag{14.1}$$

consists of the Coulomb term H_{Coul} from (12.1), the relativistic corrections $H_{\text{rel}} = H_1 + H_2 + H_3$ from Chap. 12, and the Zeeman term

$$H_Z = -\frac{e}{2mc}(L_z + 2S_z)B = -\frac{e}{2mc}(J_z + S_z)B \quad , \tag{14.2}$$

where we choose the magnetic field to point along the z-direction.

We will either work in a basis system in which the spin–orbit term $\mathbf{S} \cdot \mathbf{L}$ contained in H_{rel} is diagonalized, treating the Zeeman term perturbatively, or else the other way around, depending on the strength of the magnetic field.

We recall that, in first-order perturbation theory, H_{rel} leads to a fine-structure energy shift

$$\Delta E^0_{n,j} = \frac{mc^2(Z\alpha)^4}{2n^4}\left\{ \frac{3}{4} - \frac{n}{j + 1/2} \right\} \quad ; \tag{14.3}$$

the states which arise here are

$$|n, j = l \pm \tfrac{1}{2}, m_j, l\rangle \quad . \tag{14.4}$$

14.1.1 Weak Field

If the magnetic field is weaker than 10^5 G, the Zeeman term can be regarded as a perturbation compared to the relativistic corrections. The appropriate parent states are then the states (14.4). The fine-structure shift of the hydrogen levels due to the second term of H is given in (14.3).

Using (10.28), we find

$$\langle n, j = l \pm \tfrac{1}{2}, m_j, l | S_z | n, j = l \pm \tfrac{1}{2}, m_j, l \rangle$$
$$= \frac{\hbar}{2}(\alpha_\pm^2 - \beta_\pm^2) = \pm \frac{\hbar m_j}{2l+1} \quad . \tag{14.5}$$

Since J_z has the eigenvalue $\hbar m_j$ in the states (14.4), the Zeeman splitting in first-order perturbation theory becomes

$$\langle H_Z \rangle_{l \pm \frac{1}{2}, m_j, l} = \mu_B B m_j \left(1 \pm \frac{1}{2l+1} \right) \quad . \tag{14.6}$$

All degenerate levels are split due to the magnetic field. In contrast to the "normal" Zeeman effect, the magnitude of the splitting depends on l. The Zeeman splitting is illustrated schematically in Fig. 14.1.

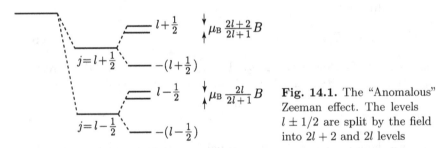

Fig. 14.1. The "Anomalous" Zeeman effect. The levels $l \pm 1/2$ are split by the field into $2l + 2$ and $2l$ levels

We note that it was permissible to use nondegenerate perturbation theory. For degenerate j, as for example in the case of the $2S_{1/2}$ and $2P_{1/2}$ levels, the two spatial wave functions (here $l = 0$ and 1) have different parity and vanishing matrix element of H_Z.

14.1.2 Strong Field, the Paschen–Back Effect

If the field is so strong that the Zeeman energy is large compared to that of the relativistic correction term, we begin with the states

$$|n, l, m_l\rangle |m_s\rangle \quad \text{with} \quad |m_s\rangle = |\uparrow\rangle \quad \text{or} \quad |\downarrow\rangle \quad ,$$

which diagonalize H_{Coul} and also H_Z.

The energy splitting of the hydrogen levels is then

$$\Delta E^Z_{n,l,m_l,m_s} = \frac{e_0\hbar}{2mc}(m_l + 2m_s)B = \mu_B(m_j + m_s)B \quad . \tag{14.7}$$

The relativistic corrections can then be taken into account within first-order perturbation theory:

$$\langle H_{\rm rel}\rangle_{n,l,m_l,m_s} = -\frac{mc^2(Z\alpha)^4}{2n^4}\left(\frac{n}{l+1/2} - \frac{3}{4}\right) + \zeta(n,l)m_l m_s \quad ,$$

$$\zeta(n,l) = \frac{Ze_0^2\hbar^2}{2m^2c^2}\left\langle\frac{1}{r^3}\right\rangle_{n,l} = \frac{mc^2(Z\alpha)^4}{2n^3l(l+1/2)(l+1)} \quad . \tag{14.8}$$

14.1.3 The Zeeman Effect for an Arbitrary Magnetic Field

After these limiting cases, we now treat the hydrogen atom for an arbitrary field. For fields lying between the two extremes we have to apply degenerate perturbation theory to the sum $H_{\rm rel} + H_Z$. That is, we must find linear combinations of the states

$$|n, j = l \pm \tfrac{1}{2}, m_j, l\rangle$$

for which the off-diagonal elements of $H_{\rm rel} + H_Z$ vanish. We start with

$$|n, j = l \pm \tfrac{1}{2}, m_j, l\rangle$$

because in these states the complicated part, $H_{\rm rel}$, is diagonal (14.3). We must also compute the matrix elements of H_Z. These vanish for differing n and l; for example, from $[\boldsymbol{L}^2, J_z + S_z] = 0$, the relation

$$(l(l+1) - l'(l'+1))\langle \ldots l|(J_z + S_z)| \ldots l'\rangle = 0$$

follows, so the matrix element of $J_z + S_z$ can differ from zero only for $l' = l$. From $[H_{\rm Coul}, J_z + S_z] = 0$, the validity of the assertion also follows for the quantum number n. From $[S_z, J_z] = 0$ one further has the relation

$$(m'_j - m_j)\langle \ldots m_j \ldots |S_z| \ldots m'_j \ldots \rangle = 0 \quad ,$$

so that all the matrix elements with differing m_j vanish. The structure of the matrix $J_z + S_z$ is characterized in Fig. 14.2 for $(n \geq 2)$ $S_{1/2}$, $P_{1/2}$, $P_{3/2}$ by dots for the non-zero matrix elements.

Using (10.28), the non-vanishing matrix elements are thus the diagonal elements

$$\langle n, l \pm \tfrac{1}{2}, m_j, l|(J_z + S_z)|n, l \pm \tfrac{1}{2}, m_j, l\rangle = \hbar m_j\left(1 \pm \frac{1}{2l+1}\right) \tag{14.9}$$

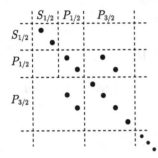

Fig. 14.2. The structure of the matrix $J_z + S_z$

and the nondiagonal elements

$$\langle n, l + \tfrac{1}{2}, m_j, l | (J_z + S_z) | n, l - \tfrac{1}{2}, m_j, l \rangle$$

$$= -\frac{\hbar}{2l+1} \sqrt{\left(l + \frac{1}{2}\right)^2 - m_j^2} \quad . \tag{14.10}$$

The states $S_{1/2}, m_j = \pm 1/2$ and $P_{3/2}, m_j = \pm 3/2$ occur only in diagonal elements and therefore for them there is no admixture due to H_Z. Consequently, for these states, and in general for

$$|n, j = l + \tfrac{1}{2}, m_j = \pm(l + \tfrac{1}{2}), l\rangle \quad ,$$

the energy shift is consequently given by (14.6), which reduces to

$$(\Delta E^Z)_{n, j = l + 1/2, m_j = \pm (l + 1/2), l} = \pm \mu_B B (l + 1) \quad . \tag{14.11}$$

For the remaining states ($|m_j| < l + 1/2$), we have to find those linear combinations

$$|\pm\rangle = \sum_{(\pm)} a_\pm |n, l \pm \tfrac{1}{2}, m_j, l\rangle \tag{14.12}$$

which diagonalize $H_{\text{rel}} + H_Z$ in the 2×2 subspaces. The coefficients $a_\pm$ and the energy shift ΔE due to $H_{\text{rel}} + H_Z$ are obtained using (14.3), (14.9), and (14.10) from the eigenvalue equation

$$\begin{pmatrix} \Delta E^0_{n, l+1/2} + \mu_B B m_j \frac{2l+2}{2l+1} - \Delta E & -\mu_B B \sqrt{(l + \tfrac{1}{2})^2 - m_j^2}/(2l+1) \\ -\mu_B B \sqrt{(l + \tfrac{1}{2})^2 - m_j^2}/(2l+1) & \Delta E^0_{n, l-1/2} + \mu_B B m_j \frac{2l}{2l+1} - \Delta E \end{pmatrix}$$

$$\times \begin{pmatrix} a_+ \\ a_- \end{pmatrix} = 0 \quad . \tag{14.13}$$

Introducing the fine-structure splitting

$$\Delta = \Delta E^0_{n, l+1/2} - \Delta E^0_{n, l-1/2} = \frac{mc^2 (\alpha Z)^4}{2n^3 l(l+1)} \quad , \tag{14.14}$$

we find as a solution of the characteristic equation of (14.13)

$$\Delta E_\pm = \Delta E^0_{n,l-1/2} + \mu_B B m_j + \frac{\Delta}{2}$$

$$\pm \sqrt{\frac{\Delta^2}{4} + \Delta \mu_B B \frac{m_j}{2l+1} + \frac{1}{4}(\mu_B B)^2} \ . \tag{14.15}$$

From this we can recover the two limiting cases:

(i) $\Delta \gg \mu_B B$:

$$\Delta E_\pm = \Delta E^0_{n,l\pm 1/2} + \mu_B B m_j \left(1 \pm \frac{1}{2l+1}\right)$$

$$\pm \frac{1}{4}\frac{\mu_B^2 B^2}{\Delta}\left(1 - \frac{4m_j^2}{(2l+1)^2}\right) \cdots \ , \tag{14.16}$$

in agreement with the sum of (14.3) and (14.6).

(ii) $\Delta \ll \mu_B B$:

$$\Delta E_\pm = \Delta E^0_{n,l-1/2} + \mu_B B\left(m_j \pm \frac{1}{2}\right) + \frac{\Delta}{2}\left(1 \pm \frac{2m_j}{2l+1}\right)$$

$$\pm \frac{\Delta^2}{4\mu_B B}\left(1 - \frac{4m_j^2}{(2l+1)^2}\right) \ . \tag{14.17}$$

Substituting $m_j = m_l \pm 1/2$, $m_s = \mp 1/2$ and using (14.3) and (14.14), one finds agreement with (14.8).

For $n = 2, l = 1, P_{3/2}$ and $P_{1/2}$, for $B \to 0$ one has

$$\Delta = \frac{mc^2(Z\alpha)^4}{32} \ , \quad \frac{\Delta E^0_{2,3/2}}{\Delta} = -\frac{1}{4} \ , \quad \frac{\Delta E^0_{2,1/2}}{\Delta} = -\frac{5}{4} \ .$$

The energy shift for all $P_{3/2}$ and $P_{1/2}$ levels is shown in Fig. 14.3.

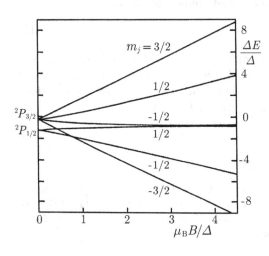

Fig. 14.3. The relative energy shift $\Delta E / \Delta$ as a function of the reduced magnetic field $\mu_B B/\Delta$ for $n = 2$ and $P_{1/2}$, $P_{3/2}$

In this treatment, the diamagnetic term $\frac{e^2}{8mc^2}[x \times B]^2$ (see Sects. 7.2 and 7.7) was not taken into account. This is justified as a rule for atoms under laboratory conditions. As mentioned in Sect. 7.2, magnetic fields such as those that prevail on the surface of neutron stars lead to significant changes in atomic structure because of the diamagnetic term.

These effects are also important at shallow donor levels in semiconductors. A singly charged defect in a semiconductor binds an electron quite in analogy to the hydrogen atom. However, the strength of Coulomb potential is decreased by the dielectric constant of the semiconductor, $V(r) = -e_0^2/(\varepsilon r)$, and the mass of the electron should be replaced by the effective mass m^*. Therefore, the Rydberg constant is replaced by Ry $= m^* e_0^4 / 2\varepsilon^2 \hbar^2$, and the cyclotron frequency by $\hbar\omega_c = e_0 \hbar B / m^* c$. In InSb (indium-antimonide), $m^* = m/77$ and $\varepsilon = 15$. The ratio $\hbar\omega_c/$Ry is thus increased by $(\varepsilon/m^*)^2 = (15 \times 77)^2 \approx 1.3 \times 10^6$ compared to the free hydrogen atom. It is not difficult to obtain a ratio of $\hbar\omega_c/$Ry ≈ 10, thus entering the regime in which the theory of hydrogen-like atoms in strong magnetic fields[1] becomes relevant under laboratory conditions.

14.2 Multielectron Atoms

For multielectron atoms, we begin with the total orbital angular momentum L and the total spin S of all electrons and their sum $J = L + S$. The Zeeman Hamiltonian is

$$H_Z = \frac{e_0}{2mc}(L_z + 2S_z)B = \frac{e_0}{2mc}(J_z + S_z)B \quad . \tag{14.18}$$

14.2.1 Weak Magnetic Field

For a weak magnetic field, the appropriate unperturbed states are

$$|J, M_J, L, S\rangle \quad . \tag{14.19}$$

These are eigenstates of J^2, J_z, L^2 and S^2, but not of S_z, and we need $\langle S_z \rangle$. We cannot fall back on the Wigner–Eckhardt theorem here, which would immediately give the ratio $\langle J_z \rangle / \langle S_z \rangle$, but rather we must make use of a short elementary computation. We begin with the identity

$$S(L \cdot S) - (L \cdot S)S = -i\hbar S \times L \quad , \tag{14.20}$$

---

[1] The hydrogen atom in a strong magnetic field is treated for example in the following papers: C. Alderich, R.L. Greene: Phys. Stat. Sol. (b) **93**, 343 (1979); H. Hasegawa: in *Physics of Solids in Intense Magnetic Fields*, ed. by E.D. Haidemenakis (Plenum, New York 1969) p. 246.

which follows from $[S_i, S_j] = i\hbar\varepsilon_{ijk}S_k$, and form its exterior product with $\boldsymbol{J}$:

$$\boldsymbol{S} \times \boldsymbol{J}(\boldsymbol{L}\cdot\boldsymbol{S}) - (\boldsymbol{L}\cdot\boldsymbol{S})\boldsymbol{S} \times \boldsymbol{J} = -i\hbar(\boldsymbol{S}\times\boldsymbol{L})\times\boldsymbol{J}$$
$$= -i\hbar\{\boldsymbol{L}(\boldsymbol{S}\cdot\boldsymbol{J}) - \boldsymbol{S}(\boldsymbol{L}\cdot\boldsymbol{J})\} = i\hbar\{-\boldsymbol{J}(\boldsymbol{S}\cdot\boldsymbol{J}) + \boldsymbol{S}\boldsymbol{J}^2\} \quad . \tag{14.21}$$

Taking the expectation value of (14.21) in the state $|J, M_J, L, S\rangle$ and using the fact that $\boldsymbol{L}\cdot\boldsymbol{S}$ is diagonal, we find that the left side vanishes and that

$$\langle\boldsymbol{S}\boldsymbol{J}^2\rangle = \langle\boldsymbol{J}(\boldsymbol{S}\cdot\boldsymbol{J})\rangle \quad . \tag{14.22}$$

With $\boldsymbol{S}\cdot\boldsymbol{J} = (\boldsymbol{J}^2 + \boldsymbol{S}^2 - \boldsymbol{L}^2)/2$,

$$\langle S_z\rangle = \langle J_z\rangle(J(J+1) + S(S+1) - L(L+1))/2J(J+1) \tag{14.23}$$

results and the energy shift becomes

$$\Delta E = \mu_B g_J M_J B \quad , \tag{14.24}$$

where

$$g_J = 1 + \frac{J(J+1) + S(S+1) - L(L+1)}{2J(J+1)} \quad , \tag{14.25}$$

is the *Landé g-factor*, which lies between 1 and 2.

Remark: This result corresponds to the classical picture that $\boldsymbol{J}$ is constant and oriented along $\boldsymbol{B}$, and $\boldsymbol{S}$ and $\boldsymbol{L}$ precess about $\boldsymbol{J}$ (Fig. 14.4):

$$\boldsymbol{\mu} = \frac{e}{2m}\boldsymbol{J}(\boldsymbol{L}\cdot\boldsymbol{J} + 2\boldsymbol{S}\cdot\boldsymbol{J})/\boldsymbol{J}^2$$
$$= \frac{e}{2m}\boldsymbol{J}(\tfrac{1}{2}(\boldsymbol{L}^2 + \boldsymbol{J}^2 - \boldsymbol{S}^2) + \boldsymbol{S}^2 + \boldsymbol{J}^2 - \boldsymbol{L}^2)/\boldsymbol{J}^2$$
$$= \frac{e}{2m}\boldsymbol{J}\left\{1 + \frac{J(J+1) + S(S+1) - L(L+1)}{2J(J+1)}\right\} \quad .$$

Fig. 14.4. Classical interpretation of the Landé g-factor

In the special case of hydrogen, $S = 1/2, L, J = L \pm 1/2$, (14.25) leads to the earlier result (14.6).

14.2.2 Strong Magnetic Field, the Paschen–Back Effect

If the Zeeman energy exceeds the relativistic splitting, the appropriate basis states are $|L, M_L, S, M_S\rangle$, in which $\boldsymbol{L}^2, L_z, \boldsymbol{S}^2$, and S_z are diagonal and H_Z is as well. The energy splitting is then

$$\Delta E = \mu_\mathrm{B} B(M_L + 2M_S) + \zeta(n, L, S)M_L M_S \quad . \tag{14.26}$$

This corresponds to the classical picture (Fig. 14.5) that $\boldsymbol{S}$ and $\boldsymbol{L}$ rotate independently of one another about $\boldsymbol{B}$, so that S, S_z, L, L_z, and J_z, but not J, remain constant. Here as well, the second term originates from the spin–orbit interaction.

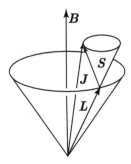

Fig. 14.5. Classical vector addition for a strong magnetic field

14.3 The Stark Effect

We now investigate the influence of an external electric field on the energy levels of the hydrogen atom. The unperturbed Hamiltonian H_Coul is given in (12.1), and the perturbation

$$H_1 = -e\boldsymbol{\mathcal{E}} \cdot \boldsymbol{x} = -e\mathcal{E}z \tag{14.27}$$

represents the interaction of the electron with the electric field $\mathcal{E}$, taken to point in the z-direction. Since the atomic electric field is of the order of magnitude $E_0/ea \approx 10^{10}$ Vm^{-1}, perturbation theory can certainly be applied for fields which are achievable in the laboratory.

We begin with a few statements, following from symmetry, concerning the matrix elements $\langle n, l, m|z|n', l', m'\rangle$ in the basis of eigenstates $|n, l, m\rangle$ of the Coulomb potential. From $[L_z, z] = 0$ it follows that

$$\langle n, l, m|[L_z, z]|n', l', m'\rangle = (m - m')\langle n, l, m|z|n', l', m'\rangle = 0 \quad ,$$

and hence one has the selection rule

$$m' = m \quad . \tag{14.28a}$$

By considering the effect of a reflection, one sees that the matrix element $\langle n, l, m|z|n', l', m' \rangle$ can only be finite for states of different parity. In connection with electric dipole radiation, it is shown in Chap. 16 that

$$l' = l \pm 1 \quad . \tag{14.28b}$$

Only when the "selection rules" (14.28a,b) are satisfied can the matrix element of z differ from zero.

14.3.1 Energy Shift of the Ground State

By (14.28b), the contribution from first-order perturbation theory vanishes for the ground state of the hydrogen atom. The second order is

$$E_1^2 = \sum_{n=2}^{\infty} e^2 \mathcal{E}^2 \frac{|\langle n, 1, 0|z|1, 0, 0 \rangle|^2}{E_1 - E_n} \quad , \tag{14.29}$$

where the selection rules (14.28a,b) have already been included, and where the E_n denote the unperturbed energy levels of H_{Coul}. Considering the typical magnitude of atomic quantities, these energy shifts must be of order $E_1^2 \approx -a^3 \mathcal{E}^2$, where a stands for the Bohr radius. The precise value is

$$E_1^2 = -\tfrac{9}{4} a^3 \mathcal{E}^2 \quad . \tag{14.30}$$

Comparison with $(-1/2) \alpha_{\mathrm{p}} \mathcal{E}^2$ immediately yields for the polarizability of the hydrogen atom in the ground state

$$\alpha_{\mathrm{p}} = \tfrac{9}{2} a^3 \quad . \tag{14.31}$$

Summarizing, we have that in the ground state there is no first-order Stark effect, but only a second-order Stark effect (first and second order in $\mathcal{E}$).

14.3.2 Excited States

The $n = 2$ states of the hydrogen atom $|2, 0, 0\rangle$, $|2, 1, 0\rangle$, $|2, 1, 1\rangle$, $|2, 1, -1\rangle$ are four-fold degenerate. We must use degenerate perturbation theory and transform to a basis in which the nondiagonal elements of z vanish.

By (14.28a) the two states $|2, 1, \pm 1\rangle$ have no off-diagonal elements with the other three states, and they remain unchanged in the new basis. Moreover, since by (14.28b)

$$\langle 2, 1, \pm 1|z|2, 1, \pm 1 \rangle = 0 \quad ,$$

they are not shifted to first order in $\mathcal{E}$ and give only a quadratic Stark effect. It remains to diagonalize the matrix formed from $|2, 0, 0\rangle$ and $|2, 1, 0\rangle$, which leads to the eigenvalue equation

$$-e\mathcal{E} \begin{pmatrix} \langle 2,0,0|z|2,0,0 \rangle & \langle 2,0,0|z|2,1,0 \rangle \\ \langle 2,1,0|z|2,0,0 \rangle & \langle 2,1,0|z|2,1,0 \rangle \end{pmatrix} \begin{pmatrix} c_1 \\ c_2 \end{pmatrix} = E^1 \begin{pmatrix} c_1 \\ c_2 \end{pmatrix} \quad . \tag{14.32}$$

The diagonal elements vanish by parity considerations (14.28b). The remaining matrix element is

$$\langle 2,0,0|z|2,1,0 \rangle = \frac{1}{8a^4} \int_0^\infty dr\, r^4 e^{-r/a} \left(1 - \frac{r}{2a}\right) \int_{-1}^{1} d\eta\, \eta^2 = -3a \quad . \tag{14.33}$$

Hence, the eigenvalue equation (14.32) becomes

$$\begin{pmatrix} E^1 & -3ea\mathcal{E} \\ -3ea\mathcal{E} & E^1 \end{pmatrix} \begin{pmatrix} c_1 \\ c_2 \end{pmatrix} = 0 \quad .$$

Its eigenvalues are

$$E^1 = \pm 3e_0 a\mathcal{E} \quad , \tag{14.34}$$

where $e = -e_0$ has been inserted, and the corresponding eigenvectors are

$$\frac{1}{\sqrt{2}} \begin{pmatrix} 1 \\ -1 \end{pmatrix} \quad \text{and} \quad \frac{1}{\sqrt{2}} \begin{pmatrix} 1 \\ 1 \end{pmatrix} \quad . \tag{14.35}$$

For these states, there exists a Stark effect of first order. The splitting in $O(\mathcal{E})$ is illustrated schematically in Fig. 14.6. A hydrogen atom in the first excited state behaves as if it had a dipole moment of magnitude $3ae_0$ orientable parallel or antiparallel to the field together with two states with no component along the field.

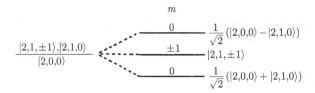

Fig. 14.6. Splitting of $n = 2$ states in first order in $\mathcal{E}$

We add a few remarks:

(i)   The perturbation (14.27) breaks the rotational invariance. One has $[\mathbf{L}^2, \mathbf{x}] \neq 0$, and therefore states with differing l are mixed.

(ii)   The fine structure need not be taken into account for field strengths greater than 10^3 V/cm, since in this case the splitting due to the electric field is larger than the fine-structure splitting. For weaker fields, one should begin with the eigenstates of $\mathbf{J}$. In linear combinations of the states $2S_{1/2}$ and $2P_{1/2}$ one then also finds a first-order Stark effect.

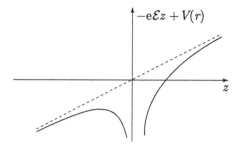

$-e\mathcal{E}z + V(r)$

z

Fig. 14.7. The potential for the Stark effect, $e = -e_0$

(iii) There is a fundamental objection to the use of perturbation theory for the Stark effect: Strictly speaking, as is evident from Fig. 14.7, the bound states are not stable, but only metastable. However, the tunneling probability is so minute that perturbation theory suffices to compute the location of these metastable levels.

(iv) Finally, we wish to draw some general conclusions from the results for the hydrogen atom. For nondegenerate levels there is no permanent dipole moment, but only an induced one. For degenerate states of different parity, permanent dipole moments can occur. In any case, exact degeneracy is present only in the Coulomb potential because of the additional symmetry generated by the Lenz vector.

On the other hand, it is clear physically that in the case of closely spaced levels of different parity, a permanent dipole moment is in effect present. This is the situation in molecules like HCl and NH_3. The ground state of these molecules is symmetric. The antisymmetric state lies barely higher in energy. From these two states, it is possible to form a linear combination with finite dipole moment d, which actually occurs if the energy $\mathcal{E}d$ is much larger than the aforementioned energy difference.

Problems

14.1 On the Stark effect: Calculate the second-order energy shift for the ground state of hydrogen E_1^2, Eqs. (14.29)–(14.30).

Hint: Instead of carrying out the summation in (14.29), derive a differential equation for the first-order perturbative correction to the wave function (method of Dalgarno and Lewis: A. Dalgarno and J. T. Lewis, Proc. Roy. Soc. (London) A**233**, 70 (1955)).

15. Molecules

15.1 Qualitative Considerations

The basic building blocks of molecules are nuclei and electrons; however, since they dissociate into atoms if energy is supplied, it seems reasonable to regard them as bound states of atoms. The determination of the energy levels and even the ground states of molecules is considerably more complicated than in the case of atoms. The electrons move in an attractive, non-rotationally symmetric potential generated by the nuclei. One should also think about to what extent the motion of the nucleus should be taken into account. Here, an important simplification occurs for the theory of molecules, due to the small ratio of the electron mass m to the nuclear mass M:[1]

$$\frac{m}{M} \approx 10^{-3} - 10^{-5} \quad . \tag{15.1}$$

Therefore the nuclei move more slowly, have a small zero-point energy, and hence are well localized. At any time, the electrons "see" effectively a static potential. The electronic wave functions are adiabatically deformed by the vibrations of the nuclei.

The nuclei of a molecule can carry out the following motions: translation, rotation, and vibration. In order to estimate the typical energies of these motions, we first consider the typical electronic energy of a valence electron, that is, an electron whose wave function extends over the whole molecule. If the linear dimension of the molecule is a, then according to the uncertainty relation and the virial theorem,

$$E_{\text{el}} \approx \frac{p^2}{2m} = \frac{(\hbar/a)^2}{2m} = \frac{\hbar^2}{2ma^2} \quad . \tag{15.2}$$

In estimating the vibrational energy, we note that the potential energy for every normal mode of oscillation (eigenoscillation of the molecule) takes the oscillator form $M\omega^2 R^2/2$, where M is roughly the nuclear mass and R is the amplitude of the oscillation. For $R \approx a$, the energy would be about E_{el}, that is

$$\frac{M\omega^2 a^2}{2} \approx \frac{\hbar^2}{2ma^2} \quad .$$

---

[1] $m_{\text{e}} = 0.911 \times 10^{-27}$ g, $m_{\text{p}} = 1.6725 \times 10^{-24}$ g.

Hence, one obtains for the vibrational frequency

$$\omega \approx \left(\frac{m}{M}\right)^{1/2} \frac{\hbar}{ma^2} \quad,$$

and the corresponding energy is

$$E_{\text{vib}} = \hbar\omega \approx \left(\frac{m}{M}\right)^{1/2} E_{\text{el}} \quad . \tag{15.3}$$

The rotational energy of the molecule is

$$E_{\text{rot}} = \frac{\hbar^2 l(l+1)}{2I} \approx \frac{\hbar^2}{Ma^2} = \frac{m}{M} E_{\text{el}} \quad, \tag{15.4}$$

where $I = Ma^2$ has been inserted for the moment of inertia. Hence, the ratios of electronic to vibrational and vibrational to rotational energy are $(m/M)^{1/2}$.

The frequencies of the transitions between electronic levels lie in the visible and ultraviolet, the vibrational transitions in the infrared, and the rotational transitions in the far infrared. From the vibrational frequency, we can also estimate the typical nuclear velocity and the amplitude of the vibrations (the mean-square deviation). The zero-point energy is composed in equal measures of kinetic and potential energy, that is,

$$\frac{P^2}{2M} = \frac{M\omega^2 R^2}{2} = \frac{\hbar\omega}{2} \quad, \quad \text{or} \quad P = \left(\frac{M}{m}\right)^{1/4} \frac{\hbar}{a} \quad .$$

Hence, the nuclear velocity becomes

$$v_{\text{N}} = \frac{P}{M} = \left(\frac{m}{M}\right)^{3/4} \frac{\hbar}{ma} = \left(\frac{m}{M}\right)^{3/4} v_{\text{el}} \tag{15.5}$$

and the vibrational amplitude

$$R = \left(\frac{\hbar}{M\omega}\right)^{1/2} = \left(\frac{m}{M}\right)^{1/4} a \quad, \tag{15.6}$$

i.e., $v_{\text{N}}/v_{\text{el}} \approx 10^{-3}$ and $R/a \approx 10^{-1}$. The velocity of the nuclei is much smaller than that of the electrons, and the deviations R from their equilibrium positions are smaller than the molecular dimension a.

15.2 The Born–Oppenheimer Approximation

The total Hamiltonian for electrons and nuclei is given by

$$H = T_e + T_N + V_{ee} + V_{eN} + V_{NN} \tag{15.7a}$$

with

$$T_e = \sum_i \frac{p_i^2}{2m} \quad , \quad T_N = \sum_N \frac{P_N^2}{2M_N} \quad . \tag{15.7b}$$

The first two terms in H stand for the kinetic energy of the electrons and nuclei, while the next three stand for the interaction of the electrons, the interaction of the electrons with the nuclei, and the interaction of the nuclei, respectively. The momenta and coordinates of the nuclei are P_N and X_N, and those of the electrons are p_i and x_i.

Before considering the total wave function of the electrons and the nuclei, let us first ignore the nuclear motion completely, i.e., we fix the coordinates X. The wave function of the electrons $\psi(x|X)$ for fixed nuclear positions X is then determined by

$$(T_e + V_{ee} + V_{eN})\psi(x|X) = E^{el}(X)\psi(x|X) \quad . \tag{15.8}$$

In the wave function $\psi(x|X)$ and in $E^{el}(X)$, the nuclear positions enter only as parameters. We introduce

$$\varepsilon(X) = V_{NN}(X) + E^{el}(X) \quad ,$$

the sum of the interaction energy of the nuclei and the energy eigenvalue of the interacting electrons in the nuclear potential.

For the wave function of the whole molecule, we make the ansatz

$$\Psi(x, X) = \psi(x|X)\Phi(X) \quad , \tag{15.9}$$

which is the product of a wave function for the nuclei and the wave function of the electrons for fixed nuclei. From $H\Psi = E\Psi$ follows

$$\psi(x|X)(T_N + V_{NN}(X) + E^{el}(X))\Phi(X) = \psi(x|X)E\Phi(X)$$
$$- \sum_N \frac{-\hbar^2}{2M_N}[\Phi(X)\nabla_X^2\psi(x|X) + 2\nabla_X\Phi(X)\nabla_X\psi(x|X)] \quad . \tag{15.10}$$

We multiply (15.10) by $\psi(x|X)^*$ and integrate over x. Ignoring the terms following from the second line, we find the *Born–Oppenheimer equation*,

$$(T_N + \varepsilon(X))\Phi(X) = E\Phi(X) \quad . \tag{15.11}$$

The Born–Oppenheimer equation, the effective Schrödinger equation for the nuclei, contains – aside from the Coulomb repulsion – the energy of the electrons, which depends on the nuclear positions. The equilibrium coordinates of the nuclei are obtained from the minima of the $\varepsilon(\boldsymbol{X})$. The nuclei arrange themselves in such a way that the sum of their Coulomb repulsion and the electronic energy attains a minimum. The oscillation frequencies of the nuclei are obtained from the second derivatives of $\varepsilon(\boldsymbol{X})$.

We must now estimate the terms neglected above. The second term in the second line of (15.10) is

$$\int d^3x\, \psi(\boldsymbol{x}|\boldsymbol{X})^* \frac{\partial}{\partial X_i} \psi(\boldsymbol{x}|\boldsymbol{X}) \sim \frac{\partial}{\partial X_i} \int d^3x\, |\psi(\boldsymbol{x}|\boldsymbol{X})|^2 = 0 \quad,$$

since bound state wave functions can always be taken as real, and the norm is independent of $\boldsymbol{X}$. The first term is estimated for the most unfavorable case of strong coupling, $\psi(\boldsymbol{x}|\boldsymbol{X}) = \psi(\boldsymbol{x} - \boldsymbol{X})$:

$$-\frac{\hbar^2}{2M}\Phi(\boldsymbol{X}) \int d^3x\, \psi(\boldsymbol{x}|\boldsymbol{X})^* \nabla_X^2 \psi(\boldsymbol{x}|\boldsymbol{X})$$

$$= -\hbar^2 \frac{\Phi(\boldsymbol{X})}{2M} \int d^3x\, \psi(\boldsymbol{x} - \boldsymbol{X})^* \nabla_x^2 \psi(\boldsymbol{x} - \boldsymbol{X})$$

$$= \frac{m}{M} E_{\text{kin}}^{\text{el}} \Phi(\boldsymbol{X}) \ll |E^{\text{el}}| \Phi(\boldsymbol{X}) \quad.$$

In the above derivation of the Born–Oppenheimer approximation, only a single electronic state $\psi(\boldsymbol{x}|\boldsymbol{X})$ was taken into account. However, one should investigate whether the admixture of other electronic states could play a role. For instance, in the evulation of the nuclear motion in the electronic ground state, matrix elements with excited states might be important. The starting point for a systematic treatment is the completely general expansion

$$\Psi(\boldsymbol{x}, \boldsymbol{X}) = \sum_\alpha \psi_\alpha(\boldsymbol{x}|\boldsymbol{X})\Phi_\alpha(\boldsymbol{X}) \quad, \tag{15.12}$$

where the states $\psi_\alpha(\boldsymbol{x}|\boldsymbol{X})$ satisfy the Schrödinger equation

$$(T_e + V_{ee} + V_{Ne})\psi_\alpha(\boldsymbol{x}|\boldsymbol{X}) = E_\alpha^{\text{el}}(\boldsymbol{X})\psi_\alpha(\boldsymbol{x}|\boldsymbol{X}) \quad. \tag{15.13}$$

The Schrödinger equation for the wave function (15.12) is

$$\sum_{\alpha'} \psi_{\alpha'}(\boldsymbol{x}|\boldsymbol{X})[T_N + V_{NN} + E_{\alpha'}^{\text{el}}(\boldsymbol{X}) - E]\Phi_{\alpha'}(\boldsymbol{X}) \tag{15.14}$$

$$= -\sum_N \sum_{\alpha'} \left(-\frac{\hbar^2}{2M_N}\right) \left(2\nabla_X\Phi_{\alpha'} \cdot \nabla_X\psi_{\alpha'}(\boldsymbol{x}|\boldsymbol{X}) + \Phi_{\alpha'} \nabla_X^2 \psi_{\alpha'}(\boldsymbol{x}|\boldsymbol{X})\right) \quad.$$

Multiplying this equation by $\psi_\alpha(\boldsymbol{x}|\boldsymbol{X})^*$ and integrating over the coordinates of the electrons, one finds

$$(T_N + V_{NN} + E_\alpha^{\text{el}}(\boldsymbol{X}) - E)\Phi_\alpha(\boldsymbol{X}) = 0 \quad. \tag{15.15}$$

Here, we have neglected the terms originating from the right side of (15.14). These are now to be estimated. The second term contains

$$\frac{1}{M} \boldsymbol{\nabla}_X^2 \psi_{\alpha'}(\boldsymbol{x}|\boldsymbol{X}) = \frac{m}{M} \frac{1}{m} \boldsymbol{\nabla}_x^2 \psi_{\alpha'}(\boldsymbol{x}|\boldsymbol{X})$$

and therefore is smaller than the electronic kinetic energy by a factor of m/M. In order to estimate the first term, we substitute, in place of the nuclear wave function, the oscillator wave function $\exp\{-(\boldsymbol{X}-\boldsymbol{X}_0)^2 M\omega/2\hbar\}$:

$$\boldsymbol{\nabla}_X \Phi_{\alpha'}(\boldsymbol{X}) \boldsymbol{\nabla}_X \psi_{\alpha'}(\boldsymbol{x}|\boldsymbol{X})$$
$$\sim (\boldsymbol{X}-\boldsymbol{X}_0)\frac{M\omega}{\hbar} \exp\left\{-\frac{(\boldsymbol{X}-\boldsymbol{X}_0)^2 M\omega}{2\hbar}\right\} \boldsymbol{\nabla}_X \psi_{\alpha'}(\boldsymbol{x}|\boldsymbol{X})$$
$$\sim \frac{M\omega}{\hbar} \Phi_{\alpha'}(\boldsymbol{X})(\boldsymbol{X}-\boldsymbol{X}_0) \boldsymbol{\nabla}_X \psi_{\alpha'}(\boldsymbol{x}|\boldsymbol{X}) \quad .$$

Let $\boldsymbol{\delta} = \boldsymbol{X} - \boldsymbol{X}_0$ be a typical nuclear displacement; it then follows that

$$(\boldsymbol{X}-\boldsymbol{X}_0)\boldsymbol{\nabla}_X \psi_{\alpha'} \approx \psi_{\alpha'}(\boldsymbol{x}|\boldsymbol{X}+\boldsymbol{\delta}) - \psi_{\alpha'}(\boldsymbol{x}|\boldsymbol{X})$$
$$\approx \psi_{\alpha'}(\boldsymbol{x}-\boldsymbol{\delta}|\boldsymbol{X}) - \psi_{\alpha'}(\boldsymbol{x}|\boldsymbol{X}) \quad .$$

Hence, these terms are of the order of magnitude

$$\frac{M\omega}{\hbar} \frac{\hbar^2}{2M} \approx \hbar\omega \approx \left(\frac{m}{M}\right)^{1/2} E_{\mathrm{el}} \quad .$$

The neglected matrix elements are smaller than the electronic energy by factors of m/M and $(m/M)^{1/2}$. They are therefore much smaller than the separations of the electronic levels and lead merely to a negligible correction.

Neglecting terms in this manner, we have recovered in (15.15) the Born–Oppenheimer equation. The nuclei move in an effective potential composed of the nuclear repulsion and the electronic energy. Hence we obtain independent Born–Oppenheimer equations for $\Phi_\alpha(\boldsymbol{X})$ corresponding to each $\psi_\alpha(\boldsymbol{x}|\boldsymbol{X})$. The energy eigenvalues of the molecule $E_{\alpha n}$ following from (15.15) depend on α and are enumerated by n, and the corresponding stationary states of the molecule are of the form

$$\Psi_{\alpha n}(\boldsymbol{x}, \boldsymbol{X}) = \psi_\alpha(\boldsymbol{x}|\boldsymbol{X}) \Phi_{\alpha n}(\boldsymbol{X}) \quad .$$

In the following, we will compute the electronic energy for fixed nuclear positions. The minima yield the molecular bound states.

15.3 The Hydrogen Molecular Ion (H_2^+)

We first consider the ionized H_2 molecule; there, a single electron moves in the attractive potential of two protons at the fixed positions $\boldsymbol{X}_A$ and $\boldsymbol{X}_B$ (Fig. 15.1). The Hamiltonian for the electron takes the form

$$H = -\frac{\hbar^2 \boldsymbol{\nabla}^2}{2m} - \frac{e^2}{|\boldsymbol{x}-\boldsymbol{X}_A|} - \frac{e^2}{|\boldsymbol{x}-\boldsymbol{X}_B|} + \frac{e^2}{|\boldsymbol{X}_A-\boldsymbol{X}_B|} \quad . \tag{15.16}$$

$\ominus^{\boldsymbol{x}}$

$\oplus$ $\oplus$

$\boldsymbol{X}_A$ $\boldsymbol{X}_B$ **Fig. 15.1.** The ionized hydrogen molecule

This problem is exactly solvable[2]; nevertheless, we will be satisfied here with a qualitative variational solution. As a variational ansatz, we take the superposition of $1s$ atomic wave functions, which are symmetric or antisymmetric by reflection symmetry:

$$\psi_\pm = C_\pm[\psi_A(\boldsymbol{x}) \pm \psi_B(\boldsymbol{x})] \quad . \tag{15.17}$$

The two $1s$ wave functions concentrated about nucleus A and B are

$$\psi_{\substack{A\\B}}(\boldsymbol{x}) = (\pi a^3)^{-1/2} \exp\{-|\boldsymbol{x} - \boldsymbol{X}_{\substack{A\\B}}|/a\} \quad . \tag{15.18}$$

The normalization constants are obtained from the overlap of the wave functions ψ_A and ψ_B:

$$1 = \int d^3x \, |\psi_\pm(\boldsymbol{x})|^2 = C_\pm^2(2 \pm 2S(R)) \tag{15.19a}$$

with

$$S(R) = \int d^3x \, \psi_A(\boldsymbol{x})\psi_B(\boldsymbol{x}) = \left(1 + \frac{R}{a} + \frac{R^2}{3a^2}\right)e^{-R/a} \quad , \tag{15.19b}$$

where $R = |\boldsymbol{X}_A - \boldsymbol{X}_B|$ is the nuclear separation and $S(R)$ represents the overlap integral. The expectation value of the Hamiltonian in the states (15.17) is

$$\begin{aligned}\langle H \rangle_\pm &= (2 \pm 2S)^{-1}(\langle A|H|A\rangle + \langle B|H|B\rangle \pm 2\langle A|H|B\rangle)\\ &= (1 \pm S)^{-1}(\langle A|H|A\rangle \pm \langle A|H|B\rangle) \quad , \end{aligned} \tag{15.20}$$

where

$$\begin{aligned}\langle A|H|A\rangle &= \int d^3x \, \psi_A(\boldsymbol{x})H\psi_A(\boldsymbol{x}) = E_1 + \frac{e^2}{R} - \int d^3x \, \psi_A^2(\boldsymbol{x})\frac{e^2}{|\boldsymbol{x} - \boldsymbol{X}_B|}\\ &= E_1 + \frac{e^2}{R}\left(1 + \frac{R}{a}\right)e^{-2R/a} \quad . \end{aligned} \tag{15.21}$$

Here, $E_1 = -1$ Ry is the ground state energy of a hydrogen atom. Furthermore,

---

[2] One can separate variables by using spheroidal coordinates. See, for instance, J.C. Slater: *Quantum Theory of Molecules and Solids. I. Electronic Structure of Molecules* (McGraw-Hill, New York 1974).

$$\langle A|H|B\rangle = \int d^3x\,\psi_A(\boldsymbol{x})H\psi_B(\boldsymbol{x})$$

$$= \left(E_1 + \frac{e^2}{R}\right)S(R) - \int d^3x\,\psi_A(\boldsymbol{x})\psi_B(\boldsymbol{x})\frac{e^2}{|\boldsymbol{x}-\boldsymbol{X}_B|} \quad ,$$

(15.22)

where the exchange integral is defined by

$$A(R) = \int d^3x\,\psi_A(\boldsymbol{x})\psi_B(\boldsymbol{x})\frac{e^2}{|\boldsymbol{x}-\boldsymbol{X}_B|} = \frac{e^2}{a}\left(1+\frac{R}{a}\right)e^{-R/a}$$

(15.23)

and

$$\int d^3x\,\psi_A^2(\boldsymbol{x})\frac{e^2}{|\boldsymbol{x}-\boldsymbol{X}_B|} = \frac{e^2}{R}\left(1 - e^{-2R/a}\left(\frac{R}{a}+1\right)\right)$$

has been used. From this, one obtains for $\varepsilon_\pm(R) \equiv \langle H\rangle_\pm$

$$\varepsilon_\pm(R) = (1\pm S)^{-1}\left[E_1 + \frac{e^2}{R}\left(1+\frac{R}{a}\right)e^{-2R/a}\right.$$

$$\left.\pm\left(E_1+\frac{e^2}{R}\right)S \mp \frac{e^2}{a}\left(1+\frac{R}{a}\right)e^{-R/a}\right] \quad .$$

(15.24)

In Fig. 15.2 $\varepsilon_\pm(R)$ is shown as a function of the separation R. Now, $\varepsilon_+(R)$ has a minimum, but $\varepsilon_-(R)$ does not. From this one can see that the symmetric wave function leads to binding and the antisymmetric wave function does not. In the region between the nuclei, $\psi_+(\boldsymbol{x})$ is larger than $\psi_-(\boldsymbol{x})$, which vanishes

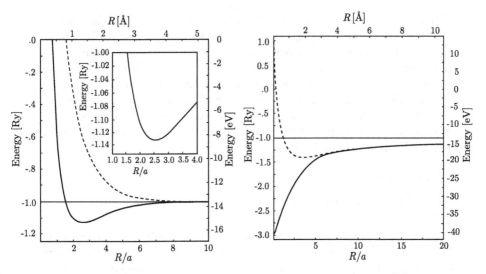

Fig. 15.2. $\varepsilon_+(R)$ (—) and $\varepsilon_-(R)$ (- - -) as functions of R; 1 Å $= 10^{-8}$ cm

Fig. 15.3. $\varepsilon_\pm(R) - e^2/R$ (notation as in Fig. 15.2)

in the plane bisecting the line connecting the nuclei. The difference in the electronic energies is explained by the fact that the potential energy there is relatively strongly attractive due to the positive superposition of the nuclear potentials. For further illustration, in Fig. 15.3 $\varepsilon_\pm(R) - e^2/R$ is shown, that is, the purely electronic energy without the Coulomb repulsion of the nuclei. The state ψ_+ is called a bonding orbital and ψ_- an antibonding orbital.

The exact ground state energy is lower than $\varepsilon_+(R)$. The values of the binding energy and the nuclear separations of H_2^+ are 2.79 eV and 1.06 Å, compared to the variational results of 1.76 eV and 1.32 Å, respectively.[3]

15.4 The Hydrogen Molecule H_2

The Hamiltonian for the two electrons of the hydrogen molecule reads (Fig. 15.4)

$$
\begin{aligned}
H = \; & -\frac{\hbar^2}{2m}\nabla_1^2 - \frac{\hbar^2}{2m}\nabla_2^2 - \frac{e^2}{|x_1 - X_A|} - \frac{e^2}{|x_1 - X_B|} \\
& - \frac{e^2}{|x_2 - X_A|} - \frac{e^2}{|x_2 - X_B|} + \frac{e^2}{|x_1 - x_2|} + \frac{e^2}{|X_A - X_B|} \; . \quad (15.25)
\end{aligned}
$$

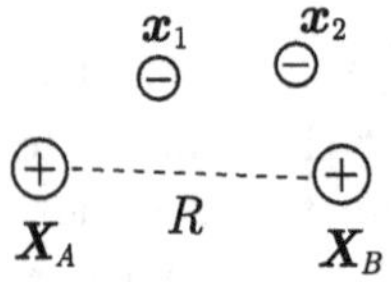

Fig. 15.4. The hydrogen molecule

From the point of view of one of the two electrons, the following picture emerges: In addition to the potential of the nuclei, there is a repulsive potential of the other electron. If the other electron is in a symmetric state, then a weaker symmetric potential results. The wave function of the electron under consideration is either symmetric or antisymmetric. Here as well, there are antisymmetric (antibonding) and symmetric (bonding) wave functions (orbitals). If a particular symmetric, molecular, one-electron state is occupied by two electrons with oppositely aligned spins, one expects for the binding energy of the hydrogen molecule E_{H_2} the result

$$
|E_{H_2}| < 2|E_{H_2^+}| \quad ,
$$

where $E_{H_2^+}$ is the binding energy of the ionized hydrogen molecule.

---

[3] The binding energy $E_{H_2^+}$ is given relative to the dissociated state, p + H, i.e.:
$E_{H_2^+} = E_{min} + 1\,Ry$, where E_{min} is the energy at the minimum (Fig. 15.2). Because of (11.15), the above calculation gives an upper bound to the binding energy. This can be improved by introducing the variational parameter Z^* into (15.18) (see Problem 15.1).

There exist two basic approximation schemes for treating the H_2 problem: One method consists in the construction of molecular orbitals, that is, one-particle wave functions for the molecule, which are occupied by electrons. The other approach is the Heitler–London method, which consists in constructing the two-particle wave functions from the one-electron wave functions of (isolated) hydrogen atoms.

The molecular orbital method

Here, the ansatz for the wave function of the two electrons is

$$\psi_s(1,2) = [\psi_A(\boldsymbol{x}_1) + \psi_B(\boldsymbol{x}_1)][\psi_A(\boldsymbol{x}_2) + \psi_B(\boldsymbol{x}_2)]$$
$$\times \chi_{\text{sing}}/2[1 + S(R)] \quad . \tag{15.26}$$

The spatial part is the product of the H_2^+ molecule wave functions $\psi_+(\boldsymbol{x}_1)$ and $\psi_+(\boldsymbol{x}_2)$. The spins are in the singlet state because of the Pauli exclusion principle. A triplet state can only be constructed by

$$[\psi_+(\boldsymbol{x}_1)\psi_-(\boldsymbol{x}_2) - \psi_-(\boldsymbol{x}_1)\psi_+(\boldsymbol{x}_2)]\chi_{\text{trip}}/\sqrt{2} \quad .$$

It contains an antibonding orbital and has higher energy.

The ansatz (15.26) has the following shortcomings: For small distances, the wave function is a product of $1s$ hydrogen wave functions instead of helium wave functions. For larger distances, a further weakness becomes apparent. For this we consider the fully expanded expression

$$\psi_s(1,2) \propto [\psi_A(\boldsymbol{x}_1)\psi_A(\boldsymbol{x}_2) + \psi_B(\boldsymbol{x}_1)\psi_B(\boldsymbol{x}_2)]$$
$$+ [\psi_A(\boldsymbol{x}_1)\psi_B(\boldsymbol{x}_2) + \psi_A(\boldsymbol{x}_2)\psi_B(\boldsymbol{x}_1)] \quad .$$

In the terms within the first pair of square brackets, both electrons are concentrated about the same atom. However, for large separations only the second pair of brackets should be present, because $H + H$ is energetically more favorable than $p + H^-$. Nevertheless, this wave function gives a reasonably accurate upper bound for the binding energy, since for this quantity only the behavior in the vicinity of the actual molecular separation of the nuclei is of relevance.

The Heitler–London method

Here, the ansatz for the singlet and triplet states

$$\psi_s(1,2) = \frac{1}{\sqrt{2(1+S^2)}}[\psi_A(\boldsymbol{x}_1)\psi_B(\boldsymbol{x}_2) + \psi_B(\boldsymbol{x}_1)\psi_A(\boldsymbol{x}_2)]\chi_{\text{sing}} , \tag{15.27a}$$

$$\psi_t(1,2) = \frac{1}{\sqrt{2(1-S^2)}}[\psi_A(\boldsymbol{x}_1)\psi_B(\boldsymbol{x}_2) - \psi_B(\boldsymbol{x}_1)\psi_A(\boldsymbol{x}_2)]\chi_{\text{trip}} \tag{15.27b}$$

is given by the symmetric and antisymmetric superposition of single-atom states. For large distances $|X_A - X_B|$, these states describe separated hydrogen atoms, which represents an improvement over the method of molecular orbitals. For small distances, the same criticism as in the method of molecular orbitals applies. The quantitative difference in the binding energies of the states (15.26) and (15.27a) is relatively small.

The expectation value of the Hamiltonian H in the Heitler–London states (15.27a,b) gives the following upper bound for the true energy:

$$\varepsilon_{\substack{s\\t}} = \langle H \rangle_{\substack{s\\t}} = (\langle AB|H|AB \rangle \pm \langle AB|H|BA \rangle)/(1 \pm S^2) \quad . \tag{15.28}$$

Here, we define

$$\langle AB|H|AB \rangle = \int d^3x_1 d^3x_2 \, \psi_A(x_1) \, \psi_B(x_2) H \psi_A(x_1) \psi_B(x_2)$$
$$= \langle BA|H|BA \rangle \quad , \tag{15.29a}$$

$$\langle AB|H|BA \rangle = \int d^3x_1 d^3x_2 \, \psi_A(x_1) \psi_B(x_2) H \psi_B(x_1) \psi_A(x_2)$$
$$= \langle BA|H|AB \rangle \quad . \tag{15.29b}$$

Using the Schrödinger equation for the $1s$ wave functions

$$\left(-\frac{\hbar^2}{2m} \nabla_{\substack{1\\2}}^2 - \frac{e^2}{|x_{\substack{1\\2}} - X_{\substack{A\\B}}|} \right) \psi_{\substack{A\\B}}(x_{\substack{1\\2}}) = E_1 \, \psi_{\substack{A\\B}}(x_{\substack{1\\2}}) \quad , \tag{15.30}$$

one can represent (15.29a) in the form

$$\langle AB|H|AB \rangle = 2E_1 + Q \quad , \tag{15.31}$$

where we have introduced the Coulomb energy Q.

$$Q = \int d^3x_1 \int d^3x_2 \, \psi_A(x_1)^2 \, \psi_B(x_2)^2$$
$$\times \left[\frac{e^2}{|x_1 - x_2|} - \frac{e^2}{|x_1 - X_B|} - \frac{e^2}{|x_2 - X_A|} \right] + \frac{e^2}{R}$$
$$= -2 \int d^3x_1 \frac{e^2}{|x_1 - X_B|} \psi_A(x_1)^2$$
$$+ \int d^3x_1 d^3x_2 \, \psi_A(x_1)^2 \frac{e^2}{|x_1 - x_2|} \psi_B(x_2)^2 + \frac{e^2}{R} \quad . \tag{15.32}$$

The terms following the second equals sign are, respectively: twice the Coulomb interaction of the electron concentrated about A with the nucleus B, which is equal to the Coulomb interaction of the electron concentrated about B with A; the Coulomb repulsion of the electrons; and the Coulomb repulsion of the nuclei.

For (15.29b), one obtains using (15.25) and (15.30)

$$\langle AB|H|BA\rangle = S^2\, 2E_1 + A \quad . \tag{15.33}$$

The exchange energy A is defined by

$$
\begin{aligned}
A &= S^2 \frac{e^2}{R} + \int d^3x_1 d^3x_2\, \psi_A(\boldsymbol{x}_1)\psi_B(\boldsymbol{x}_2)\psi_A(\boldsymbol{x}_2)\psi_B(\boldsymbol{x}_1) \\
&\quad \times \left[\frac{e^2}{|\boldsymbol{x}_1 - \boldsymbol{x}_2|} - \frac{e^2}{|\boldsymbol{x}_1 - \boldsymbol{X}_B|} - \frac{e^2}{|\boldsymbol{x}_2 - \boldsymbol{X}_A|} \right] \\
&= S^2 \frac{e^2}{R} + \int d^3x_1 d^3x_2\, \psi_A(\boldsymbol{x}_1)\psi_B(\boldsymbol{x}_2)\frac{e^2}{|\boldsymbol{x}_1 - \boldsymbol{x}_2|}\psi_B(\boldsymbol{x}_1)\psi_A(\boldsymbol{x}_2) \\
&\quad - 2S \int d^3x_1 \frac{e^2\psi_A(\boldsymbol{x}_1)\psi_B(\boldsymbol{x}_1)}{|\boldsymbol{x}_1 - \boldsymbol{X}_A|} \quad .
\end{aligned} \tag{15.34}
$$

It is a measure of the square of the overlap of the wave functions, weighted by the potential energies. The exchange energy results from the interplay of quantum mechanics (Pauli principle) and the Coulomb interaction.

We now substitute (15.29–15.34) into (15.28) and obtain for the energies in the singlet and triplet states

$$\varepsilon_{\substack{s \\ t}} = 2E_1 + \frac{Q \pm A}{1 \pm S^2} \quad . \tag{15.35}$$

The Coulomb and exchange energy depend on the nuclear separation. The Coulomb energy Q is positive and everywhere small. Except for very small distances, the exchange energy is negative and exceeds the Coulomb energy; thus, the singlet state is binding, but not the triplet state. Physically, this results from the fact that for the singlet function the spatial probability density is large in between the two nuclei, whereas in the triplet state it has a node in the midplane bisecting the line connecting the nuclei. Although the Coulomb repulsion of the electrons in the singlet state is larger, this is over-compensated by the attraction of the charge distribution in the internuclear region by the nuclei to the left and right. Quantitatively, this computation is unsatisfactory; for the nuclear distance, one finds $R_0 = 0.8 \times 10^{-8}$ cm in place of the experimental value 0.7395×10^{-8} cm. Nevertheless, it shows the origin of the homopolar bonding of the hydrogen molecule.

Remark: The explicit expressions for Q and A are[4]

$$Q = \frac{e^2}{a\varrho}e^{-2\varrho}\left(1 + \frac{5}{8}\varrho - \frac{3}{4}\varrho^2 - \frac{1}{6}\varrho^3\right) \quad , \tag{15.36}$$

[4] Y. Sugiura: Z. Phys. **45**, 484 (1927)

$$A = \frac{e^2}{a} \left\{ \frac{S^2}{\varrho} \left[1 + \frac{6}{5}(C + \ln \varrho) \right] - e^{-2\varrho} \left(\frac{11}{8} + \frac{103}{20} \varrho + \frac{49}{15} \varrho^2 + \frac{11}{15} \varrho^3 \right) \right.$$
$$\left. + \frac{6M}{5\varrho} [M \, \mathrm{Ei}(-4\varrho) - 2S \, \mathrm{Ei}(-2\varrho)] \right\} \quad ,$$

$$C = 0.57722 \quad ,$$

$$\mathrm{Ei}(x) = - \int_{-x}^{\infty} \frac{e^{-\xi}}{\xi} d\xi \quad ,$$

$$M = (1 - \varrho + \tfrac{1}{3}\varrho^2)e^{\varrho} \quad ,$$

$$S = \left(1 + \varrho + \frac{\varrho^2}{3} \right) e^{-\varrho} \quad .$$

At this point, of course, a further discussion of covalent (homopolar) bonding might be warranted. However, since this discussion could only be qualitative, we will refrain from carrying it further. To conclude this section, we give a summary of the orders of magnitude of binding energies occurring in solids for the various bonding types:

Homopolar = covalent bonding	10	eV
Ionic bonding	10	eV
Van der Waals bonding	0.1	eV
Metallic bonding	1–5	eV
Hydrogen bonding	0.1	eV

15.5 Energy Levels of a Two-Atom Molecule: Vibrational and Rotational Levels

In the two preceding sections, we determined the effective potential energy of the nuclei and its minimum. Of course, one can just as well compute the excited states of the nuclei from the Born–Oppenheimer equation. In the following, we would like to determine the oscillatory and rotational states of two-atom molecules such as HCl. In this case, the Born–Oppenheimer equation (15.11) is a two-particle Schrödinger equation, which as in Sect. 6.4 can be reduced to a one-particle Schrödinger equation by introducing center-of-mass and relative coordinates and separating off the center-of-mass part. We denote the relative coordinate of nucleus 1 and 2 by $\boldsymbol{x} = \boldsymbol{X}_1 - \boldsymbol{X}_2$. The Born–Oppenheimer equation (15.11) then yields

$$\left[-\frac{\hbar^2}{2m}\Delta + \varepsilon(r) \right] \psi(\boldsymbol{x}) = E\psi(\boldsymbol{x}) \quad . \tag{15.37}$$

Here, we have introduced the reduced mass $m = M_1 M_2/(M_1 + M_2)$, and we recall that the effective potential energy depends only on the separation

$r = |\boldsymbol{x}|$ of the nuclei. In (15.37) we are again confronted with a rotationally symmetric problem.

With the ansatz $\psi(\boldsymbol{x}) = R_{nl}(r)Y_{lm}(\vartheta, \varphi)$ one obtains

$$\left[-\frac{\hbar^2}{2m} \left(\frac{d^2}{dr^2} + \frac{2}{r} \frac{d}{dr} \right) + \varepsilon(r) + \frac{\hbar^2 l(l+1)}{2mr^2} \right] R_{nl}(r) = ER_{nl}(r) \quad . \quad (15.38)$$

Here,

$$V_{\text{eff}}(r) = \varepsilon(r) + \frac{\hbar^2 l(l+1)}{2mr^2} \tag{15.39}$$

enters as a new effective potential. As in previous chapters, the substitution

$$R_{nl}(r) = \frac{u_{nl}(r)}{r} \tag{15.40}$$

leads to

$$\left[-\frac{\hbar^2}{2m} \frac{d^2}{dr^2} + V_{\text{eff}}(r) \right] u_{nl}(r) = Eu_{nl}(r) \quad . \tag{15.41}$$

For small l, V_{eff} has a minimum depending on the angular momentum quantum number l, which we denote by r_l. In the neighborhood of this minimum, we can expand V_{eff} in a Taylor series,

$$V_{\text{eff}} = V_{\text{eff}}(r_l) + \tfrac{1}{2} m\omega_l^2 (r - r_l)^2 + \ldots \quad , \tag{15.42}$$

where $m\omega_l^2 = (d^2 V_{\text{eff}}/dr^2)|_{r_l}$. For small displacements, we can cut off (15.42) after the harmonic term, and we obtain from (15.41) after introduction of $x = r - r_l$,

$$\left[-\frac{\hbar^2}{2m} \frac{d^2}{dx^2} + \varepsilon(r_l) + \frac{\hbar^2 l(l+1)}{2mr_l^2} + \frac{m\omega_l^2}{2} x^2 \right] u = Eu \quad , \tag{15.43}$$

which is the Schrödinger equation of a harmonic oscillator. Its energy eigenvalues are

$$E = \varepsilon(r_l) + \frac{\hbar^2 l(l+1)}{2I_l} + \hbar\omega_l \left(n + \frac{1}{2} \right) \quad , \tag{15.44}$$

in which the effective moment of inertia $I_l = mr_l^2$ enters. The corresponding stationary states are

$$u_{nl} = A_n H_n \left(\frac{x}{x_{0l}} \right) \exp \left\{ -\frac{1}{2} \left(\frac{x}{x_{0l}} \right)^2 \right\} \tag{15.45}$$

with $x_{0l} = (\hbar/m\omega_l)^{1/2}$. Although the wave functions u_{nl} do not satisfy the condition $u(r = 0) = 0$, the accuracy of the energy eigenvalues is not impaired, since the exponent of $u(0) \propto \exp\{-(r_l/x_{0l})^2/2\}$ contains the square of the ratio of nuclear separation to oscillation amplitude (see (15.6)) and is therefore exponentially small.

The energy eigenvalues (15.44) consist of contributions from the effective electronic energy, the rotational energy, and the vibrational energy.

The rotational levels correspond to a wavelength of $\lambda = 0.1 - 1\,\mathrm{cm}$ and lie in the far infrared and microwave regions, whereas the vibrational levels lie at a wavelength of $\lambda = 2 \times 10^{-3} - 3 \times 10^{-3}\,\mathrm{cm}$ in the infrared. The quantum nature of these excitations manifests itself in a macroscopic property of gases, the specific heat. The classical specific heat is $7k_B/2$ (three translational, two rotational, and two vibrational degrees of freedom). This value is only found at high temperatures. Experimentally, and on the basis of quantum statistics, one finds that when the temperature is lowered, the vibrations freeze out first at $10^3\,\mathrm{K}$ followed then by the rotations.

15.6 The van der Waals Force

This is important in the noble gases (He, Ne, Ar, Kr, Xe) and in compounds of large molecules. The charge distribution of the noble gases is spherically symmetric, but not static; for this reason, an interaction of fluctuating dipole moments occurs.

Qualitatively, we can imagine how the van der Waals interaction arises between two atoms in the following manner: The fluctuating dipole moment of the first atom induces a dipole moment in the second atom (Fig. 15.5). The potential of the first atom is $\boldsymbol{p}_1 \cdot \boldsymbol{x}/r^3$, and the electric field of the first atom is

$$\frac{1}{R^3}\left[-\boldsymbol{p}_1 + 3\frac{(\boldsymbol{p}_1 \cdot \boldsymbol{R})\boldsymbol{R}}{R^2} \right] \quad .$$

This causes an induced dipole moment in the second atom,

$$\boldsymbol{p}_2 = \frac{\alpha_2}{R^3}\left[-\boldsymbol{p}_1 + 3\frac{(\boldsymbol{p}_1 \cdot \boldsymbol{R})\boldsymbol{R}}{R^2} \right]$$

(α_2 is the polarizability). The interaction energy of the two dipole moments is

$$\begin{aligned}
W &= \frac{1}{R^3}\left[\boldsymbol{p}_1 \cdot \boldsymbol{p}_2 - 3\frac{(\boldsymbol{p}_1 \cdot \boldsymbol{R})(\boldsymbol{p}_2 \cdot \boldsymbol{R})}{R^2} \right] \\
&= -\frac{\alpha_2}{R^6}p_{1i}(\delta_{ij} + 3\delta_{i1}\delta_{j1})p_{1j} < 0 \quad .
\end{aligned}$$

This yields an attractive interaction $\sim -e^2 r_0^5/R^6$.

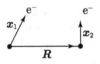

Fig. 15.5. The van der Waals interaction of two hydrogen atoms: separation of the nuclei $\boldsymbol{R}$, electrons e^-

We now turn to the quantum mechanical theory of the van der Waals interaction. We formulate this for two hydrogen atoms, and we denote the coordinates and momenta of the two electrons by $\boldsymbol{x}_1(\boldsymbol{x}_2)$ and $\boldsymbol{p}_1(\boldsymbol{p}_2)$ and the separation vector of the two nuclei by $\boldsymbol{R}$. The Hamiltonian of the two electrons is

$$H = H_1(\boldsymbol{x}_1) + H_2(\boldsymbol{x}_2) + W(\boldsymbol{x}_1, \boldsymbol{x}_2, \boldsymbol{R}) \quad , \tag{15.46}$$

where

$$H_{\frac{1}{2}} = \frac{1}{2m} p_{\frac{1}{2}}^2 - \frac{e_0^2}{|\boldsymbol{x}_{\frac{1}{2}}|} \tag{15.47a}$$

and

$$W(\boldsymbol{x}_1, \boldsymbol{x}_2, \boldsymbol{R}) = \frac{e_0^2}{R} + \frac{e_0^2}{|\boldsymbol{R} + \boldsymbol{x}_2 - \boldsymbol{x}_1|} - \frac{e_0^2}{|\boldsymbol{R} + \boldsymbol{x}_2|} - \frac{e_0^2}{|\boldsymbol{R} - \boldsymbol{x}_1|} \quad . \tag{15.47b}$$

The expansion

$$(1 + x)^{-1/2} = 1 - \tfrac{1}{2}x + \tfrac{3}{8}x^2 + \ldots$$

yields for large distances the dipole interaction

$$W(\boldsymbol{x}_1, \boldsymbol{x}_2, \boldsymbol{R}) = \frac{e_0^2}{R^3} \left(\boldsymbol{x}_1 \cdot \boldsymbol{x}_2 - 3\frac{(\boldsymbol{x}_1 \cdot \boldsymbol{R})(\boldsymbol{x}_2 \cdot \boldsymbol{R})}{R^2} \right) \quad . \tag{15.47b'}$$

The influence of W will now be determined by perturbation theory. The hydrogen eigenstates $|n_1\rangle$ and $|n_2\rangle$ satisfy

$$H_1|n_1\rangle = E_{n_1}|n_1\rangle \ , \quad H_2|n_2\rangle = E_{n_2}|n_2\rangle \quad . \tag{15.48}$$

The change in the ground state energy is

$$\Delta E(R) = \langle 00|W|00\rangle + {\sum_{n_1 n_2}}' \frac{|\langle 00|W|n_1 n_2\rangle|^2}{E_{00}^0 - E_{n_1 n_2}^0} \quad . \tag{15.49}$$

Here, $E_{n_1 n_2}^0 = E_{n_1} + E_{n_2}$, and the prime on the summation symbol indicates that the sum is not taken over $n_1 = n_2 = 0$. For atoms without a dipole moment, such as hydrogen atoms in the ground state or noble gases, the first term vanishes. The second term is always negative; we estimate it for hydrogen atoms. Choosing $\boldsymbol{R}$ in the x-direction, we can write the second term in the form $-e_0^4 A/R^6$ with

$$A = {\sum_{n_1 n_2}}' \frac{|\langle 00|w|n_1 n_2\rangle|^2}{E_{n_1 n_2}^0 - E_{00}^0} \tag{15.50}$$

and $w = (-2x_1 x_2 + y_1 y_2 + z_1 z_2)$.

The ground state wave function of the hydrogen atom is spherically symmetric, as in the noble gases. Thus,

$$\langle 00|w|0n_2\rangle = \langle 00|w|n_1 0\rangle = 0 \quad .$$

In the intermediate state, both electrons must go into an excited state. The energy denominator

$$E^0_{n_1 n_2} - E^0_{00} = (-\tfrac{1}{2} + 2)\,\text{Ry} \ldots (0 + 2)\,\text{Ry}$$

therefore varies between $+3/2$ Ry and $+2$ Ry. Therefore, it follows approximately that

$$A \approx \frac{a}{e_0^2} {\sum_{n_1 n_2}}' |\langle 00|w|n_1 n_2\rangle|^2 = \frac{a}{e_0^2} \langle 00|w^2|00\rangle \quad ,$$

where a is the Bohr radius. Using

$$\langle 0|x^2|0\rangle = \frac{1}{3}\langle 0|r^2|0\rangle = \frac{4\pi}{3}\int dr\, r^4 \frac{e^{-2r/a}}{4\pi a^3} = a^2 \quad ,$$

$$\langle 00|w^2|00\rangle = 6a^4 \quad ,$$

one then finds from (15.49) the attractive *van der Waals interaction*

$$V(R) = -\frac{6e_0^2 a^5}{R^6}$$

$$\sim -\frac{e_0^2}{a}\frac{\alpha_1 \alpha_2}{R^6} \quad , \quad \alpha_i \sim a^3 \text{ (polarizability)} \quad . \tag{15.51}$$

The precise evaluation of the sum (15.50) by London and Eisenschitz gave in place of the factor 6 in (15.51) the value 6.47. The van der Waals force is weaker than the covalent binding forces by a factor of 1000. But it is responsible for the formation of crystals of the noble gases and the binding of large molecules.

The calculation presented here is based on the static Coulomb potential. In fact, the electric interaction occurs via the exchange of photons, and thus it is not instantaneous but propagates with the speed of light c. The propagation time of the light is $\tau_{\text{light}} = 2R/c$, and the characteristic time for one revolution of the electron in the atom is $\tau = 1/me^4$. For small distances,

$$\tau_{\text{light}} \ll \tau \quad , \quad R \ll \lambda \left(\approx \frac{e^2 a}{\hbar c} \sim 10^{-6}\,\text{cm} \right) \quad ,$$

and thus $V(R) \propto -1/R^6$, as above.

For large distances $(R \gg \lambda)$, retardation effects become important, and instead of (15.51) one finds

$$V(R) = -\frac{23}{4\pi}\hbar c \frac{\alpha_1 \alpha_2}{R^7} \quad , \tag{15.52}$$

i.e., R^7 replaces R^6. (In other words, in (15.51) the energy of the atoms e_0^2/a is replaced by $\hbar c/R$, the energy of photons of wavelength R.)[5]

Further reading:

J.C. Slater: *Quantum Theory of Molecules and Solids, Vol. I, Electronic Structure of Molecules* (McGraw-Hill, New York 1963)

M. Weissbluth: *Atoms and Molecules* (Academic Press, New York 1978)

Problems

15.1 Calculate the ground state energy of the H_2^+ molecule by introducing the variational parameter Z^* into the electron wave function

$$\psi = \left(\frac{Z^*}{\pi a^3}\right)^{1/2} e^{-Z^* r/a} \quad .$$

15.2 Calculate the integrals S and A – (15.19b), (15.23) – introduced for the H_2^+ molecule.

15.3 Show that the He_2^{3+} molecule does not exist.

15.4 Consider Problem 3.18 as a simple model for a molecule consisting of two nuclei at the positions $x = +a$ and $x = -a$ and an electron. Assume additionally that a repulsive potential of the form $V_{NN}(a) = \varepsilon \lambda/(2a)$ ($\varepsilon > 0$) is present between the two nuclei. The total energy of the system as a function of the separation of the nuclei is given by the sum of the energy of the electron and the potential of the nuclei. Show that the state of even parity is stable for sufficiently small ε. Hint: Show qualitatively that the total energy as a function of the separation has a minimum.

15.5 Solve the preceding exercise by a variational ansatz analogous to the H_2^+ problem treated in Sect. 15.3. Note: As an ansatz, use the sum and difference of the solutions for the individual δ-potentials and determine which functions yield the smaller expectation value of the energy

$$\left(\frac{d}{dx} \operatorname{sgn} x = 2\delta(x)\right) \quad .$$

15.6 The interaction of the nuclei in a two-atom molecule can be described by

$$V(r) = V_0 \left(\frac{a^2}{r^2} - \frac{2a}{r}\right) \quad , \quad V_0 > 0 \text{ and } a > 0 \quad .$$

Solve the Schrödinger equation for the bound states and find the energy eigenvalues. Think about analogies to the hydrogen atom. Note: As in the H-atom, set $\psi = u/\varrho$,

---

[5] H.B.G. Casimir, D. Polder: Phys. Rev. **73**, 360 (1948); E.A. Power: *Advances in Chemical Physics XII*, ed. by J.O. Hirschfelder (Interscience, London 1967)

$\varrho = r/a$ and factor out the behavior of the solution for $\varrho \to \infty$ and $\varrho \to 0$ in u. The resulting differential equation takes the form

$$\left(z\frac{d^2}{dz^2} + (c - z)\frac{d}{dz} - a \right) F = 0 \quad ,$$

and the solutions regular at the origin are the confluent hypergeometric functions $F(a, c, z)$.

15.7 Compare the rotational levels with the vibrational levels for a $^1\text{H}-^{35}\text{Cl}$ molecule with the potential

$$V(r) = 4\varepsilon \left[\left(\frac{d}{r}\right)^{12} - \left(\frac{d}{r}\right)^{6} \right] \quad , \quad \varepsilon = 3.1 \times 10^{-12}\,\text{eV} \quad , \quad d = 3.3\,\text{Å} \quad .$$

Hint: Since the equilibrium separation r_l depends only weakly on the angular momentum, one can expand the effective potential about r_0. It is advantageous to compute $b = \hbar^2/md^2$ and to express the potential in terms of b and $x_0 = r_0/d$. Represent both the rotational and the vibrational energy as functions of the quantities x_0, b, ε, and l. Give the energies in eV, Hz, and K.

15.8 The Hamiltonian for a rigid body is given by

$$H = \sum_{\alpha=1}^{3} \frac{1}{2I'_\alpha} \left(L'_\alpha \right)^2 \quad ,$$

where L'_α are the components of the angular momentum in the corotating frame of reference. Determine the eigenvalues for a two-atom molecule. Hint: $[L'_\alpha, L'_\beta] = -i\hbar\varepsilon_{\alpha\beta\gamma} L_\gamma$. Introduce $L_\alpha = -L_\alpha'$ and use rotational symmetry about the axis.

15.9 Solve the time independent Schrödinger equation for the Kronig–Penney model

$$V(x) = \lambda \sum_{n} \delta(x - na) \quad .$$

Here, n takes values over all integers, and a is the lattice constant.

16. Time Dependent Phenomena

16.1 The Heisenberg Picture
for a Time Dependent Hamiltonian

Up to now we have studied problems which could be represented by a time independent Hamiltonian. Once the stationary states had been found, the quantum mechanical initial value problem was solved by

$$|\psi, t\rangle = e^{-iHt/\hbar}|\psi, 0\rangle = \sum_n e^{-iE_n t/\hbar} \langle n|\psi, 0\rangle |n\rangle \quad . \tag{16.1}$$

If the Hamiltonian has, in addition to a time independent part H_0, a time dependent part $V(t)$,

$$H(t) = H_0 + V(t) \quad , \tag{16.2}$$

we can either try to solve the Schrödinger equation

$$i\hbar \frac{\partial}{\partial t}|\psi, t\rangle = H(t)|\psi, t\rangle \tag{16.3}$$

or else the Heisenberg equations to be discussed shortly. The solution of the Schrödinger equation (16.3) can be represented formally by

$$|\psi, t\rangle = U(t, t_0)|\psi, t_0\rangle \quad . \tag{16.4}$$

The quantity $U(t, t_0)$ expresses the time development of an arbitrary initial state $|\psi, t_0\rangle$. Substitution into the Schrödinger equation gives the following equation determining $U(t, t_0)$:

$$i\hbar \frac{\partial}{\partial t} U(t, t_0) = H(t)U(t, t_0) \quad , \tag{16.5}$$

which is to be solved subject to the initial condition

$$U(t_0, t_0) = 1 \quad . \tag{16.6}$$

For time independent H, we find immediately

$$U(t, t_0) = \exp\left\{-\frac{i}{\hbar}H(t - t_0)\right\} \quad ,$$

a result which is known from Sect. 8.5 and is equivalent to (16.1). For infinitesimal time differences, the Schrödinger equation implies

$$|\psi, t + dt\rangle = \left(1 - \frac{i}{\hbar}H(t)dt\right)|\psi, t\rangle \tag{16.7}$$

and thus generally

$$U(t + dt, t) = 1 - \frac{i}{\hbar}H(t)\,dt \quad . \tag{16.8}$$

We can now represent $U(t, t_0)$ as a product of infinitesimal time development operators (16.8):

$$U(t, t_0) = \exp\left\{-\frac{i}{\hbar}H(t_0 + \Delta(n-1))\Delta\right\} \dots \exp\left\{-\frac{i}{\hbar}H(t_0)\Delta\right\} \quad , \tag{16.9}$$

where the time interval $(t - t_0)$ is decomposed into n infinitesimal subintervals $\Delta = (t - t_0)/n$, $n \to \infty$. From (16.9), one immediately sees that $U(t, t_0)$ is unitary:

$$U(t, t_0)^\dagger U(t, t_0) = U(t, t_0)U(t, t_0)^\dagger = 1 \quad . \tag{16.10}$$

We now define the Heisenberg operator $O_H(t)$ corresponding to the Schrödinger operator O by

$$O_H(t) = U(t, t_0)^\dagger\, O\, U(t, t_0) \quad . \tag{16.11}$$

Since $U(t, t_0)$ satisfies the same differential equation as $\exp\{-iHt/\hbar\}$ in the time independent case, just as in Sect. 8.5 the equation of motion (Heisenberg equation) follows:

$$\frac{d}{dt}O_H(t) = \frac{i}{\hbar}[H_H(t), O_H(t)] + \frac{\partial}{\partial t}O_H(t) \quad . \tag{16.12}$$

The last term in (16.12) arises only for operators

$$O_H(t) = O(\boldsymbol{x}_H(t), \boldsymbol{p}_H(t), \boldsymbol{S}_H(t), \dots ; t) \tag{16.13}$$

depending explicitly on time (last argument in (16.13)). The Heisenberg state is defined by

$$|\psi\rangle_H = U(t, t_0)^\dagger|\psi, t\rangle \quad . \tag{16.14}$$

Thus, $|\psi\rangle_H = |\psi, t_0\rangle$ and is therefore independent of the time:

$$\frac{d}{dt}|\psi\rangle_H = 0 \quad . \tag{16.15}$$

Just as in (8.59), the matrix elements of operators are given by

$$\langle\psi, t|O|\varphi, t\rangle = \langle\psi|_H O_H(t)|\varphi\rangle_H \quad . \tag{16.16}$$

Remark: We note that (16.9) can also be written compactly in the form

$$U(t, t_0) = T \exp\left\{ -\frac{i}{\hbar} \int_{t_0}^{t} dt' H(t') \right\} \quad , \tag{16.9'}$$

where T is the time-ordering operator, which orders subsequent factors from right to left consecutively in time, e.g.:

$$T\big(H(t_1)H(t_2)\big) = \Theta(t_1 - t_2)H(t_1)H(t_2) + \Theta(t_2 - t_1)H(t_2)H(t_1) \quad .$$

Then (16.9) may be rewritten as

$$
\begin{aligned}
U(t, t_0) &= T\big\{ e^{-\frac{i}{\hbar} H(t_0 + \Delta(n-1))\Delta} \cdots e^{-\frac{i}{\hbar} H(t_0)\Delta} \big\} \\
&= T e^{-\frac{i}{\hbar}\{ H(t_0 + \Delta(n-1)) + \cdots + H(t_0) \}\Delta} \quad .
\end{aligned}
$$

Here we have used the fact that the operators $H\big(t_0 + \Delta(n-1)\big)$, $H\big(t_0 + \Delta(n-2)\big)$, ... can be treated as commuting within a time ordered product and that for commuting operators $e^A e^B = e^{A+B}$. Thus, in the limit $\Delta \to 0$ one obtains (16.9').

For time dependent $H(t)$, the solution of the Schrödinger equation or of the Heisenberg equations is only possible in special cases, e.g., for harmonic oscillators or low-dimensional systems such as the motion of a spin of magnitude $1/2$ in a magnetic field. As a rule, however, one is forced to use approximation methods.

16.2 The Sudden Approximation

As an illustration of the domain of applicability of this approximation method, we give the following example: An atomic nucleus emits an α-particle. The electrons must adjust to the new nuclear charge, which is two units smaller. Here, the characteristic time for the restructuring of the electronic cloud is much larger than the duration of the α-decay. Thus, the electrons still find themselves immediately after the decay in the same state as before it. For example, the question arises as to the probability that the electrons will be in the ground state afterwards. The general problem can be posed as follows: Let the Hamiltonian (16.2) for $t \leq 0$ be independent of time, $H = H_0$. Suppose it is changed within the "switching time" τ_s to $H = H_0 + V$ and that it thereafter remains time independent (Fig. 16.1). Suppose that τ_s is much shorter than the characteristic time τ_{ch} for the restructuring of the system, i.e., $\tau_s \ll \tau_{ch}$. In this sense, the change is "sudden". How large for example is the probability that afterwards the system is in an excited state of the new Hamiltonian if it was originally in the ground state?

Let the (original) stationary states of H_0 be

$$|n^0, t\rangle = e^{-iE_n^0 t/\hbar} |n^0\rangle \quad ,$$

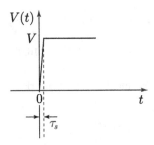

Fig. 16.1. Sudden change of the time dependent part of the Hamiltonian (16.2)

and the stationary states of $H_0 + V$ (afterwards) be

$$|\overline{n}, t\rangle = e^{-i\overline{E}_{\overline{n}} t/\hbar} |\overline{n}\rangle \quad .$$

If the system was in the state $|\psi, t\rangle$ before the change, then by the assumption $\tau_s \ll \tau_{ch}$, it remains in this state just after the change. The further time development satisfies

$$|\overline{\psi}, t\rangle = \sum_{\overline{n}} e^{-i\overline{E}_{\overline{n}} t/\hbar} |\overline{n}\rangle \langle \overline{n} | \psi, 0\rangle \quad . \tag{16.17}$$

The probability that a transition from $|\psi\rangle$ to $|\overline{n}\rangle$ takes place is

$$P_{\psi \to \overline{n}} = |\langle \overline{n} | \psi, 0 \rangle|^2 \quad . \tag{16.18}$$

The sudden approximation is only applicable for transitions between discrete states. In transitions into the continuum, the condition $\tau_s \ll \tau_{ch}$ is not fulfilled.

16.3 Time Dependent Perturbation Theory

16.3.1 Perturbative Expansion

We now suppose that the time dependent part $V(t)$ in the Hamiltonian (16.2) is small compared to H_0 and can be regarded as a perturbation. Suppose also that $V(t) = 0$ for times $t \leq t_0$.

For $t \leq t_0$, let the system be in the state $|\psi^0, t\rangle$ satisfying the Schrödinger equation

$$i\hbar \frac{\partial}{\partial t} |\psi^0, t\rangle = H_0 |\psi^0, t\rangle \quad . \tag{16.19}$$

After the perturbation has been switched on, the state evolves into $|\psi, t\rangle$, which must satisfy the new Schrödinger equation

$$i\hbar \frac{\partial}{\partial t} |\psi, t\rangle = [H_0 + V(t)] |\psi, t\rangle \tag{16.20}$$

and the initial condition

$$|\psi, t\rangle = |\psi^0, t\rangle \text{ for } t \le t_0 \quad . \tag{16.21}$$

For the systematic development of perturbation theory, it is convenient to separate off that part of the time evolution which occurs due to H_0 and which is supposed to be known from the outset. For this purpose, we transform to the *interaction picture* (also called the Dirac picture). The states in the interaction picture $|\psi, t\rangle_{\mathrm{I}}$ are defined by (Sect. 8.5.3)

$$|\psi, t\rangle_{\mathrm{I}} = e^{iH_0 t/\hbar} |\psi, t\rangle \quad . \tag{16.22}$$

The equations of motion in the interaction picture are obtained by differentiating (16.22) with respect to time and using the Schrödinger equation (16.20):

$$i\hbar \frac{\partial}{\partial t} |\psi, t\rangle_{\mathrm{I}} = -H_0 |\psi, t\rangle_{\mathrm{I}} + e^{iH_0 t/\hbar} [H_0 + V(t)] |\psi, t\rangle \quad ,$$

or

$$i\hbar \frac{\partial}{\partial t} |\psi, t\rangle_{\mathrm{I}} = V_{\mathrm{I}}(t) |\psi, t\rangle_{\mathrm{I}} \quad . \tag{16.23}$$

Here, the perturbation operator in the interaction picture is defined by

$$V_{\mathrm{I}}(t) = e^{iH_0 t/\hbar} V(t) e^{-iH_0 t/\hbar} \quad . \tag{16.24}$$

Instead of (16.20), we can solve the Schrödinger equation in the interaction picture (16.23). The time integration yields

$$|\psi, t\rangle_{\mathrm{I}} = |\psi, t_0\rangle_{\mathrm{I}} + \frac{1}{i\hbar} \int_{t_0}^{t} dt' \, V_{\mathrm{I}}(t') |\psi, t'\rangle_{\mathrm{I}} \quad , \tag{16.25}$$

an integral equation which gives by means of iterative substitution the following series expansion in $V_{\mathrm{I}}(t)$:

$$|\psi, t\rangle_{\mathrm{I}} = |\psi, t_0\rangle_{\mathrm{I}} + \frac{1}{i\hbar} \int_{t_0}^{t} dt' \, V_{\mathrm{I}}(t') |\psi, t_0\rangle_{\mathrm{I}}$$

$$+ \frac{1}{(i\hbar)^2} \int_{t_0}^{t} dt' \int_{t_0}^{t'} dt'' \, V_{\mathrm{I}}(t') V_{\mathrm{I}}(t'') |\psi, t_0\rangle_{\mathrm{I}} + \dots \quad . \tag{16.26}$$

This expansion is known as the Neumann series. In principle, it allows one to calculate the state up to arbitrary order in the perturbation, provided the assumptions of perturbation theory are satisfied. We will restrict ourselves here to those applications in which the series can be truncated after the first term.

16.3.2 First-Order Transitions

Let the system be initially in the eigenstate

$$|m, t\rangle = e^{-iH_0t/\hbar}|m\rangle = e^{-iE_mt/\hbar}|m\rangle$$

of the unperturbed Hamiltonian H_0. We seek the probability of finding the system at the time t in the state

$$|n, t\rangle = e^{-iH_0t/\hbar}|n\rangle \doteq e^{-iE_nt/\hbar}|n\rangle$$

after the action of $V(t)$. The probability amplitude for this transition is

$$\langle n, t|\psi, t\rangle = \langle n|e^{iH_0t/\hbar}|\psi, t\rangle = \langle n|\psi, t\rangle_I \quad . \tag{16.27}$$

Substituting the initial state

$$|\psi^0, t\rangle_I = e^{iH_0t/\hbar}|m, t\rangle = |m\rangle$$

into (16.26), we find to first order in $V_I(t)$

$$|\psi, t\rangle_I = |m\rangle + \frac{1}{i\hbar} \int_{t_0}^{t} dt' V_I(t')|m\rangle \quad . \tag{16.28}$$

Hence, the transition amplitude becomes

$$\langle n, t|\psi, t\rangle = \delta_{n,m} + \frac{1}{i\hbar} \int_{t_0}^{t} dt' \langle n|V_I(t')|m\rangle$$

$$= \delta_{n,m} + \frac{1}{i\hbar} \int_{t_0}^{t} dt' e^{i(E_n - E_m)t'/\hbar} \langle n|V(t')|m\rangle \quad . \tag{16.29}$$

The transition probability $P_{mn}(t)$ for the transition of $|m\rangle$ to another orthogonal state $|n\rangle$ is the square of the modulus of this expression:

$$P_{mn}(t) = |\langle n, t|\psi, t\rangle|^2 = \left| \frac{1}{\hbar} \int_{t_0}^{t} dt' \, e^{i(E_n - E_m)t'/\hbar} \langle n|V(t')|m\rangle \right|^2 \quad . \tag{16.30}$$

16.3.3 Transitions into a Continuous Spectrum, the Golden Rule

We now apply (16.30) to transitions into a continuous spectrum of final states. Examples of this are:

(a) Scattering: Here, the momentum k of a particle goes over to k'. First-order perturbation theory leads to the Born approximation of scattering theory.
(b) α-decay: Here as well, the final states, i.e., the momenta of the α-particles, lie in a continuum.
(c) Optical transitions: An excited state makes a transition to a lower state by emitting a photon, whose momentum varies continuously.

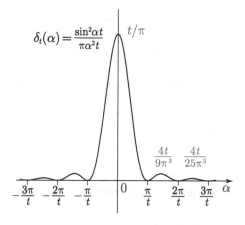

Fig. 16.2. The function $\delta_t(\alpha)$

We will take up these concrete examples again later, but for now let us formulate the general theory of such transitions.

We consider first a perturbation that is switched on at the time $t = 0$ and subsequently remains unchanged:

$$V(t) = V\,\Theta(t) \quad . \tag{16.31}$$

We cannot utilize the sudden approximation here, since transitions to states within a continuum occur in an arbitrarily short time.

Substituting (16.31) into (16.30), we find

$$
\begin{aligned}
P_{mn}(t) &= \frac{1}{\hbar^2}\left| \int_0^t dt'\,\mathrm{e}^{\mathrm{i}(E_n - E_m)t'/\hbar}\langle n|V|m\rangle \right|^2 \\
&= \frac{1}{\hbar^2}\left| \frac{\mathrm{e}^{\mathrm{i}\omega_{nm}t} - 1}{\omega_{nm}}\langle n|V|m\rangle \right|^2 \\
&= \frac{1}{\hbar^2}\left[\frac{\sin\left(\omega_{nm}t/2\right)}{\omega_{nm}/2} \right]^2 |\langle n|V|m\rangle|^2
\end{aligned}
\tag{16.32}
$$

with

$$\omega_{nm} = (E_n - E_m)/\hbar \quad . \tag{16.33}$$

In order to evaluate (16.32) further, we consider the sequence of functions

$$\delta_t(\alpha) = \frac{\sin^2\alpha t}{\pi\alpha^2 t} \quad , \tag{16.34a}$$

which is illustrated in Fig. 16.2 as a function of α. It has the properties

$$
\delta_t(\alpha)
\begin{cases}
= t/\pi & \text{for } \alpha = 0 \\
\leq 1/\pi\alpha^2 t & \text{for } \alpha \neq 0
\end{cases}
\quad , \tag{16.34b}
$$

and $\int_{-\infty}^{\infty} dy \, \sin^2 y/y^2 = \pi$. For a test function $F(\alpha)$, one has

$$\lim_{t \to \infty} \int_{-\infty}^{\infty} d\alpha \, \delta_t(\alpha) F(\alpha) = \lim_{t \to \infty} \int_{-\infty}^{\infty} dy \, \frac{\sin^2 y}{\pi y^2} \, F\left(\frac{y}{t}\right) = F(0) \quad , \quad (16.34c)$$

and thus the sequence of functions $\delta_t(\alpha)$ is a representation of the δ-function:

$$\lim_{t \to \infty} \delta_t(\alpha) = \delta(\alpha) \quad . \tag{16.34d}$$

For long times, we thus obtain from (16.32)

$$P_{mn}(t) = t \frac{2\pi}{\hbar} \delta(E_n - E_m) |\langle n|V|m \rangle|^2 \quad . \tag{16.35}$$

From this, one obtains the transition rate, that is, the transition probability per unit time,

$$\Gamma_{mn} = \frac{2\pi}{\hbar} \delta(E_n - E_m) |\langle n|V|m \rangle|^2 \quad . \tag{16.36}$$

Since we are treating transitions into states of a continuous spectrum, the transition rate to a group of states is of interest; in scattering, for example, to all of the states with wave numbers in a certain angular region. We suppose that the matrix element for all of these final states is equal and introduce the density of states $\varrho(E_n)$. The quantity $\varrho(E_n)dE_n$ gives the number of states in the interval dE_n. The transition rate into this set of states is then

$$\sum_n \Gamma_{mn} = \int dE_n \, \varrho(E_n) \Gamma_{mn} = \varrho(E_m) \frac{2\pi}{\hbar} |\langle n|V|m \rangle|^2 \quad . \tag{16.37}$$

The energy of the final state $|n\rangle$ entering the matrix element of (16.37) must be equal to E_m.

The formulae (16.36, 16.37) and their analogs for periodically varying potentials $V(t)$ were derived by Pauli in 1928, and, because of the multitude of applications, Fermi coined the phrase "golden rule"[1].

We now add a few remarks concerning the validity of the golden rule. For this, we must go back to our original representation of the δ-function $\delta_t[(E_n - E_m)/2\hbar]$ in (16.34a). For every finite t, the width of this function is $4\pi\hbar/t$. In order for this function to be replaceable by a δ-function, the width of the energy distribution of the final states ΔE must be much larger than $2\pi\hbar/t$ (Fig. 16.3). The second condition is that many states must lie within this δ-like function, because only then can we characterize this set of states by a density of states. Denoting the separation of the energy levels by $\delta\varepsilon$, we thus have the condition

$$\Delta E \gg \frac{2\pi\hbar}{t} \gg \delta\varepsilon \quad , \quad \text{or} \quad \frac{2\pi\hbar}{\Delta E} \ll t \ll \frac{2\pi\hbar}{\delta\varepsilon} \quad .$$

---

[1] W. Pauli: Über das H-Theorem vom Anwachsen der Entropie vom Standpunkt der neuen Quantenmechanik. In *Probleme der modernen Physik, Arnold Sommerfeld zum 60. Geburtstag* (Hirzel, Leipzig 1928); E. Fermi: *Nuclear Physics* (Univ. Chicago Press, Chicago 1950) p. 148.

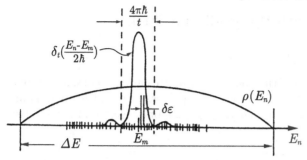

Fig. 16.3. The function $\delta_t((E_n - E_m)/2\hbar)$, the state density $\varrho(E_n)$, and the separation of the final states

16.3.4 Periodic Perturbations

We now extend the golden rule to the case where the perturbation which has been switched on at $t = 0$ varies periodically with time. This can occur for example in the case of a periodic external electric field. In general,

$$V(t) = \Theta(t)(Fe^{-i\omega t} + F^\dagger e^{i\omega t}) \quad , \tag{16.38}$$

where F is an operator. For the coupling to the electromagnetic field, which will be treated later, V is given in (16.43). The transition matrix element is then (Fig. 16.4)

$$\langle n, t|\psi, t\rangle = \frac{1}{i\hbar} \int_0^t dt' \left[e^{i(\omega_{nm}-\omega)t'} \langle n|F|m\rangle + e^{i(\omega_{nm}+\omega)t'} \langle n|F^\dagger|m\rangle \right] \quad . \tag{16.39}$$

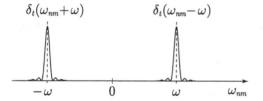

Fig. 16.4. The frequency dependence of the transition matrix element, schematic

Because the two δ-like functions do not overlap, the cross terms in the square of the modulus of (16.39) do not contribute:

$$|\langle n, t|\psi, t\rangle|^2 = t\frac{2\pi}{\hbar^2}[\delta(\omega_{nm} - \omega)|\langle n|F|m\rangle|^2 + \delta(\omega_{nm} + \omega)|\langle n|F^\dagger|m\rangle|^2] \quad .$$

Hence, the transition rate becomes

$$\begin{aligned}
\Gamma_{mn} &= \frac{2\pi}{\hbar}[\delta(E_n - E_m - \hbar\omega)|\langle n|F|m\rangle|^2 \\
&\quad + \delta(E_n - E_m + \hbar\omega)|\langle n|F^\dagger|m\rangle|^2] \quad .
\end{aligned} \tag{16.40}$$

16.4 Interaction with the Radiation Field

16.4.1 The Hamiltonian

An important application of time dependent perturbation theory, and in particular of the golden rule, is the theory of processes in which photons are emitted and absorbed. The familiar Hamiltonian of an electron in an electromagnetic field is

$$H = \frac{1}{2m}\left[\boldsymbol{p} - \frac{e}{c}\boldsymbol{A}(\boldsymbol{x},t)\right]^2 + e\varPhi(\boldsymbol{x},t) + V(\boldsymbol{x}) \quad , \tag{16.41}$$

disregarding the spin in this section. For multielectron atoms[2], this yields the unperturbed Hamiltonian

$$H_0 = \sum_i \left(\frac{1}{2m}\boldsymbol{p}_i^2 + V(\boldsymbol{x}_i)\right) \tag{16.42}$$

as well as the the time dependent interaction term

$$V(t) = \sum_i \left(-\frac{e}{2mc}\{\boldsymbol{p}_i, \boldsymbol{A}(\boldsymbol{x}_i,t)\} + \frac{e^2}{2mc^2}\boldsymbol{A}^2(\boldsymbol{x}_i,t) + e\varPhi(\boldsymbol{x}_i,t)\right), \tag{16.43}$$

where we sum over the electrons. Defining the particle number density and current density[3] by

$$\varrho(\boldsymbol{x}) = \sum_i \delta(\boldsymbol{x} - \boldsymbol{x}_i) \tag{16.44a}$$

and

$$\boldsymbol{j}(\boldsymbol{x}) = \frac{1}{2}\sum_i \left\{\frac{\boldsymbol{p}_i}{m}, \delta(\boldsymbol{x} - \boldsymbol{x}_i)\right\} \quad , \tag{16.44b}$$

we can also write $V(t)$ in the form

$$V(t) = \int d^3x \left[-\frac{e}{c}\boldsymbol{j}(\boldsymbol{x})\cdot\boldsymbol{A}(\boldsymbol{x},t) + \frac{e^2}{2mc^2}\varrho(\boldsymbol{x})\boldsymbol{A}^2(\boldsymbol{x},t) + e\varrho(\boldsymbol{x})\varPhi(\boldsymbol{x},t)\right] \quad . \tag{16.45}$$

---

[2] The index i labels the electrons.
[3] The electric current density is

$$\boldsymbol{J}(\boldsymbol{x}) = -c\frac{\delta H}{\delta \boldsymbol{A}} = \frac{e}{2m}\sum_i \left\{\left[\boldsymbol{p}_i - \frac{e}{c}\boldsymbol{A}(\boldsymbol{x}_i,t)\right], \delta(\boldsymbol{x} - \boldsymbol{x}_i)\right\} \quad .$$

16.4.2 Quantization of the Radiation Field

Electromagnetic waves are oscillations of the electromagnetic field. We will quantize them in analogy to the one-dimensional harmonic oscillator. For a one-dimensional oscillator q, we know from Sect. 3.1 that the Hamiltonian

$$H = \frac{m\dot{q}^2}{2} + \frac{m\omega^2}{2}q^2 \tag{16.46a}$$

can, by means of the transformation

$$q = \sqrt{\frac{\hbar}{2m\omega}}(ae^{-i\omega t} + a^\dagger e^{i\omega t}) \quad , \tag{16.46b}$$

$$[a, a^\dagger] = 1 \quad , \tag{16.46c}$$

be brought into the form

$$H = \hbar\omega(a^\dagger a + \tfrac{1}{2}) \quad . \tag{16.46d}$$

For the free radiation field, we use the transverse gauge div $A = 0$, also designated the Coulomb gauge. It then follows, because of the absence of sources, that $\Phi = 0$ and

$$E = -\frac{1}{c}\dot{A} \quad , \quad B = \nabla \times A \quad . \tag{16.47a}$$

Because of the absence of external currents, A satisfies the free wave equation. We can now expand the radiation field in a Fourier series

$$A(x,t) = \sum_k A_k(t)e^{ik \cdot x} \quad . \tag{16.47b}$$

The individual oscillation amplitudes $A_k(t)$ will be decomposed as in (16.46b) into sums of annihilation and creation operators. For this, it remains only to find out what to insert for m and ω. We can read off this information from the energy of the radiation field, which is given by

$$H_{\text{rad}} = \frac{1}{8\pi} \int d^3x \, (E^2 + B^2) = \frac{V}{8\pi} \sum_k \left(\frac{1}{c^2}|\dot{A}_k|^2 + |k \times A_k|^2\right) . \tag{16.48}$$

Comparison of (16.48) with (16.46a) shows that ck should be substituted for ω and $1/4\pi c^2$ for m. Finally, representing the vector field A_k by the two polarization vectors $\varepsilon_{k,\lambda}$ ($\lambda = 1, 2$) orthogonal to k and to each other, we obtain

$$A(x,t) = \sum_{k,\lambda} \sqrt{\frac{2\pi\hbar c}{kV}} \left(a_{k,\lambda}\varepsilon_{k,\lambda}e^{ik \cdot x - i\omega_k t} + a^\dagger_{k,\lambda}\varepsilon^*_{k,\lambda}e^{-ik \cdot x + i\omega_k t}\right) \quad .$$

$$\tag{16.49}$$

Here, we have introduced the frequency $\omega_k = ck$ and the volume V, on whose boundaries we assume periodic boundary conditions, thus fixing the values of k.

In analogy to the oscillator, we demand

$$[a_{\boldsymbol{k},\lambda}, a^{\dagger}_{\boldsymbol{k}',\lambda'}] = \delta_{\boldsymbol{k},\boldsymbol{k}'}\delta_{\lambda,\lambda'} \quad , \quad [a_{\boldsymbol{k},\lambda}, a_{\boldsymbol{k}',\lambda'}] = [a^{\dagger}_{\boldsymbol{k},\lambda}, a^{\dagger}_{\boldsymbol{k}',\lambda'}] = 0 \quad . \quad (16.50)$$

Then, $a^{\dagger}_{\boldsymbol{k},\lambda}(a_{\boldsymbol{k},\lambda})$ creates (annihilates) a quantum of wave number $\boldsymbol{k}$ and polarization λ. Substituting the expansion (16.49) of $\boldsymbol{A}(\boldsymbol{x},t)$ into (16.48), one then obtains for the Hamiltonian of the free radiation field[4]

$$H_{\text{rad}} = \sum_{\boldsymbol{k},\lambda} \hbar c k \left(a^{\dagger}_{\boldsymbol{k},\lambda} a_{\boldsymbol{k},\lambda} + \tfrac{1}{2} \right) \quad . \tag{16.51a}$$

The operators

$$\hat{n}_{\boldsymbol{k},\lambda} = a^{\dagger}_{\boldsymbol{k},\lambda} a_{\boldsymbol{k},\lambda}$$

have eigenvalues $n_{\boldsymbol{k},\lambda} = 0, 1, 2, \ldots$, and their eigenstates are of the form

$$|n_{\boldsymbol{k},\lambda}\rangle = \frac{1}{\sqrt{n_{\boldsymbol{k},\lambda}!}} \left(a^{\dagger}_{\boldsymbol{k},\lambda} \right)^{n_{\boldsymbol{k},\lambda}} |0\rangle \quad ,$$

where $|0\rangle$ is the vacuum state, the state without photons. The eigenstates of the Hamiltonian H_{rad} are direct products of the above states

$$|\ldots, n_{\boldsymbol{k},\lambda}, \ldots\rangle = \prod_{\boldsymbol{k}_i} \prod_{\lambda_i} |n_{\boldsymbol{k}_i,\lambda_i}\rangle$$

with energy

$$\sum_{\boldsymbol{k}} \sum_{\lambda} \hbar \omega_{\boldsymbol{k}} \left(n_{\boldsymbol{k},\lambda} + \frac{1}{2} \right) \quad .$$

The operator $a_{\boldsymbol{k},\lambda}$ reduces the occupation number $n_{\boldsymbol{k},\lambda}$ of the mode $\boldsymbol{k}, \lambda$ by one, while all other occupation numbers remain unchanged,

$$a_{\boldsymbol{k},\lambda} |\ldots, n_{\boldsymbol{k},\lambda}, \ldots\rangle = \sqrt{n_{\boldsymbol{k},\lambda}} |\ldots, n_{\boldsymbol{k},\lambda} - 1, \ldots\rangle \quad .$$

Accordingly, the energy is lowered by $\hbar\omega_{\boldsymbol{k}}$. Thus we refer to $a_{\boldsymbol{k},\lambda}$ as an annihilation (or destruction) operator, which annihilates a photon in the mode $\boldsymbol{k}, \lambda$, i.e., with momentum $\hbar\boldsymbol{k}$, polarization vector $\boldsymbol{\varepsilon}_{\boldsymbol{k},\lambda}$ and energy $\hbar\omega_{\boldsymbol{k}}$. Correspondingly, $a^{\dagger}_{\boldsymbol{k},\lambda}$ is the creation operator of photons with these quantum numbers:

$$a^{\dagger}_{\boldsymbol{k},\lambda} |\ldots, n_{\boldsymbol{k},\lambda} \ldots\rangle = \sqrt{n_{\boldsymbol{k},\lambda} + 1} |\ldots, n_{\boldsymbol{k},\lambda} + 1, \ldots\rangle \quad .$$

---

[4] The sum of the zero point energies in (16.51a) diverges. In quantum field theory the products in (16.48) are defined in normal ordered form, i.e. all creation operators are placed to the left of the annihilation operators. Then the Hamiltonian (16.51a) no longer contains any zero point energies, while otherwise remaining unchanged (see F. Schwabl, *Advanced Quantum Mechanics*, 3rd ed., Springer, Berlin, Heidelberg 2005).

The total Hamiltonian of matter coupled to the quantized radiation field reads

$$H = H_0 + H_{\text{rad}} + H_1 \quad . \tag{16.51b}$$

Here H_0 and H_{rad} are the Hamiltonians of the electrons and of the free radiation field, given by (16.42) and (16.51a), respectively. The interaction Hamiltonian following from (16.45) reads

$$H_1 = \int d^3x \left[-\frac{e}{c} j(x) \cdot A(x) + \frac{e^2}{2mc^2} \varrho(x) A^2(x) \right] \quad . \tag{16.51c}$$

In the Schrödinger picture, $j(x)$ and $\varrho(x)$ are given in (16.44a,b) and the field operator $A(x)$ by (16.49), with t replaced by $t = 0$. In the interaction picture, $A(x)$ is to be replaced by $A(x,t)$ of (16.49), and $(\varrho(x), j(x))$ by $e^{iH_0t/\hbar}(\varrho(x), j(x)) e^{-iH_0t/\hbar}$.

The first term of the interaction Hamiltonian (16.51c) gives rise to processes in which one photon is annihilated or created. The second term leads to the annihilation (creation) of two photons and also to the simultaneous annihilation and creation of a photon.

16.4.3 Spontaneous Emission

We now consider the spontaneous emission of a photon by an atom; an atomic electron makes a transition from its initial state $|m\rangle$ to the state $|n\rangle$ and emits a photon of wave number k and polarization λ. Let the radiation field be initially in the ground state – also known as the vacuum state – which we denote by $|0\rangle$. The initial state of the complete system is then $|0\rangle|m\rangle$, and its final state is $a_{k,\lambda}^\dagger |0\rangle|n\rangle$. The perturbative operator inducing the transition is given by the first term in (16.45) (or (16.51c)), where we substitute (16.49); it is of the form (16.38). Because of $a_{k',\lambda'}|0\rangle = 0$ and the oscillator (Bose) commutation relations, only the second term contributes in the golden rule for the transition rate (16.40):

$$\Gamma_{m \to n,k,\lambda} = \frac{(2\pi)^2 e^2}{kc} \delta(E_m - E_n - \hbar ck)$$
$$\times \left| \langle n| \int d^3x j(x) \cdot \varepsilon_{k,\lambda}^* \frac{e^{-ik \cdot x}}{\sqrt{V}} |m\rangle \right|^2 \quad . \tag{16.52}$$

From this we can determine the power dP_λ radiated into the solid angle $d\Omega$. For this, we recall that in a volume element d^3k of k-space, there are $d^3k \, V/(2\pi)^3$ states, and we write $d^3k = k^2 dk \, d\Omega$. Introducing now the Fourier transform of the current density

$$j_k = \int d^3x \, j(x) e^{-ik \cdot x} \quad , \tag{16.53}$$

we finally find

$$dP_\lambda = \sum_{k \in d\Omega} \hbar c k \Gamma_{m \to n, k, \lambda}$$

$$= d\Omega \int \frac{dk k^3 \hbar}{(2\pi)^3} \frac{(2\pi)^2}{k} e^2 |\langle n| \boldsymbol{j_k} \cdot \boldsymbol{\varepsilon}^*_{k,\lambda} |m\rangle|^2 \delta(E_m - E_n - \hbar c k) \quad,$$

(16.54)

$$\frac{dP_\lambda}{d\Omega} = \frac{\omega^2 e^2}{2\pi c^3} |\langle n| \boldsymbol{j_k} \cdot \boldsymbol{\varepsilon}^*_{k,\lambda} |m\rangle|^2 \quad .$$

(16.55)

Because of energy conservation, implied by the δ-function in (16.54), the frequency is $\omega = (E_m - E_n)/\hbar$, and the wave number $\boldsymbol{k}$ in the matrix element is restricted to $|\boldsymbol{k}| = \omega/c$.

Remarks: Before evaluating this quantum mechanical formula, we would like to gain an overview of the orders of magnitude occurring in it. The energy of a photon emitted in atomic transitions is $\hbar\omega \sim mc^2 (Z\alpha)^2 \sim (Ze)^2/a$, and therefore the wavelength is $\lambda = \lambda/(2\pi) = c/\omega \approx a/(Z^2\alpha)$, where a is the Bohr radius, α the fine-structure constant (see Sect. 6.3), and $\tau \approx 2\pi a/Zv \approx 2\pi a/(Z^2\alpha c) \simeq 2\pi/\omega$ the classical period. In order to determine the lifetime ΔT or the duration of the optical transition, we take from *classical* electrodynamics the power emitted by an accelerated charge e in all spatial directions:

$$P = \int dP = \frac{2e^2}{3c^3} \ddot{x}^2 \quad .$$

For atomic dimensions, we can estimate this as

$$P \sim \frac{\omega^4 e^2}{c^3} \left(\frac{a}{Z}\right)^2 \sim \omega \frac{1}{\lambda^3} \frac{e^2}{a} \frac{a^3}{Z^2} \sim \hbar\omega^2 \alpha^3 Z^2 \quad .$$

In order of magnitude, the power must also be equal to the ratio of $\hbar\omega$ to ΔT, i.e., $P \approx \hbar\omega/\Delta T$, and therefore

$$\Delta T \approx \frac{1}{Z^2} \omega^{-1} \alpha^{-3} \sim \omega^{-1} \alpha^{-3} \quad .$$

(16.56)

With this, we can also estimate the coherence length of the emitted light wave $\Delta L = c \Delta T$. For visible light, $1/\lambda = 1.3 \times 10^4 - 2.8 \times 10^4$ cm^{-1}, i.e., $\omega \approx 10^{15}$ Hz, and therefore $\Delta T \approx 10^{-9}$ s. This corresponds to a coherent wave train of length $\Delta L \approx 10 - 20$ cm.

We now return to the expression (16.55) for the power radiated per unit solid angle in the direction of $\boldsymbol{k}$. Since for atoms

$$\boldsymbol{k} \cdot \boldsymbol{x} \approx ka = a/\lambda \approx \alpha \ll 1 \quad ,$$

(16.57)

we may expand the exponential function of (16.53):

$$\langle n| \boldsymbol{j_k} |m\rangle = \langle n| \int d^3x \, (1 - \mathrm{i}\boldsymbol{k} \cdot \boldsymbol{x} + \tfrac{1}{2}(\mathrm{i}\boldsymbol{k} \cdot \boldsymbol{x})^2 + \ldots) \boldsymbol{j}(\boldsymbol{x}) |m\rangle$$

$$= \langle n| \boldsymbol{j_0} |m\rangle - \mathrm{i} \langle n| \int d^3x \, (\boldsymbol{k} \cdot \boldsymbol{x}) \boldsymbol{j}(\boldsymbol{x}) |m\rangle + \ldots \quad .$$

(16.58)

The terms in (16.58) give, respectively, electric dipole transitions, magnetic dipole transitions, electric quadrupole transitions, etc.

16.4.4 Electric Dipole ($E1$) Transitions

These come from the terms of zeroth order in (16.58). The Fourier transform of the current density at $\boldsymbol{k} = 0$ is, according to (16.53) and (16.44b),

$$\boldsymbol{j}_0 = \frac{1}{m}\boldsymbol{P} \quad , \tag{16.59}$$

where

$$\boldsymbol{P} = \sum_i \boldsymbol{p}_i \tag{16.60}$$

is the total momentum. The equation of motion for the center of mass

$$\boldsymbol{X} = \sum_i \boldsymbol{x}_i \tag{16.61}$$

is just

$$\frac{1}{m}\boldsymbol{P} = \frac{\mathrm{i}}{\hbar}[H_0, \boldsymbol{X}] \quad , \tag{16.62}$$

by means of which the matrix element required in (16.58) can be obtained:

$$\begin{aligned}
\langle n|\boldsymbol{j}_0|m\rangle &= \frac{\mathrm{i}}{\hbar}\langle n|[H_0, \boldsymbol{X}]|m\rangle \\
&= \frac{\mathrm{i}}{\hbar}(E_n - E_m)\langle n|\boldsymbol{X}|m\rangle \quad .
\end{aligned} \tag{16.63}$$

We define the dipole matrix element

$$\boldsymbol{d}_{nm} = \langle n|\boldsymbol{X}|m\rangle \tag{16.64}$$

and finally obtain for the radiated power of photons with polarization λ

$$\frac{dP_\lambda}{d\Omega} = \frac{\omega^4 e^2}{2\pi c^3}|\boldsymbol{d}_{nm} \cdot \boldsymbol{\varepsilon}^*_{\boldsymbol{k},\lambda}|^2 \quad . \tag{16.65}$$

The radiated power is proportional to the fourth power of the frequency $\omega = (E_n - E_m)/\hbar$ of the radiated light. The amplitude of the emitted light is proportional to the projection of $\boldsymbol{d}_{nm}$ onto $\boldsymbol{\varepsilon}^*_{\boldsymbol{k},\lambda}$, and the intensity is proportional to its square. The matrix element $e\boldsymbol{d}_{nm}$ of the dipole operator $e\boldsymbol{X}$ determines which electric dipole transitions are possible.

16.4.5 Selection Rules for Electric Dipole ($E1$) Transitions

Statements concerning vanishing or possibly nonvanishing matrix elements $\boldsymbol{d}_{nm}$ are called selection rules. In order to derive them, we begin with

$$[L_z, Z] = 0 \quad , \quad [L_z, X \pm \mathrm{i}Y] = \pm(X \pm \mathrm{i}Y)\hbar \quad , \tag{16.66a,b}$$

where

$$L = \sum_i (\boldsymbol{x}_i \times \boldsymbol{p}_i) \tag{16.67}$$

is the total angular momentum and X, Y, and Z are the components of the center of mass defined above. We assume that the initial (final) states are eigenstates of $\boldsymbol{L}^2$ and L_z with quantum numbers $l(l')$ and $m(m')$. It then follows from (16.66a,b) that

$$\langle l', m'|Z|l, m \rangle (m' - m) = 0 \quad , \tag{16.68a}$$

$$\langle l', m'|(X + iY)|l, m \rangle (m' - m - 1) = 0 \quad , \tag{16.68b}$$

$$\langle l', m'|(X - iY)|l, m \rangle (m' - m + 1) = 0 \quad . \tag{16.68c}$$

Electric dipole transitions are only possible if one of the dipole matrix elements is finite, that is, according to (16.68a–c), if one of the selection rules

$$m' = \begin{cases} m \\ m \pm 1 \end{cases} \tag{16.69}$$

is satisfied.

By means of the commutator

$$[\boldsymbol{L}^2, [\boldsymbol{L}^2, \boldsymbol{X}]] = 2\hbar^2 \{\boldsymbol{X}, \boldsymbol{L}^2\}$$

one can show in complete analogy that

$$\langle l', m'|\boldsymbol{X}|l, m \rangle (l + l')(l + l' + 2)[(l - l')^2 - 1] = 0 \quad .$$

The last factor yields the selection rule

$$l' = l \pm 1 \quad . \tag{16.70}$$

Since l' and l are both nonnegative, the third factor cannot vanish, and the second can only vanish for $l' = l = 0$. However, this selection rule cannot be satisfied, since the states with $l' = l = 0$ are independent of direction, and therefore these matrix elements of $\boldsymbol{X}$ vanish. Formally, one easily shows this using the parity operator P (see (5.35)),

$$\langle 0|\boldsymbol{X}|0 \rangle = \langle 0|P\boldsymbol{X}P|0 \rangle = -\langle 0|\boldsymbol{X}P^2|0 \rangle = -\langle 0|\boldsymbol{X}|0 \rangle \quad ,$$

that is, $\langle 0|\boldsymbol{X}|0 \rangle = 0$.

We now discuss the polarization of the emitted light. For transitions with m unchanged, $m' = m$, (16.68) yields $\boldsymbol{d}_{mn} \sim \boldsymbol{e}_z$. For such transitions, the emitted light is linearly polarized according to (16.65) in the $\boldsymbol{k} - \boldsymbol{e}_z$-plane (see Fig. 16.5a). There is no radiation in the z-direction.

For electric dipole transitions with $m' = m \pm 1$, the matrix elements of $X \pm iY$ are finite, and, by (16.68b,c),

$$\langle l', m'|(X \mp iY)|l, m \rangle = 0 \quad , \text{ i.e., } \langle l', m'|Y|l, m \rangle = \mp i\langle l', m'|X|l, m \rangle \quad .$$

Moreover, since by (16.68a) the Z-component vanishes, we find

$$d \sim \begin{pmatrix} 1 \\ \mp i \\ 0 \end{pmatrix} . \tag{16.71}$$

If the wave vector k of the emitted photon points in the z-direction, then by $\varepsilon_{r(l)} = \varepsilon_1 \mp i\varepsilon_2$ it is right (left) circularly polarized, and its helicity is negative (positive), corresponding to spin $-\hbar$ $(+\hbar)$ in the direction of k, which it must carry off because of the conservation of the z-component of angular momentum. On the other hand, light emitted in the "equatorial" xy-plane is linearly polarized. For all other k-directions it is elliptically polarized (see Fig. 16.5b).

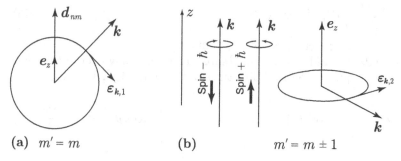

(a) $m' = m$ (b) $m' = m \pm 1$

Fig. 16.5a,b. Polarization for electric dipole transitions

The selection rules considerably restrict the optical transitions. For example, an H atom cannot make a direct transition from one of the Rydberg states with large principal quantum number n into the ground state if in addition $l = n - 1$, but only stepwise by means of dipole transitions from one level to the next. The same thing happens in the capture of muons, which first enter a highly excited state, i.e., $l \approx n$ (see Fig. 16.6).

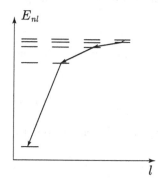

Fig. 16.6. Transitions of $n, n-1$ states

16.4.6 The Lifetime for Electric Dipole Transitions

We are now in a position to compute the transition probability per unit time and determine the lifetime of an excited state. The transition probability per unit time with a photon emitted into the solid angle $d\Omega$ is given by

$$dw_\lambda = \sum_{k \in d\Omega} \Gamma_{m \to n, k, \lambda} \quad . \tag{16.72}$$

In comparison with the radiated power of Eq. (16.54) the factor $\hbar c k$ is absent. Hence, for electric dipole transitions the same steps leading to Eq. (16.55) give

$$
\begin{aligned}
dw_\lambda &= \frac{\omega^3 e^2}{2\pi c^3 \hbar} |d_{nm} \cdot \varepsilon_{k,\lambda}|^2 d\Omega \\
&= \frac{\omega^3 e^2}{2\pi c^3 \hbar} |d_{nm}|^2 \cos^2 \theta_\lambda d\Omega \quad .
\end{aligned}
\tag{16.73}
$$

Here θ_λ is the angle enclosed by d_{nm} and $\varepsilon_{k,\lambda}^*$. From Fig. 16.7 one sees $\cos \theta_1 = \sin \vartheta \cos \varphi$ and $\cos \theta_2 = \sin \vartheta \sin \varphi$. Thus summing over both polarizations the integration over all directions of the emitted photon gives

$$\int d\Omega (\cos^2 \theta_1 + \cos^2 \theta_2) = \int d\Omega \sin^2 \vartheta = 2\pi \int_{-1}^{1} d(\cos \vartheta) \sin^2 \vartheta = \frac{8\pi}{3}.$$

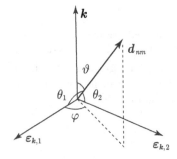

Fig. 16.7. The orientation of d_{nm} and the polarization vectors

Then one finds from (16.73) the total transition probability per unit time from a state m into a state n

$$w = \frac{4\omega^3 e^2}{3c^3 \hbar} |d_{nm}|^2 \quad . \tag{16.74}$$

In analogy to (3.75a) concerning the α-decay the lifetime τ is related to w by

$$\frac{1}{\tau} = \sum_n w \quad , \tag{16.75}$$

where the sum extends over all allowed final states. For instance the transition from the hydrogenic state nlm into $n'l'm'$, where $m' = m, m \pm 1$, gives a contribution

$$\sum_{m'=0\pm1} w(nlm, n'l'm') = \frac{4\omega^3 e^2}{3c^3\hbar} \left\{ \begin{matrix} \frac{l+1}{2l+1} \\ \frac{l}{2l+1} \end{matrix} \right\} \left| \int_0^\infty dr\, r^3 R_{n'l'}(r) R_{nl}(r) \right|^2$$

$$\text{for } l' = \begin{cases} l+1 \\ l-1 \end{cases} . \qquad (16.76)$$

In particular, this transition rate may be computed easily for the transition of the hydrogenic $2p$ into the $1s$ state leading to the lifetime $\tau(2p \to 1s) = 1.6 \times 10^{-9}$ sec.

16.4.7 Electric Quadrupole and Magnetic Dipole Transitions

The electric quadrupole transitions $(E2)$ and the magnetic dipole transitions $(M1)$ result from the second term in (16.58) for the expansion of the current density. We first symmetrize with respect to the wave number and polarization

$$-i \int d^3x\, (\boldsymbol{k} \cdot \boldsymbol{x})(\boldsymbol{j}(\boldsymbol{x}) \cdot \boldsymbol{\varepsilon}_{\boldsymbol{k},\lambda}^*)$$

$$= -i \int d^3x \Big[\tfrac{1}{2}\{(\boldsymbol{k} \cdot \boldsymbol{x})(\boldsymbol{j}(\boldsymbol{x}) \cdot \boldsymbol{\varepsilon}_{\boldsymbol{k},\lambda}^*) - (\boldsymbol{\varepsilon}_{\boldsymbol{k},\lambda}^* \cdot \boldsymbol{x})(\boldsymbol{j}(\boldsymbol{x}) \cdot \boldsymbol{k})\}$$

$$+ \tfrac{1}{2}\{(\boldsymbol{k} \cdot \boldsymbol{x})(\boldsymbol{j}(\boldsymbol{x}) \cdot \boldsymbol{\varepsilon}_{\boldsymbol{k},\lambda}^*) + (\boldsymbol{\varepsilon}_{\boldsymbol{k},\lambda}^* \cdot \boldsymbol{x})(\boldsymbol{j}(\boldsymbol{x}) \cdot \boldsymbol{k})\} \Big] \quad . \qquad (16.77)$$

The first term can also be brought into the form

$$-i\tfrac{1}{2} \int d^3x (\boldsymbol{k} \times \boldsymbol{\varepsilon}_{\boldsymbol{k},\lambda}^*)(\boldsymbol{x} \times \boldsymbol{j}(\boldsymbol{x})) = -\frac{i}{2m}(\boldsymbol{k} \times \boldsymbol{\varepsilon}_{\boldsymbol{k},\lambda}^*) \cdot \boldsymbol{L} \quad . \qquad (16.78)$$

Because of the proportionality to the orbital angular momentum and thus to the orbital part of the magnetic moment operator, the resulting transitions are called *magnetic dipole transitions* $(M1)$. The matrix element

$$\frac{1}{2m}\langle l', m' | (\boldsymbol{k} \times \boldsymbol{\varepsilon}_{\boldsymbol{k},\lambda}^*) \cdot \boldsymbol{L} | l, m\rangle$$

is only different from zero if the selection rules

$$m' - m = \begin{cases} 0 \\ \pm 1 \end{cases} \quad \text{and } l' - l = 0 \qquad (16.79)$$

for magnetic dipole transitions are fulfilled. The former follows from the known properties of the angular momentum, and the latter follows from $[\boldsymbol{L}^2, \boldsymbol{L}] = 0$. For the second term in (16.77), one finds using (16.44b)[5]

---

[5] The index i again labels the electrons. Einstein's summation convention is used for the indices j and l, which label the Cartesian vector components.

$$\frac{1}{2}\varepsilon^*_{\boldsymbol{k},\lambda l}k_j \int d^3x \left(j_l(\boldsymbol{x})x_j + x_l j_j(\boldsymbol{x})\right)$$

$$= \frac{1}{2}\varepsilon^*_{\boldsymbol{k},\lambda l}k_j \frac{1}{2m}\sum_i \left(\{p_{il}, x_{ij}\} + \{x_{il}, p_{ij}\}\right) \quad . \tag{16.80}$$

Using

$$[x_j x_l, H_0] = \frac{\mathrm{i}\hbar}{m}(x_j p_l + p_j x_l) \quad , \tag{16.81}$$

one therefore finds the matrix element for *quadrupole transitions*

$$\frac{1}{2\mathrm{i}}\frac{E_m - E_n}{\hbar}\langle n|\sum_i (\boldsymbol{k}\cdot\boldsymbol{x}_i)(\varepsilon^*_{\boldsymbol{k},\lambda}\cdot\boldsymbol{x}_i)|m\rangle \quad .$$

Here, $x_l x_j$ is the quadrupole moment operator. From (16.58), because of the additional factor $\boldsymbol{k}\cdot\boldsymbol{x}$, it is evident that the amplitudes of the $M1$ and $E2$ transitions are smaller than those of the $E1$ transitions by $ka \approx \alpha = 1/137$. Of course, one also sees this immediately from the matrix elements. The higher-multipole transitions, corresponding to the term of higher order in $\boldsymbol{k}\cdot\boldsymbol{x} \sim \alpha$ in (16.58), are even weaker and are thus only noticeable if the lower-order transitions are forbidden due to selection rules.

Since the perturbation Hamiltonian in the form (16.45) or (16.51c) does not contain the spin of the electrons, the transitions discussed up till now conserve the spin, i.e., they satisfy the additional selection rule

$$\Delta S = S' - S = 0 \quad . \tag{16.82}$$

According to our discussion in Chap. 5, however, there is also a coupling of the radiation field to the spin. By (9.28) and (9.25), we have for this the additional perturbation operator

$$-\frac{ge}{2mc}\sum_i \boldsymbol{S}_i \cdot \boldsymbol{B}(\boldsymbol{x}_i, t)$$

$$= -\frac{ge}{2mc}\sum_i \boldsymbol{S}_i \cdot \boldsymbol{\nabla} \times \left\{ \sum_{\boldsymbol{k},\lambda}\sqrt{\frac{2\pi\hbar c}{k}}\left[a_{\boldsymbol{k},\lambda}\varepsilon_{\boldsymbol{k},\lambda}\frac{\mathrm{e}^{\mathrm{i}(\boldsymbol{k}\cdot\boldsymbol{x}_i - \omega_k t)}}{\sqrt{V}} + \mathrm{h.c.}\right] \right\}$$

$$= -\frac{ge}{2mc}\sum_i \boldsymbol{S}_i \cdot \mathrm{i}\boldsymbol{k} \times \left\{ \sum_{\boldsymbol{k},\lambda}\sqrt{\frac{2\pi\hbar c}{k}}\left[a_{\boldsymbol{k},\lambda}\varepsilon_{\boldsymbol{k},\lambda}\frac{\mathrm{e}^{\mathrm{i}(\boldsymbol{k}\cdot\boldsymbol{x}_i - \omega_k t)}}{\sqrt{V}} - \mathrm{h.c.}\right] \right\} \quad , \tag{16.83}$$

where we use $\boldsymbol{A}(\boldsymbol{x}, t)$ in the representation (16.49). As one sees, this also leads to a matrix element smaller than the matrix element of the electric dipole transition by a factor of $\hbar k/p \approx ka \approx \alpha$. For the spin-dependent dipole transitions, the selection rule (16.82) no longer holds. One might think that (16.83) would give a transition for He from the triplet state $1s2s\ {}^3S_1$ into the singlet state $1s^2\ {}^1S_0$. However, the spatial part of the matrix element is zero, since $|100\rangle|100\rangle$ and $\sum_i \exp\{i\boldsymbol{k}\cdot\boldsymbol{x}_i\}$ are symmetric and $(|100\rangle|2lm\rangle - |2lm\rangle|100\rangle)$ is antisymmetric.

16.4.8 Absorption and Induced Emission

As our final example of the interaction of atomic electrons with the radiation field, we briefly study the absorption of photons and induced emission.

Let us first consider the absorption of a photon with wave number and polarization $(\boldsymbol{k}, \lambda)$. We assume that in the initial state there are just $n_{\boldsymbol{k},\lambda}$ photons of the kind $(\boldsymbol{k}, \lambda)$ present. Then the final state consists of $n_{\boldsymbol{k},\lambda} - 1$ such photons. Let the initial state of the atom be $|n\rangle$ and the final (higher) state $|m\rangle$: The process described results from the first term in (16.51c) and the annihilation operator contained therein. The transition amplitude is given by the matrix element

$$\langle m|\langle n_{\boldsymbol{k},\lambda} - 1|H_1|n_{\boldsymbol{k},\lambda}\rangle|n\rangle = -e\sqrt{\frac{2\pi\hbar n_{\boldsymbol{k},\lambda}}{V\omega_{\boldsymbol{k}}}}\langle m|\boldsymbol{\varepsilon}_{\boldsymbol{k},\lambda}\cdot\boldsymbol{j}_{-\boldsymbol{k}}|n\rangle \quad . \tag{16.84}$$

If this is inserted into the golden rule, one finds for the absorption rate (absorption per unit time)

$$\Gamma^{\text{abs}}_{n\to m} = n_{\boldsymbol{k},\lambda}\frac{(2\pi)^2e^2}{Vck}\delta(E_m - E_n - \hbar ck)|\langle m|\boldsymbol{\varepsilon}_{\boldsymbol{k},\lambda}\cdot\boldsymbol{j}_{-\boldsymbol{k}}|n\rangle|^2 \quad , \tag{16.85}$$

in full analogy to (16.52). The absorption rate increases linearly with the number of incident photons.

Next we discuss the emission process. Now the initial state of the atom is $|m\rangle$ and the final state is $|n\rangle$. Again there are $n_{\boldsymbol{k},\lambda}$ photons incident on the atom, but the final state contains $n_{\boldsymbol{k},\lambda} + 1$ such photons. The relevant matrix element now comes from the creation operator contained in the first term of (16.51c)

$$\langle n|\langle n_{\boldsymbol{k},\lambda} + 1|H_1|n_{\boldsymbol{k},\lambda}\rangle|m\rangle$$
$$= -e\sqrt{\frac{2\pi\hbar(n_{\boldsymbol{k},\lambda} + 1)}{V\omega_{\boldsymbol{k}}}}\langle n|\boldsymbol{\varepsilon}^*_{\boldsymbol{k},\lambda}\cdot\boldsymbol{j}_{\boldsymbol{k}}|m\rangle \quad . \tag{16.86}$$

Employing the golden rule again, one finds for the emission rate

$$\Gamma^{\text{emiss}}_{m\to n} = (n_{\boldsymbol{k},\lambda} + 1)\frac{(2\pi)^2e^2}{Vck}\delta(E_n + \hbar ck - E_m)|\langle n|\boldsymbol{\varepsilon}^*_{\boldsymbol{k},\lambda}\cdot\boldsymbol{j}_{\boldsymbol{k}}|m\rangle|^2 . \tag{16.87}$$

First we note that the dependence on the atomic matrix elements is the same in (16.85) and (16.87)

$$|\langle n|\boldsymbol{\varepsilon}^*_{\boldsymbol{k},\lambda}\cdot\boldsymbol{j}_{\boldsymbol{k}}|m\rangle|^2 = |\langle m|\boldsymbol{\varepsilon}_{\boldsymbol{k},\lambda}\cdot\boldsymbol{j}_{-\boldsymbol{k}}|n\rangle|^2 \quad .$$

If there are no photons present initially, i.e. $n_{\boldsymbol{k},\lambda} = 0$, the emission rate (16.87) reduces to the rate for spontaneous emission (16.52) evaluated in Sect. 16.4.3.

The $n_{\boldsymbol{k},\lambda}$-dependent contribution to $\Gamma^{\mathrm{emiss}}_{m \to n}$ is the rate of induced (or stimulated) emission and is equal in magnitude to the absorption rate $\Gamma^{\mathrm{abs}}_{n \to m}$. This equality is referred to as detailed balancing. In the quantum field theoretic description of the radiation field developed in this section, spontaneous emission and induced emission are just facets of one and the same theoretical treatment.

Equations (16.85), (16.87) are the basis for the relation between the famous "A and B coefficients" of Einstein. Suppose that the atomic levels E_n and E_m are thermally populated according to the Boltzmann factors $e^{-E_n/kT}$ and $e^{-E_m/kT}$, respectively. Then the condition that the emission (gain) and absorption (loss) processes balance each other leads to the Planck radiation law (1.2) for black-body radiation at temperature T.

Problems

16.1 A hydrogen atom is located in a homogeneous electric field $\boldsymbol{E} = (0, 0, E(t))$ with

$$E(t) = \frac{B\tau}{\pi e} \frac{1}{\tau^2 + t^2} \quad .$$

Let the atom be in the ground state at $t = -\infty$. Calculate the probability of finding the atom at the time $t = \infty$ in the 2p-state.

16.2 An electrically charged linear harmonic oscillator in the ground state is suddenly acted upon by a homogeneous electric field E, constant in time from then on. Determine the probability of exciting the particle into the nth state by means of the "sudden approximation". Hint: The potential corresponding to the electric field takes the form $\varphi(x) = -eEx$. Determine first the wave functions for the harmonic oscillator under the influence of this potential. The matrix elements occurring in the transition probability can be computed with the help of the generating function for the Hermite polynomials.

16.3 In β-decay, the nuclear charge number Z of a $(Z - 1)$-times ionized atom changes suddenly to $Z+1$; the effect on the electron wave function can be described with the help of the "sudden approximation". Using the wave functions for an electron in the Coulomb potential of the nucleus, calculate the probabilities for the transition of the electron into the 2s- and 3s-states, provided that the electron was in the ground state before the β-decay.

16.4 A magnetic field $(0, 0, B_z)$ which is constant in time acts on a spin of magnitude $\hbar/2$. In addition, the spin is acted upon by a transverse impulse $(\delta(t)\lambda B_x, 0, 0)$. Solve the Pauli equation

$$i\hbar\dot{\Psi} = H\Psi$$

for this problem, where

$$H = -\frac{\hbar e}{2mc} (B_z\sigma_z + \lambda\delta(t)B_x\sigma_x) \equiv H_0 + V(t) \quad .$$

Hint: Transform to the interaction picture

$$\Psi_I(t) = e^{iH_0 t/\hbar}\Psi(t) \quad ,$$

and solve the resulting equation of motion. By transforming back, obtain the desired wave function.

16.5 A harmonic oscillator

$$H = \frac{p^2}{2m} + \frac{m}{2}\omega^2 x^2$$

is acted upon by an external force described by the potential $V = -xf(t)$, where

$$f(t) = \begin{cases} 0 & -\infty \le t \le t_1 \\ D\cos\Omega t & t_1 \le t \le t_2 \\ 0 & t \ge t_2 \end{cases} \quad .$$

Determine the probability for a transition from the ground state $(t < t_1)$ to the nth excited state $(t > t_2)$. Hint: Use the Heisenberg representation and introduce for x and p creation and annihilation operators.

16.6 Prove the two equations following (16.69), which lead to the selection rule (16.70).

16.7 An electron moves in the (one-dimensional) potential $V(x) = -\lambda\delta(x)$. At the time $t = 0$, the strength of the potential changes suddenly to the value μ $(\lambda, \mu > 0)$. Using the sudden approximation, calculate the probability of a transition from the old ground state to the new ground state. Consider the special case $\mu = \lambda/2$ and discuss $\mu/\lambda \gg 1$ and $\mu/\lambda \ll 1$.

16.8 Galilei transformation: Consider the Schrödinger equation in two inertial systems S and S', whose coordinates are related by

$$x = x' + vt \quad ,$$
$$t = t' \quad .$$

Let the solution of the wave equation in the system S be $\psi(x, t)$. Show that the solution in the system S' is given by

$$\psi' = \psi \exp\left[\frac{i}{\hbar}\left(-m v \cdot x + \frac{m v^2 t}{2}\right)\right]$$
$$= \psi(x' + vt', t') \exp\left[\frac{i}{\hbar}\left(-m v \cdot x' - \frac{m v^2 t'}{2}\right)\right] \quad .$$

Hint: Transform the Schrödinger equation in S' to coordinates x and t; the result then follows in analogy to the gauge transformation.

16.9 The nucleus of a hydrogen atom previously at rest experiences a sudden jolt by which it obtains a velocity v. Let the orbital electron be in the ground state before the collision. What is the probability for the electron to make a transition to the nth excited state due to this collision? Hint: Use the result of the previous problem and expand up to first order in v.

16.10 Prove for a particle in one dimension the oscillator strength sum rule (Thomas–Reiche–Kuhn sum rule)

$$\sum_n (E_n - E_a)|\langle n|x|a\rangle|^2 = \frac{\hbar^2}{2m} \quad .$$

Hint: Consider $\langle a|[\dot{x}, x]|a\rangle$ and compute this expression by direct use of $[p, x] = -i\hbar$ on the one hand and by substitution of $\dot{x} = i[H, x]/\hbar$ on the other hand.

16.11 (a) Prove, assuming potentials depending only on coordinates, the f-sum-rule for N particles,

$$\big[\, [H, \varrho_q], \varrho_q \,\big] = -N q^2 \hbar^2 / m \quad ,$$

where $\varrho_q = \sum_{i=1}^{N} e^{-i q \cdot x_i}$ is the particle density operator in the momentum representation.

(b) Consider the limit $q \to 0$ and, by forming the expectation value in the state $|a\rangle$, derive the Thomas–Reiche–Kuhn sum rule

$$\sum_{i,k} \sum_n (E_n - E_a)\langle a|q \cdot x_i|n\rangle\langle n|q \cdot x_k|a\rangle = N\frac{q^2 \hbar^2}{2m} \quad .$$

Compare with the special case in Problem 16.10.

17. The Central Potential II

17.1 The Schrödinger Equation for a Spherically Symmetric Square Well

In this chapter we will study the spherically symmetric square well (Fig. 17.1),

$$V(r) = \begin{cases} -V_0 & r < a \\ 0 & r > a \end{cases} \quad . \tag{17.1}$$

Firstly, it serves as a simple model of a short-range potential as required in nuclear physics, and secondly, in the course of the discussion we will encounter some of the mathematical preparation for the scattering theory of the next chapter.

The Schrödinger equation for the radial component of the wave function is

$$\left\{ -\frac{\hbar^2}{2m} \left[\frac{d^2}{dr^2} + \frac{2}{r}\frac{d}{dr} - \frac{l(l+1)}{r^2} \right] + V(r) \right\} R(r) = ER(r) \quad . \tag{17.2}$$

The potential is piecewise constant. We first determine the solution in an interval of constant potential with $E > V$ and define

$$k = \sqrt{2m(E - V)}/\hbar \quad . \tag{17.3}$$

Hence (17.2) becomes

$$\left[\frac{d^2}{dr^2} + \frac{2}{r}\frac{d}{dr} - \frac{l(l+1)}{r^2} + k^2 \right] R(r) = 0 \quad , \tag{17.4}$$

and, after substitution of $\varrho = kr$,

$$\left[\frac{d^2}{d\varrho^2} + \frac{2}{\varrho}\frac{d}{d\varrho} - \frac{l(l+1)}{\varrho^2} + 1 \right] R(\varrho) = 0 \quad . \tag{17.4'}$$

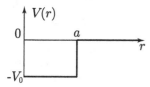

Fig. 17.1. A spherical square well potential

17.2 Spherical Bessel Functions

The differential equation (17.4′) can be solved by elementary methods and leads to the spherical Bessel functions. We first consider the case $l = 0$,

$$\frac{d^2}{d\varrho^2}(\varrho R_0) + \varrho R_0 = 0 \quad , \tag{17.5}$$

and obtain two linearly independent solutions, one regular at $\varrho = 0$,

$$R_0 = \frac{\sin \varrho}{\varrho} \quad , \tag{17.6a}$$

and one singular at $\varrho = 0$,

$$R_0 = -\frac{\cos \varrho}{\varrho} \quad . \tag{17.6b}$$

For $l \neq 0$, considering the behavior found in (17.6a) at small distances, we introduce the substitution

$$R_l = \varrho^l \chi_l \tag{17.7}$$

and obtain from (17.4′) the differential equation

$$\chi_l'' + \frac{2(l+1)}{\varrho}\chi_l' + \chi_l = 0 \quad . \tag{17.8}$$

We now attempt to find a recursion relation for the solutions. If χ_l is a solution of the differential equation (17.8), what is the equation satisfied by $\chi = (1/\varrho)\chi_l'$?

To this end, we differentiate the differential equation (17.8) once, giving

$$\chi_l''' + \frac{2(l+1)}{\varrho}\chi_l'' + \left[1 - \frac{2(l+1)}{\varrho^2}\right]\chi_l' = 0$$

and obtain

$$(\varrho\chi)'' + \frac{2(l+1)}{\varrho}(\varrho\chi)' + \left[1 - \frac{2(l+1)}{\varrho^2}\right]\varrho\chi = 0 \quad ,$$

from which we find

$$\chi'' + \frac{2(l+2)}{\varrho}\chi' + \chi = 0 \quad .$$

This is the differential equation for $l + 1$! Hence,

$$\chi_{l+1} = \frac{1}{\varrho}\frac{d}{d\varrho}\chi_l \tag{17.9}$$

and

$$\chi_l = \left(\frac{1}{\varrho}\frac{d}{d\varrho}\right)^l \chi_0 \quad , \tag{17.10}$$

where χ_0 is given in in (17.6a) and (17.6b).

We summarize the solutions of the differential equation (17.4′) in the following definitions. Starting from (17.6a), we define the *spherical Bessel functions*

$$j_l(\varrho) = (-\varrho)^l \left(\frac{1}{\varrho}\frac{d}{d\varrho}\right)^l \frac{\sin \varrho}{\varrho} \tag{17.11a}$$

and starting from (17.6b) the *spherical Neumann functions*

$$n_l(\varrho) = -(-\varrho)^l \left(\frac{1}{\varrho}\frac{d}{d\varrho}\right)^l \frac{\cos \varrho}{\varrho} \quad . \tag{17.11b}$$

The factor $(-1)^l$ is conventional. The lowest spherical Bessel and Neumann functions are

$$j_0(\varrho) = \frac{\sin \varrho}{\varrho} \quad , \qquad n_0(\varrho) = -\frac{\cos \varrho}{\varrho} \quad ,$$

$$j_1(\varrho) = \frac{\sin \varrho}{\varrho^2} - \frac{\cos \varrho}{\varrho} \quad , \quad n_1(\varrho) = -\frac{\cos \varrho}{\varrho^2} - \frac{\sin \varrho}{\varrho} \quad , \tag{17.12}$$

$$j_2(\varrho) = \left(\frac{3}{\varrho^3} - \frac{1}{\varrho}\right)\sin \varrho - \frac{3}{\varrho^2}\cos \varrho \quad ,$$

$$n_2(\varrho) = -\left(\frac{3}{\varrho^3} - \frac{1}{\varrho}\right)\cos \varrho - \frac{3}{\varrho^2}\sin \varrho \quad .$$

For $\varrho \to 0$, these functions behave like

$$j_l(\varrho) = \frac{\varrho^l}{1 \times 3 \times 5 \times \ldots \times (2l+1)} \quad ,$$

$$n_l(\varrho) = -\frac{1 \times 3 \times 5 \times \ldots \times (2l-1)}{\varrho^{l+1}} \quad , \tag{17.13}$$

whereas the asymptotic behavior for $\varrho \gg 1$, $\varrho \gg l$ is given by

$$j_l(\varrho) \simeq \frac{1}{\varrho}\sin\left(\varrho - \frac{l\pi}{2}\right) \quad , \quad n_l(\varrho) \simeq -\frac{1}{\varrho}\cos\left(\varrho - \frac{l\pi}{2}\right) \quad . \tag{17.14}$$

The *spherical Hankel functions* are linear combinations of the spherical Bessel and Neumann functions. The first Hankel function is defined by

$$h_l^{(1)}(\varrho) = j_l(\varrho) + i n_l(\varrho) \tag{17.15a}$$

and the second by

$$h_l^{(2)}(\varrho) = h_l^{(1)}(\varrho)^* \quad . \tag{17.15b}$$

Inversion gives

$$j_l(\varrho) = \tfrac{1}{2}(h_l^{(1)} + h_l^{(2)}) = \operatorname{Re} h_l^{(1)} \quad, \tag{17.16a}$$

$$n_l(\varrho) = \frac{1}{2i}(h_l^{(1)} - h_l^{(2)}) = \operatorname{Im} h_l^{(1)} \quad. \tag{17.16b}$$

In comparison to plane waves, j_l, n_l, and $h_l^{(1)}$ correspond to $\sin kx, \cos kx$, and $\exp\{ikx\}$.

The lowest Hankel functions are

$$h_0^{(1)}(\varrho) = \frac{e^{i\varrho}}{i\varrho} \quad,$$

$$h_1^{(1)}(\varrho) = -\frac{e^{i\varrho}}{\varrho}\left(1 + \frac{i}{\varrho}\right) \quad, \tag{17.17}$$

$$h_2^{(1)}(\varrho) = \frac{ie^{i\varrho}}{\varrho}\left(1 + \frac{3i}{\varrho} - \frac{3}{\varrho^2}\right) \quad.$$

The asymptotic behavior is given by

$$h_l^{(1)}(\varrho) \simeq -\frac{i}{\varrho}e^{i(\varrho - l\pi/2)} \quad. \tag{17.18}$$

In Chap. 18, we will use the abbreviated notation $h_l \equiv h_l^{(1)}$ together with (17.15b).

17.3 Bound States of the Spherical Potential Well

If the energy lies in the interval $-V_0 < E < 0$, the radial Schrödinger equation becomes

$$\left[\frac{d^2}{dr^2} + \frac{2}{r}\frac{d}{dr} - \frac{l(l+1)}{r^2} + \binom{q^2}{-\kappa^2}\right]R = 0 \quad, \tag{17.19}$$

where

$$q = \sqrt{2m(V_0 + E)}/\hbar \quad, \quad \kappa = \sqrt{2m(-E)}/\hbar \tag{17.20}$$

have been introduced inside and outside the well. The solution regular at the origin is

$$R(r) = A\, j_l(qr) \quad, \quad 0 \leq r \leq a \quad. \tag{17.21a}$$

Outside the well, due to the normalizability of the wave function, only the solution which decreases exponentially for $r \to \infty$ is admissible,

$$R(r) = B h_l^{(1)}(i\kappa r) \quad, \quad a < r \quad. \tag{17.21b}$$

The continuity conditions give

$$A j_l(qa) = B h_l^{(1)}(i\kappa a) \quad , \tag{17.22a}$$

$$A j_l'(qa)q = B h_l^{(1)'}(i\kappa a)i\kappa \quad . \tag{17.22b}$$

After division of (17.22b) by (17.22a), we obtain for the logarithmic derivatives

$$q \frac{d \log j_l}{d\varrho}\bigg|_{qa} = i\kappa \frac{d \log h_l^{(1)}}{d\varrho}\bigg|_{i\kappa a} \quad . \tag{17.23}$$

For $l = 0$, the solution is

$$u(r) = rR(r) = \begin{cases} A \sin qr & \text{for } r < a \\ B e^{-\kappa r} & \text{for } r > a \end{cases} \quad . \tag{17.24}$$

The continuity of $R(r)$ and $R'(r)$ implies that $u(r)$ and $u'(r)$ must also be continuous; hence, one immediately obtains the continuity condition

$$\cot qa = -\frac{(2m|E|)^{1/2}}{\hbar q} \quad . \tag{17.25}$$

Consistent with our general considerations in Sect. 6.2, the energy eigenvalues following from (17.25) coincide with those of the odd states of the one-dimensional potential well, Sect. 3.4.2. In order for a bound state to exist, the potential must have, according to Eq. (3.91), a minimum strength of

$$V_{0 \min} = \frac{\pi^2}{8} \frac{\hbar^2}{ma^2} \quad . \tag{17.26}$$

One bound state exists for $\pi/2 = 1.57 < \sqrt{(2mV_0a^2)/\hbar^2} < 3\pi/2$, two bound states exist for $3\pi/2 = 4.71 < \sqrt{(2mV_0a^2)/\hbar^2} < 5\pi/2 = 7.85$, etc., see Fig. 17.2.

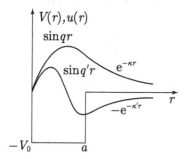

Fig. 17.2. A square well potential $V(r)$ (*thin*) possessing two s-bound states $u(r)$ (*thick*)

17.4 The Limiting Case of a Deep Potential Well

We now assume $qa \gg l$. One can then use the asymptotic formula (17.14) on the left-hand side of the transcendental equation (17.23), and using

$$\frac{d}{d\varrho} \frac{\sin (\varrho - l\pi/2)}{\varrho} = -\frac{1}{\varrho^2} \sin \left(\varrho - \frac{l\pi}{2} \right) + \frac{\cos (\varrho - l\pi/2)}{\varrho} \quad ,$$

one obtains the condition

$$-\frac{1}{a} + q \cot \left(qa - \frac{l\pi}{2} \right) = \mathrm{i}\kappa \frac{d \log h_l^{(1)}(\varrho)}{d\varrho} \bigg|_{\mathrm{i}\kappa a} \quad .$$

The right-hand side of this equation does not depend on V_0. Thus, for $V_0 \gg |E|$, the factor multiplying q must be close to zero, i.e., $\cot (qa - l\pi/2) \approx 0$, in order that in the limit $V_0 \to \infty$ the term $q \cot (qa - l\pi/2)$ remain finite. Hence, for this limiting case,

$$qa - \frac{l\pi}{2} \approx (n + \tfrac{1}{2})\pi \quad . \tag{17.27}$$

For large q, or $|E| \ll V_0$, we can expand q, in (17.20), in terms of E. If we truncate after the linear term,

$$q \approx \left(\frac{2mV_0}{\hbar^2} \right)^{1/2} \left(1 + \frac{E}{2V_0} \right) \quad ,$$

we obtain from (17.27)

$$\frac{E}{2V_0} = -1 + \pi \left(\frac{\hbar^2}{2mV_0 a^2} \right)^{1/2} \left(n + \frac{1}{2} + \frac{l}{2} \right) \quad . \tag{17.28}$$

One sees immediately that in this limit the states with equal $n + l/2$ are degenerate. Formula (17.28) holds only for large n, since only then is the assumption $qa \gg l$ satisfied. However, in order to obtain a rough overview, we can calculate (17.28) even for small n, as is summarized in Table 17.1. From this, the following order of the energy levels would result: $1S$; $1P$; $1D$, $2S$; $1F$, $2P$; $1G$, $2D$, $3S$; .... Degenerate eigenvalues are separated only by commas.

Table 17.1. Values of $(n + l/2)$

l	0	1	2	3
n				
1	1	3/2	2	5/2
2	2	5/2	3	7/2
3	3	7/2	4	9/2

For the *infinitely deep* potential well

$$V(r) = \begin{cases} 0 & r < a \\ \infty & r > a \end{cases} \quad , \tag{17.29}$$

the stationary solutions of the Schrödinger equation simplify to

$$R(r) = \begin{cases} Aj_l(qr) & \text{for} \quad r < a \\ 0 & \text{for} \quad r > a \end{cases} \quad , \tag{17.30}$$

where

$$q = \left(\frac{2mE}{\hbar^2} \right)^{1/2} \quad . \tag{17.31}$$

The continuity condition is then given by the transcendental equation

$$j_l(qa) = 0 \quad , \tag{17.32}$$

i.e., the wave function vanishes at the infinitely high barrier. The permissible values of qa and thus the energy eigenvalues $(\hbar q)^2/2m$ are obtained from the zeros of the Bessel functions. These are known, and we list the lowest ones in Table 17.2. The quantity N labels the sequence of zeros; it also gives the number of nodes for this wave number and corresponds to the radial quantum number of the Coulomb wave functions.

Table 17.2. Zeros of the Bessel functions, qa

l N	S 0	P 1	D 2	F 3	G 4	H 5
1	3.14	4.49	5.76	6.99	8.18	9.36
2	6.28	7.73	9.10	10.42		
3	9.42					

We label the eigenvalues for each l consecutively by the radial quantum number $N = 1, 2, \ldots$. The following sequence of energies results:

$$1S, \ 1P, \ 1D, \ 2S, \ 1F, \ 2P, \ 1G, \ 2D, \ 1H, \ 3S, \ 2F \quad .$$

Atomic nuclei consist of nucleons, that is, protons and neutrons. In the *shell model* of nuclear structure, one assumes that every nucleon moves in a short range, rotationally symmetric potential generated by the other nucleons. Let us assume as the simplest model a spin independent, deep potential well. Since nucleons are fermions, the occupation number of every energy level can reach $2 \times (2l + 1)$. A shell for protons (neutrons) is filled on the basis of the sequence of energy levels found in Table 17.1 for proton (neutron) number 2, $8 \ (= 2 + 2 \times 3)$, $18 \ (= 2 + 2 \times 3 + 2 \times 5)$, $20 \ (= 18 + 2)$, $34 \ (= 20 + 2 \times 7)$, 40, 58, 68, 90, 106. These numbers are called "*magic numbers*". Nuclei with magic proton number and neutron number N are called doubly magic nuclei.

These are especially stable. In comparison to their neighboring nuclei, they have an anomalously small tendency to bind a further nucleon, and their excited states lie anomalously high.

In nature, things are of course more complicated due to the effective spin–orbit interaction of the nucleons. The levels are shifted for this reason and the true magic numbers are[1]

2, 8, 20, 28, 50, 82 and 126   .

With the nomenclature $_Z^A\text{El}_N$, $A = Z + N$, $_2^4\text{He}_2$, $_8^{16}\text{O}_8$, $_{20}^{40}\text{Ca}_{20}$, $_{82}^{208}\text{Pb}_{126}$, are doubly magic. The ^4He nucleus (identical to the α-particle) cannot bind an additional nucleon. The next stable nuclei are $_3^6\text{Li}_3$, $_3^7\text{Li}_4$.

17.5 Continuum Solutions for the Potential Well

As preparation for the scattering theory of the next chapter, we also study the continuum states of the spherical potential well. For $E > 0$, (17.3) and (17.4) yield the wave function

$$R_l(r) = \begin{cases} Aj_l(qr) & r < a \\ Bj_l(kr) + Cn_l(kr) & r > a \end{cases} , \qquad (17.33)$$

where

$$k = \sqrt{2mE}/\hbar \quad \text{and} \quad q = \sqrt{2m(E + V_0)}/\hbar \quad . \qquad (17.34)$$

The matching condition at a is

$$q\frac{dj_l/d\varrho}{j_l}\bigg|_{\varrho=qa} = k\left[\frac{Bdj_l/d\varrho + Cdn_l/d\varrho}{Bj_l + Cn_l}\right]_{\varrho=ka} . \qquad (17.35)$$

This yields the ratio C/B. Asymptotically, one has

$$R_l(r) = \frac{B}{kr}\left[\sin\left(kr - \frac{l\pi}{2}\right) - \frac{C}{B}\cos\left(kr - \frac{l\pi}{2}\right)\right] .$$

Introducing for the amplitude ratio the notation $C/B = -\tan\delta_l(k)$, we find for the asymptotic form of $R_l(r)$

$$R_l(r) = \frac{B}{\cos\delta_l(k)}\frac{1}{kr}\sin\left(kr - \frac{l\pi}{2} + \delta_l(k)\right) . \qquad (17.36)$$

In comparison to a free spherical wave without potential, this is a phase-shifted spherical wave.

---

[1] E. Segré: *Nuclei and Particles*, 2nd edn. (Benjamin, New York, Amsterdam 1977) p. 281

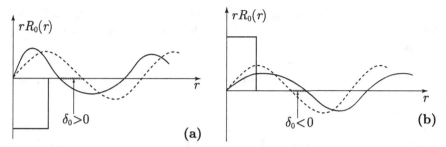

Fig. 17.3a,b. Wave functions $u(r) = rR_0$. (a) For attractive potential $V < 0$; $\delta_0 > 0$, the wave function is pulled in. (b) For a repulsive potential, $V > 0, \delta_0 < 0$, the wave function is pushed out; (- - -): indicates the wave function for $V = 0$

For $l = 0$, one can easily calculate the phase shift δ_0. From (17.35) one obtains

$$q \cot qa = k\frac{B \cos ka + C \sin ka}{B \sin ka - C \cos ka} = k\frac{\cos ka \cos \delta_0 - \sin ka \sin \delta_0}{\sin ka \cos \delta_0 + \cos ka \sin \delta_0}$$

$$= k \cot (ka + \delta_0) \quad ,$$

and hence

$$\delta_0 = \arctan\left(\frac{k}{q} \tan qa\right) - ka \quad . \tag{17.37}$$

From (17.37) one obtains the phase shift as a function of energy and potential strength. The sign of the phase shift can be understood even without computation by physical considerations. For negative potential, the wave number inside of the potential is raised and the wave function in the exterior region is pulled in, i.e., $\delta_0 > 0$. For a repulsive potential, the wave number is decreased in the interior, and the wave function is pushed out in comparison with the force-free wave function. This is shown in Fig. 17.3 for s-waves.

17.6 Expansion of Plane Waves in Spherical Harmonics

We now derive the expansion of plane waves in spherical harmonics, which is important for the theory of scattering and diffraction. It is of course possible to expand plane waves in spherical solutions of the free Schrödinger equation:

$$e^{i\boldsymbol{k} \cdot \boldsymbol{x}} = \sum_{l=0}^{\infty} \sum_{m=-l}^{l} c_{lm}(\boldsymbol{k}) \, j_l(kr) Y_{lm}(\vartheta, \varphi) \tag{17.38}$$

with expansion coefficients $c_{lm}(\boldsymbol{k})$ to be determined. We first specialize to the case $\boldsymbol{k} \parallel \boldsymbol{e}_z$, i.e. $\boldsymbol{k} \cdot \boldsymbol{x} = k \cos \vartheta$. Since the left-hand side then no longer contains φ, only

$$Y_{l0}(\vartheta, \varphi) = \left(\frac{2l+1}{4\pi} \right)^{1/2} P_l(\cos \vartheta) \tag{17.39}$$

occurs on the right:

$$\mathrm{e}^{\mathrm{i}kr \cos \vartheta} = \sum_{l=0}^{\infty} \left(\frac{2l+1}{4\pi} \right)^{1/2} A_l \, j_l(kr) P_l(\cos \vartheta) \quad . \tag{17.40}$$

Using the orthogonality relation for the Legendre polynomials, which follow from (5.28),

$$\int_{-1}^{1} d \cos \vartheta \, P_l(\cos \vartheta) P_{l'}(\cos \vartheta) = \frac{2\delta_{ll'}}{2l+1} \quad , \tag{17.41}$$

and multiplying (17.40) by $P_l(\cos \vartheta)$, and integrating over ϑ, we obtain

$$A_l j_l(kr) = \tfrac{1}{2} [4\pi(2l+1)]^{1/2} \int_{-1}^{1} dz \, P_l(z) \mathrm{e}^{\mathrm{i}krz} \quad . \tag{17.42}$$

Since A_l is independent of r, we can go to the limiting case of small r and compute the leading term for $kr \to 0$. The right-hand side becomes, upon repeated use of (5.24) and (17.41),

$$\int_{-1}^{1} dz \, P_l(z) \mathrm{e}^{\mathrm{i}krz} = \int_{-1}^{1} dz \, P_l(z) \left[\ldots + \frac{(\mathrm{i}krz)^l}{l!} + \frac{(\mathrm{i}krz)^{l+1}}{(l+1)!} + \ldots \right]$$

$$= (\mathrm{i}kr)^l \frac{2^l l!}{(2l)!} \int_{-1}^{1} dz \, P_l(z) P_l(z) + O((kr)^{l+1})$$

$$= \frac{(\mathrm{i}kr)^l \, 2^{l+1} \, l!}{(2l+1)!} + O((kr)^{l+1}) \quad .$$

With (17.13), the left-hand side of (17.42) is in the limit of small kr

$$A_l \frac{2^l l!}{(2l+1)!} (kr)^l \quad ,$$

and hence A_l becomes

$$A_l = \mathrm{i}^l [4\pi(2l+1)]^{1/2} \quad . \tag{17.43}$$

This yields the expansion

$$\mathrm{e}^{\mathrm{i}kr \cos \vartheta} = \sum_{l=0}^{\infty} \mathrm{i}^l (2l+1) j_l(kr) P_l(\cos \vartheta) \quad . \tag{17.44}$$

As a side result, we obtain the integral representation for $j_l(x)$,

$$j_l(x) = (-\mathrm{i})^l \tfrac{1}{2} \int_{-1}^{1} dz \, P_l(z) \mathrm{e}^{\mathrm{i}xz} \quad . \tag{17.45}$$

For an arbitrary direction $\boldsymbol{k}$, one substitutes the addition theorem for the spherical harmonics, Eq. (5.32),

$$P_l(\cos \vartheta) = \frac{4\pi}{2l+1} \sum_{m=-l}^{l} Y_{lm}(\Omega_{\boldsymbol{k}})^* Y_{lm}(\Omega_{\boldsymbol{x}})$$

into (17.44) and finds for a *general plane wave*

$$e^{i\boldsymbol{k} \cdot \boldsymbol{x}} = 4\pi \sum_{l=0}^{\infty} \sum_{m=-l}^{l} i^l j_l(kr) Y_{lm}(\Omega_{\boldsymbol{k}})^* Y_{lm}(\Omega_{\boldsymbol{x}}) \quad . \tag{17.46}$$

Here, the polar angles of $\boldsymbol{x}$ are written as $\Omega_{\boldsymbol{x}} = (\vartheta, \varphi)$ and likewise for $\boldsymbol{k}$: $\Omega_{\boldsymbol{k}} = (\vartheta_{\boldsymbol{k}}, \varphi_{\boldsymbol{k}})$.

Relation to the Bessel functions

We claim that

$$j_l(\varrho) = \sqrt{\frac{\pi}{2\varrho}} J_{l+1/2}(\varrho) \quad , \tag{17.47}$$

where $J_{l+1/2}$ is the *Bessel function* of index $l + 1/2$.

Substituting this and the derivatives

$$j_l' = \sqrt{\frac{\pi}{2}} \left(-\frac{1}{2}\varrho^{-3/2} J_{l+1/2} + \varrho^{-1/2} J_{l+1/2}' \right) \quad ,$$

$$j_l'' = \sqrt{\frac{\pi}{2}} \left(\frac{3}{4}\varrho^{-5/2} J_{l+1/2} - \varrho^{-3/2} J_{l+1/2}' + \varrho^{-1/2} J_{l+1/2}'' \right)$$

into the differential equation (17.4$'$) for j_l, one finds

$$J_{l+1/2}'' + \frac{1}{\varrho} J_{l+1/2}' + \left[1 - \frac{(l+1/2)^2}{\varrho^2} \right] J_{l+1/2} = 0 \quad . \tag{17.48}$$

This is the differential equation for the Bessel functions, (11.32), whereby the assertion (17.47) has been shown. The Bessel functions occur in cylindrically symmetric problems.

The following *orthogonality and completeness relations* hold:

$$\int dr\, r^2 \, j_l(kr) j_l(k'r) = \frac{2\pi}{(2k)^2} \delta(k - k') \quad , \tag{17.49}$$

$$\int dr\, r^2 d\Omega\, j_l(kr) Y_{lm}^*(\Omega)\, j_{l'}(k'r) Y_{l'm'}(\Omega) = \frac{2\pi}{(2k)^2} \delta(k-k') \delta_{ll'} \delta_{mm'} \quad , \tag{17.50}$$

$$\int dk\, k^2 \sum_{l=0}^{\infty} \sum_{m=-l}^{l} \psi_{lm}(r,\Omega;k)\psi_{lm}(r',\Omega';k)^*$$

$$= \frac{1}{r^2 \sin\vartheta}\delta(r-r')\delta(\vartheta-\vartheta')\delta(\varphi-\varphi') \quad , \tag{17.51}$$

where

$$\psi_{lm}(r,\Omega;k) = \frac{2k}{\sqrt{2\pi}} j_l(kr) Y_{lm}(\Omega) \quad . \tag{17.52}$$

Problems

17.1 Determine the stationary states and the energy eigenvalues for the three-dimensional spherical harmonic oscillator

$$V(r) = \frac{m\omega^2}{2} x^2 \quad .$$

(a) by using creation and annihilation operators,

(b) by going over to polar coordinates.

17.2 Investigate the bound states of the δ-shell potential

$$V(r) = -\lambda \frac{\hbar^2}{2m}\delta(r-a) \quad .$$

It is useful to introduce the dimensionless variables $y = r/a$, $\xi = ka$, and $g = \lambda a$. It turns out that there is at most one bound state for each l.

(a) Determine the s-wave function. Show that a bound state exists only for $g > 1$.

(b) Show that there is at most one bound state corresponding to each l.

(c) Show for general l that the minimum strength of the potential for the existence of a bound state is $g = 2l + 1$.

17.3 A particle moves under the influence of the nonlocal, separable potential

$$V(\boldsymbol{x},\boldsymbol{x}') = \lambda\varrho(|\boldsymbol{x}|)\,\varrho(|\boldsymbol{x}'|) \quad ,$$

where $\int d^3x\, \varrho(|\boldsymbol{x}|) = 1$ and $\int d^3x\, |\varrho(|\boldsymbol{x}|)|^2 < \infty$ are assumed. The potential term in the Schrödinger equation then has the form $\int d^3x'\, V(\boldsymbol{x},\boldsymbol{x}')\psi(\boldsymbol{x}')$, so that the time independent Schrödinger equation becomes

$$\frac{-\hbar^2}{2m}\nabla^2\psi(\boldsymbol{x}) + \int d^3x'\, V(\boldsymbol{x},\boldsymbol{x}')\,\psi(\boldsymbol{x}') = E\psi(\boldsymbol{x}) \quad .$$

Determine the bound states and the condition for the existence of a bound state, and show that at most one bound state exists. Hint: It is useful to go over to the momentum representation.

18. Scattering Theory

To gain information about the structure of matter, ranging from elementary particles to solids, one studies the scattering of particles (electrons, neutrons, He atoms, photons) (Fig. 18.1).

The particle incident on the target will be represented by a wave packet. This must be large compared to the dimensions of the scatterer, but small compared to the remaining spatial dimensions, in order not to overlap simultaneously with the scatterer and the detector. Its width in momentum space must be sufficiently narrow, so that the spreading during the experiment is negligible, Eq. (2.108). After the interaction, the wave function consists of an unscattered part propagating in the forward direction as well as a scattered part. In what follows, we consider a single scattering center described by a potential $V(\boldsymbol{x})$, which is taken to be situated at the origin of coordinates.

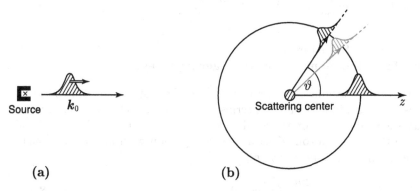

(a)   **(b)**

Fig. 18.1. The scattering of a wave packet: (a) before, (b) after the scattering

18.1 Scattering of a Wave Packet and Stationary States

18.1.1 The Wave Packet

Suppose that the incident wave packet leaves the source, far to the left of the potential, at time t_0 and is represented by

$$\psi_0(\boldsymbol{x}, t_0) = \int \frac{d^3k}{(2\pi)^3} e^{i\boldsymbol{k} \cdot \boldsymbol{x}} a_{\boldsymbol{k}} \quad . \tag{18.1}$$

Suppose the maximum of a_k lies at k_0, so that the wave packet moves with the velocity $v = \hbar k_0/m$ towards the target. We seek the wave function $\psi(x, t)$ for later times, in particular, after the interaction with the scattering center.

Let $\psi_k(x)$ be the exact eigenstates of the Hamiltonian for the potential $V(x)$ with energy

$$E_k = \frac{\hbar^2 k^2}{2m} \geq 0 \quad , \tag{18.2a}$$

$$\left[-\frac{\hbar^2}{2m}\nabla^2 + V(x) \right] \psi_k(x) = E_k \psi_k(x) \quad . \tag{18.2b}$$

Instead of plane waves, we can also expand $\psi_0(x, t_0)$ using the eigenstates $\psi_k(x)$ as:

$$\psi_0(x, t_0) = \int \frac{d^3 k}{(2\pi)^3} \psi_k(x) A_k \quad . \tag{18.3}$$

Here, we have introduced new expansion coefficients A_k, which we will determine later. In expansion (18.3), only states corresponding to a wave incident from the left and an outgoing scattered wave appear. No bound states occur, since these fall off exponentially.

Starting with (18.3), we immediately find the time evolution of $\psi_0(x, t_0)$

$$\psi(x, t) = \int \frac{d^3 k}{(2\pi)^3} \psi_k(x) A_k e^{-iE_k(t-t_0)/\hbar}. \tag{18.4}$$

18.1.2 Formal Solution
of the Time Independent Schrödinger Equation

We must now determine the general structure of the stationary states $\psi_k(x)$. To this end, we add and subtract terms in (18.2b) in such a way as to interchange the second term on the left-hand side and the term on the right-hand side. We then are confronted with a wave equation with an inhomogeneity (source term) dependent on $\psi_k(x)$:

$$(\nabla^2 + k^2)\psi_k(x) = \frac{2m}{\hbar^2} V(x)\psi_k(x) \quad . \tag{18.2b$'$}$$

In order to obtain a formal solution, or, more precisely, in order to transform to an integral equation, we make use of the retarded *Green's function* $G_+(x)$ of the free Schrödinger equation (wave equation, see also Appendix A.3)

$$(\nabla^2 + k^2)G_+(x) = \delta^{(3)}(x) \quad , \tag{18.5}$$

and hence obtain from (18.2b$'$)

$$\psi_k(x) = e^{ik \cdot x} + \frac{2m}{\hbar^2} \int d^3 x' G_+(x - x')V(x')\psi_k(x') \quad . \tag{18.6}$$

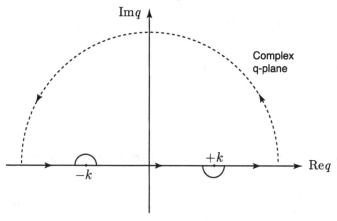

Fig. 18.2. The path of integration for the retarded Green's function

The first term on the right-hand side is a solution of the free Schrödinger equation. Thus, $\psi_k(x)$ is the sum of an incident plane wave and a scattered wave. We now compute $G_+(x)$:

$$G_+(x) = \int \frac{d^3q}{(2\pi)^3} \frac{e^{iq \cdot x}}{k^2 - q^2} = -\frac{1}{4\pi^2 ir} \int_{-\infty}^{+\infty} \frac{dq\, q e^{iqr}}{q^2 - k^2} \quad . \tag{18.7}$$

The expression after the first equality is verified by applying the operator $(\nabla^2 + k^2)$ to (18.5) or by Fourier transformation of (18.5). The poles of the integrand lie in the complex q-plane at $q = \pm k$. The path of integration is shown in Fig. 18.2. Since $r > 0$, one can close the path in the upper half-plane. As discussed in more detail in Appendix A.3, the path along the real q-axis in Fig. 18.2 is chosen in such a way that, in accordance with the physical situation, an outgoing spherical wave results. Indeed, by means of the residue theorem, we immediately obtain

$$G_+(x) = -\frac{1}{4\pi} \frac{e^{ikr}}{r} \quad . \tag{18.7'}$$

The quantity $G_+(x - x')$ represents an outgoing spherical wave emitted from x'.

Substituting (18.7') into (18.6), we find

$$\psi_k(x) = e^{ik \cdot x} - \frac{m}{2\pi\hbar^2} \int d^3x' \frac{e^{ik|x-x'|}}{|x - x'|} V(x')\psi_k(x') \quad , \tag{18.8}$$

an integral equation for $\psi_k(x)$, from which general conclusions concerning the structure of the scattering phenomenon can be drawn. On considering the time dependence, by inserting the factor $\exp\{-iE_k t/\hbar\}$, one sees that (18.8) is the sum of a plane wave incident from the left and a wave moving outward. The detectors are located far from the scatterer, i.e., $|\boldsymbol{x}| \gg |\boldsymbol{x}'|$, so that

$$k|\boldsymbol{x} - \boldsymbol{x}'| = k\sqrt{\boldsymbol{x}^2 - 2\boldsymbol{x} \cdot \boldsymbol{x}' + \boldsymbol{x}'^2} \approx kr - k\frac{\boldsymbol{x}}{r} \cdot \boldsymbol{x}' = kr - \boldsymbol{k}' \cdot \boldsymbol{x}' \quad ,$$

where

$$\boldsymbol{k}' = k\frac{\boldsymbol{x}}{r} \quad .$$

Thus, far from the scatterer, by (18.8), the stationary solution $\psi_{\boldsymbol{k}}(\boldsymbol{x})$ takes the form

$$\psi_{\boldsymbol{k}}(\boldsymbol{x}) = e^{i\boldsymbol{k} \cdot \boldsymbol{x}} + \frac{e^{ikr}}{r} f_{\boldsymbol{k}}(\vartheta, \varphi) \quad , \tag{18.9}$$

where

$$f_{\boldsymbol{k}}(\vartheta, \varphi) = -\frac{m}{2\pi\hbar^2} \int d^3x' e^{-i\boldsymbol{k}' \cdot \boldsymbol{x}'} V(\boldsymbol{x}')\psi_{\boldsymbol{k}}(\boldsymbol{x}') \tag{18.10}$$

has been defined as the *scattering amplitude* It depends only on the direction $\boldsymbol{x}/r$, i.e. ϑ and φ, but not on the distance and has the dimensions of a length. Computing the current density for the wave function $\exp\{ikr\}/r$:

$$\boldsymbol{e}_r \frac{\hbar}{m} \operatorname{Im}\left(\psi^* \frac{\partial}{\partial r}\psi\right) = \frac{\hbar k}{mr^2}\boldsymbol{e}_r \quad ,$$

one sees that the second term in (18.9) is an outgoing spherical wave. Equation (18.9) constitutes the general form of the stationary scattering solutions.

18.1.3 Asymptotic Behavior of the Wave Packet

We can now continue with the calculation of the time evolution of the *wave packet* (18.4). We first determine the coefficients $A_{\boldsymbol{k}}$ entering the expansion (18.3) of the wave packet in terms of the exact eigenfunctions $\psi_{\boldsymbol{k}}(\boldsymbol{x})$ of the scatterer. For this, we express $\exp\{i\boldsymbol{k} \cdot \boldsymbol{x}\}$ in (18.1) in terms of the solution (18.8):

$$\psi_0(\boldsymbol{x}, t_0) = \int \frac{d^3k}{(2\pi)^3} a_{\boldsymbol{k}}\left[\psi_{\boldsymbol{k}}(\boldsymbol{x}) + \frac{m}{2\pi\hbar^2}\int d^3x' \frac{e^{ik|\boldsymbol{x}-\boldsymbol{x}'|}}{|\boldsymbol{x} - \boldsymbol{x}'|} V(\boldsymbol{x}')\psi_{\boldsymbol{k}}(\boldsymbol{x}')\right] \quad . \tag{18.11}$$

Since $k_0 \gg |\boldsymbol{k} - \boldsymbol{k}_0|$, we can write approximately

$$k = \sqrt{(\boldsymbol{k}_0 + \boldsymbol{k} - \boldsymbol{k}_0)^2} \approx \sqrt{k_0^2 + 2\boldsymbol{k}_0 \cdot (\boldsymbol{k} - \boldsymbol{k}_0)} \approx \hat{\boldsymbol{k}}_0 \cdot \boldsymbol{k} \quad ,$$

where $\hat{\boldsymbol{k}}_0 = \boldsymbol{k}_0/k_0$.

The second term in (18.11) contains the k-integral

$$\int \frac{d^3k}{(2\pi)^3} a_k e^{ik|x-x'|} \psi_k(x') \approx \int \frac{d^3k}{(2\pi)^3} a_k e^{ik\cdot\hat{k}_0|x-x'|} \psi_{k_0}(x')$$
$$= \psi_0(\hat{k}_0|x-x'|, t_0)\psi_{k_0}(x') = 0 \quad .$$

The first factor vanishes, since the vector $\hat{k}_0|x-x'|$ is to the right of the potential, while the wave packet is localized at the source at time t_0. Hence, by (18.11) and (18.3),

$$\psi_0(x, t_0) = \int \frac{d^3k}{(2\pi)^3} a_k \psi_k(x) \quad \text{and} \quad A_k = a_k \quad . \tag{18.12}$$

Substitution into (18.4) gives for the wave packet

$$\psi(x, t) = \int \frac{d^3k}{(2\pi)^3} a_k \psi_k(x) e^{-iE_k(t-t_0)/\hbar} \quad . \tag{18.13}$$

For large distances from the scatterer, we can substitute the asymptotic formula (18.9) into (18.13),

$$\psi(x, t) = \psi_0(x, t)$$
$$+ \int \frac{d^3k}{(2\pi)^3} a_k \frac{\exp\{i(kr - E_k(t-t_0))/\hbar\}}{r} f_k(\vartheta, \varphi) \quad , \tag{18.14}$$

where the first term

$$\psi_0(x, t) = \int \frac{d^3k}{(2\pi)^3} a_k \exp\{ik\cdot x - iE_k(t-t_0)/\hbar\}$$
$$\approx \psi_0(x - v(t-t_0), t_0) \tag{18.15}$$

is identical to the wave packet which would result at time t in the absence of the scatterer. Recall Eq. (2.105). Again using the above approximate expression for $k \approx k_0 \cdot k$, the second term in (18.14) can be rearranged , so that using $f_k(\vartheta, \varphi) \approx f_{k_0}(\vartheta, \varphi)$ we gain altogether

$$\psi(x, t) = \psi_0(x, t) + \frac{f_{k_0}(\vartheta, \varphi)}{r} \psi_0(\hat{k}_0 r, t) \quad . \tag{18.16}$$

In the radial direction, $\psi_0(\hat{k}_0 r, t)$ has the the same shape as the incident wave packet. The wave function after the scattering is a linear *superposition* of the transmitted wave packet and the scattered wave, deflected to all solid angles according to $f_{k_0}(\vartheta, \varphi)/r$ (Fig. 18.1). Equation (18.16) does not hold in the following circumstances

(a) if narrow scattering resonances are present, causing a strong deformation of the wave packet (see Sect. 3.7);
(b) for the long-range Coulomb potential, for which a different r-dependence results (see Sect. 18.11.2).

18.2 The Scattering Cross Section

The *differential* scattering cross section gives the number of particles scattered into the element of solid angle $d\Omega$ around Ω, divided by $d\Omega$ and by the number of incident particles per cm^2:

$$\frac{d\sigma}{d\Omega} = \frac{dN(\Omega)}{N_{\text{in}}d\Omega} \quad . \tag{18.17}$$

Here, N_{in} signifies the number of incident particles and $dN(\Omega)$ the number of particles scattered into the element of solid angle $d\Omega$,

$$N_{\text{in}} = \int_{-\infty}^{\infty} dt\, j_{\text{in}} \quad , \quad dN(\Omega) = \int_{-\infty}^{\infty} dt\, j_r r^2 d\Omega \quad , \tag{18.18}$$

where j_{in} is the incident current density,

$$j_{\text{in}} = \frac{\hbar}{2mi}(\psi_0^* \boldsymbol{\nabla}\psi_0 - \psi_0 \boldsymbol{\nabla}\psi_0^*) \approx \frac{\hbar k_0}{m}|\psi_0(\boldsymbol{x},t)|^2 \quad . \tag{18.19}$$

As a consequence,

$$N_{\text{in}} = \frac{\hbar k_0}{m} \int_{-\infty}^{\infty} dt\, |\psi_0(\boldsymbol{x}_{\text{source}},t)|^2 \quad . \tag{18.20}$$

The outgoing radial component of the current density is[1]

$$\begin{aligned}
j_r &= \frac{\hbar}{m}\, \text{Im}\left(\frac{f^*}{r}\psi_0(\hat{\boldsymbol{k}}_0 r, t)^* \frac{\partial}{\partial r}\frac{f}{r}\psi_0(\hat{\boldsymbol{k}}_0 r, t)\right) \\
&= \frac{\hbar k_0}{m}\frac{|f_{\boldsymbol{k}_0}(\Omega)|^2}{r^2}|\psi_0(\hat{\boldsymbol{k}}_0 r, t)|^2 \quad ,
\end{aligned} \tag{18.21}$$

and therefore

$$dN(\Omega) = \int_{-\infty}^{\infty} dt\, j_r d\Omega\, r^2 = |f_{\boldsymbol{k}_0}(\Omega)|^2 d\Omega \frac{\hbar k_0}{m}\int_{-\infty}^{\infty} dt\, |\psi_0(\hat{\boldsymbol{k}}_0 r, t)|^2. \tag{18.22}$$

Neglecting the spreading of the wave packet, one sees that the two integrals in (18.20) and (18.22) are equal, and it follows from (18.17) that

$$\frac{d\sigma}{d\Omega} = |f_{\boldsymbol{k}_0}(\vartheta, \varphi)|^2 \quad . \tag{18.23}$$

---

[1]
$$\begin{aligned}
\frac{\partial}{\partial r}\psi_0(\hat{\boldsymbol{k}}_0 r, t) &= \frac{\partial}{\partial r}\int \frac{d^3 k}{(2\pi)^3} \exp\{i\boldsymbol{k} \cdot \hat{\boldsymbol{k}}_0 r\}a_{\boldsymbol{k}} \exp\{-iE_k(t - t_0)/\hbar\} \\
&= \int \frac{d^3 k}{(2\pi)^3}\, i\boldsymbol{k} \cdot \hat{\boldsymbol{k}}_0 \exp\{i\boldsymbol{k} \cdot \hat{\boldsymbol{k}}_0 r\}a_{\boldsymbol{k}} \exp\{-iE_k(t - t_0)/\hbar\} \\
&= ik_0\psi_0(\hat{\boldsymbol{k}}_0 r, t) \quad .
\end{aligned}$$

The *total scattering cross section* σ is given by the integral of (18.23) over all angles:

$$\sigma = \int d\Omega |f_{\boldsymbol{k}_0}(\vartheta, \varphi)|^2 \quad . \tag{18.24}$$

Remark: The derivation of the differential scattering cross section is shorter for a stationary solution. The incident plane wave has the current density

$$I = \left| \frac{\hbar}{2m} (e^{-i\boldsymbol{k}\cdot\boldsymbol{x}} \nabla e^{i\boldsymbol{k}\cdot\boldsymbol{x}} - e^{i\boldsymbol{k}\cdot\boldsymbol{x}} \nabla e^{-i\boldsymbol{k}\cdot\boldsymbol{x}}) \right| = \left| \frac{\hbar\boldsymbol{k}}{m} \right| \quad . \tag{18.25a}$$

The radially outgoing current density, determined after (18.10), is

$$j_r = \frac{\hbar 2ik}{2mi} \frac{1}{r^2} |f_{\boldsymbol{k}}(\vartheta, \varphi)|^2 = \frac{1}{r^2} \frac{\hbar k}{m} |f_{\boldsymbol{k}}(\vartheta, \varphi)|^2 \tag{18.25b}$$

and from the ratio of $r^2 j_r$ and I one finds

$$\frac{d\sigma}{d\Omega} = |f_{\boldsymbol{k}}(\vartheta, \varphi)|^2 \quad . \tag{18.25c}$$

18.3 Partial Waves

We now suppose that the potential $V(\boldsymbol{x}) = v(r)$ is *spherically symmetric*. To start with, the formal scattering solution $\psi_{\boldsymbol{k}}(\boldsymbol{x})$ of (18.9) will be expanded in spherical harmonics. Our goal in this section is to investigate stationary solutions which are also eigenfunctions of the angular momentum (partial waves). We will characterize the asymptotic behavior of these partial waves by phase shifts. Comparison of the formal scattering solution (18.28) with the equivalent expansion in partial waves allows one to represent the scattering amplitude and the scattering cross section in terms of the phase shifts.

We first recall the expansion (17.44) of a plane wave in spherical harmonics,

$$e^{i\boldsymbol{k}\cdot\boldsymbol{x}} = \sum_{l=0}^{\infty} i^l (2l+1) j_l(kr) P_l(\cos\vartheta) \quad , \tag{18.26}$$

where we assume that the incident plane wave is in the z-direction (Fig. 18.3).

Due to the spherical symmetry of $v(r)$ the scattering is cylindrically symmetric. By rotational invariance about $\boldsymbol{e}_z$, the scattering amplitude (18.10) and the wave function $\psi_{\boldsymbol{k}}(\boldsymbol{x})$ (18.9) are independent of the azimuthal angle φ:

$$f_{\boldsymbol{k}}(\vartheta, \varphi) = f_k(\vartheta) = \sum_{l=0}^{\infty} (2l+1) f_l P_l(\cos\vartheta) \quad . \tag{18.27}$$

The expansion coefficients f_l are called partial wave scattering amplitudes.

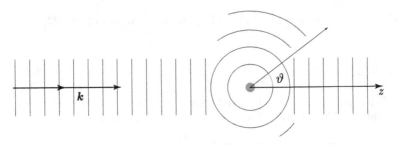

Fig. 18.3. Scattering: the plane wave and the spherical wave

Inserting (18.26) and (18.27) in (18.9), one finds by means of (17.14) for large r

$$\psi_{\boldsymbol{k}}(\boldsymbol{x}) \simeq \sum_{l=0}^{\infty} \frac{(2l+1)}{kr} P_l(\cos \vartheta)$$

$$\times \left[\frac{\mathrm{i}^l}{2\mathrm{i}} \left(\mathrm{e}^{\mathrm{i}(kr-l\pi/2)} - \mathrm{e}^{-\mathrm{i}(kr-l\pi/2)} \right) + k f_l \mathrm{e}^{\mathrm{i}kr} \right] \quad . \tag{18.28}$$

According to chapter 6 the stationary states for a potential $v(r)$ take the form $R_l(r) Y_{lm}(\vartheta, \varphi)$, where the radial wave functions obey the radial Schrödinger equation (6.9)

$$\left[\frac{d}{dr^2} + k^2 + \frac{l(l+1)}{r^2} \right] r R_l(r) = \frac{2m}{t_1^2} v(r) r R_l(r) \quad . \tag{18.29}$$

The stationary states $\psi_{\boldsymbol{k}}(\boldsymbol{x})$ can be expanded in terms of these radial eigenfunctions and spherical harmonics as:

$$\psi_{\boldsymbol{k}}(\boldsymbol{x}) = \sum_{l=0}^{\infty} \mathrm{i}^l (2l+1) R_l(r) P_l(\cos \vartheta) \quad . \tag{18.30}$$

Because of the rotational symmetry emphasized above (φ-independence), only $Y_{10} \sim P_l(\cos \vartheta)$ appears in (18.30). Separating off $(2l+1)$ in (18.27) and (18.30) will simplify later formulas. One refers to (18.30) as the *partial-wave expansion*. We first study the individual partial waves for large r. At distances r, for which $v(r) = 0$ or at least $v(r) < 1/r^2$, the radial solution $R_l(r)$ behaves like

$$R_l(r) = B_l \left(h_l^{(2)}(kr) + S_l(E) h_l^{(1)}(kr) \right) \quad , \tag{18.31}$$

i.e. it is a superposition of spherical Hankel functions, Eqs. (17.15a)–(17.18). These have the asymptotic behaviour

$$h_l^{(1)} \sim -\mathrm{i}\mathrm{e}^{\mathrm{i}(kr-l\pi)}/kr \quad \text{and} \quad h_l^{(2)} \sim \mathrm{i}\mathrm{e}^{-\mathrm{i}(kr-l\pi)}/kr \quad . \tag{18.32}$$

The first term in (18.31) corresponds to an incident spherical wave, the second to an outgoing spherical wave. The coefficients B_l and $S_l(E)$ have now to be determined.

According to (18.9) or (18.28), $\psi_{\boldsymbol{k}}$ consists of a plane wave and an outgoing spherical wave. The plane wave is, by (18.26), the sum of incident and outgoing spherical partial waves. The incident partial waves in (18.31) must coincide with those in (18.28); this fixes the amplitude B_l as

$$B_l = \tfrac{1}{2} \quad . \tag{18.33}$$

The quantity $S_l(E)$ is an amplitude dependent on k or equivalently on the energy. (In the framework of general scattering theory, the $S_l(E)$ are eigenvalues of the S-matrix.)

Potential scattering is elastic. Due to the conservation of the probability density, for each individual partial wave, the radial component of the current density must satisfy $j_r = 0$. If we imagine a spherical shell of radius r, there must be as many particles passing through this shell outward as inward, which implies $|S_l(E)| = 1$, a property one refers to as unitarity.

Formally, this follows from

$$j_r = \frac{\hbar}{m} \operatorname{Im} \left(R_l^* \frac{\partial}{\partial r} R_l \right) = \frac{\hbar k}{m} \operatorname{Im}(h_l h_l^{*\prime} + |S_l(E)|^2 h_l^* h_l' + 2 \operatorname{Re}(h_l S_l h_l'))$$

$$\sim -\frac{\hbar}{mkr^2}(1 - |S_l(E)|^2) \tag{18.34}$$

on using the asymptotic formulae (18.32), the abbreviation $h_l \equiv h_l^{(1)}$ and one leaves out the factor $|P_l(\cos \vartheta)|^2 \geq 0$.

Thus, $S_l(E)$ must be of the form

$$S_l(E) = \mathrm{e}^{2\mathrm{i}\delta_l(E)} \quad . \tag{18.35}$$

Thus, the partial waves of (18.31) take the form

$$R_l(r) = \frac{1}{2} \left(h_l^{(2)}(kr) + \mathrm{e}^{2\mathrm{i}\delta_l(E)} h_l^{(1)}(kr) \right) \quad . \tag{18.36a}$$

Expressing R_l in terms of Bessel and Neumann functions, we obtain

$$R_l(r) = \mathrm{e}^{\mathrm{i}\delta_l} (j_l(kr) \cos \delta_l - n_l(kr) \sin \delta_l) \quad . \tag{18.36b}$$

At large separations, the influence of the potential is manifest in the phase shifts $\delta_l(E)$ of the outgoing waves. These must be found by solving the Schrödinger equation (18.29), subject to the boundary condition $(rR_l(r))|_{r=0} = 0$.

We can now express the scattering amplitude in terms of the phase shifts. From (18.30), one obtains with the help of (18.36a) and (18.32) the asymptotic expression for the partial wave expansion of $\psi_{\boldsymbol{k}}(\boldsymbol{x})$:

$$\psi_{\boldsymbol{k}}(\boldsymbol{x}) = \sum \frac{\mathrm{i}^l (2l + 1)}{kr2\mathrm{i}} \left(\mathrm{e}^{\mathrm{i}(kr - l\pi/2 + 2\delta_l)} - \mathrm{e}^{-\mathrm{i}(kr - l\pi/2)} \right) \quad . \tag{18.36c}$$

The comparison of the formal scattering solution (18.28) with (18.36c) yields the following representation of the partial wave scattering amplitudes in terms of the phase shifts:

$$f_l = \frac{e^{2i\delta_l} - 1}{2ik} = \frac{e^{i\delta_l} \sin \delta_l}{k} \quad . \tag{18.37}$$

Substituting this into the scattering amplitude (18.27)

$$f_k(\vartheta) = \frac{1}{k} \sum_{l=0}^{\infty} (2l + 1) e^{i\delta_l} \sin \delta_l \, P_l(\cos \vartheta) \quad , \tag{18.38}$$

we finally find for the differential scattering cross section

$$\frac{d\sigma}{d\Omega} = \frac{1}{k^2} \sum_{l,l'} (2l + 1)(2l' + 1) e^{i(\delta_l - \delta_{l'})} \sin \delta_l \, \sin \delta_{l'} \, P_l(\cos \vartheta) P_{l'}(\cos \vartheta) \quad . \tag{18.39}$$

Whereas there are interference terms of the various partial waves in the differential scattering cross section, in the total scattering cross section

$$\sigma = \int d\Omega \frac{d\sigma}{d\Omega} = \sum_{l=0}^{\infty} \sigma_l \tag{18.40}$$

the contributions of the partial waves

$$\sigma_l = \frac{4\pi}{k^2} (2l + 1) \sin^2 \delta_l \tag{18.41}$$

are additive.

Remarks:

(i)   The contribution of a partial wave to the total scattering cross section σ is

$$\frac{4\pi}{k^2} (2l + 1) \sin^2 \delta_l \leq \frac{4\pi}{k^2} (2l + 1) \quad .$$

The equal sign holds for $\delta_l = (n + 1/2)\pi$.

(ii)   In the sum (18.40), only l with $l \lesssim ka$ contribute, where a is the range of the potential. Classically, one sees this condition immediately from the fact that scattering only occurs if the impact parameter d is smaller than the range of the potential, i.e., $d < a$ and $L = pd = \hbar kd$. *Quantum mechanically,* the argument is as follows: For $r > a$, only the centrifugal potential $\hbar^2 l(l + 1)/2mr^2$ acts. For the energy $E = \hbar^2 k^2/2m$, the classical turning radius is $r_{cl} = \sqrt{l(l + 1)}/k$. For $r < r_{cl}$, the wave function falls off exponentially. If $r_{cl} > a$, then the particle "feels" nothing from the potential. The particle is thus only scattered for $r_{cl} \leq a$, i.e., $\sqrt{l(l + 1)} \approx l \leq ka$.

18.4 The Optical Theorem

Taking the imaginary part of (18.38)

$$\text{Im } f_k(\vartheta) = \frac{1}{k} \sum_{l=0}^{\infty} (2l+1) \sin^2 \delta_l \, P_l(\cos \vartheta) \quad,$$

comparing with (18.40), and using $P_l(1) = 1$, we obtain the *optical theorem*:

$$\sigma = \frac{4\pi}{k} \text{ Im } f_k(0) \quad. \tag{18.42}$$

This relationship between the total scattering cross section and the imaginary part of the scattering amplitude in the forward direction is a consequence of the conservation of probability density.

Remark: To show this for a wave packet, we compute the radial current density

$$j_r = \frac{\hbar}{m} \text{ Im } \left(\psi^* \frac{\partial \psi}{\partial r} \right) \tag{18.43}$$

after scattering. Starting from (18.16), we use[2]

$$\frac{\partial \psi_0(\boldsymbol{x}, t)}{\partial r} \approx \mathrm{i} k_0 \cdot \frac{\boldsymbol{x}}{r} \psi_0(\boldsymbol{x}, t) \quad, \quad \frac{\partial \psi_0(\hat{\boldsymbol{k}}_0 r, t)}{\partial r} \approx \mathrm{i} k_0 \psi_0(\hat{\boldsymbol{k}}_0 r, t) \quad.$$

The derivative of the $1/r$-factor can be neglected. The radial current density

$$j_r = j_{r,0} + j_{r,\text{scat}} + j_{r,\text{int}} \tag{18.44a}$$

consists of the *transmitted* current density

$$j_{r,0} = \frac{\hbar k_0}{m} \cdot \frac{\boldsymbol{x}}{r} |\psi_0(\boldsymbol{x}, t)|^2 \quad, \tag{18.44b}$$

the *scattered* current density

$$j_{r,\text{scat}} = \frac{\hbar k_0}{m} \frac{1}{r^2} |\psi_0(\hat{\boldsymbol{k}}_0 r, t)|^2 |f|^2 \quad, \tag{18.44c}$$

and the *interference* current density of scattered and transmitted wave functions

$$j_{r,\text{int}} = \frac{\hbar}{m} \text{ Im } \left[\mathrm{i} k_0 \cdot \frac{\boldsymbol{x}}{r} \psi_0(\boldsymbol{x}, t) \frac{f^*}{r} \psi_0^*(\hat{\boldsymbol{k}}_0 r, t) + \frac{f}{r} \mathrm{i} k_0 \psi_0(\hat{\boldsymbol{k}}_0 r, t) \psi_0^*(\boldsymbol{x}, t) \right]$$

$$\approx \frac{\hbar k_0}{m} 2 \text{ Re } \left[\frac{f_{k_0}(\vartheta, \varphi)}{r} \psi_0(\hat{\boldsymbol{k}}_0 r, t) \psi_0^*(\boldsymbol{x}, t) \right] \quad. \tag{18.44d}$$

---

[2]
$$\int d^3 k \, \mathrm{e}^{\mathrm{i}\boldsymbol{k} \cdot \boldsymbol{x}} \psi(\boldsymbol{k}, t) \mathrm{i} k \cdot \frac{\boldsymbol{x}}{r} \approx \mathrm{i} k_0 \cdot \frac{\boldsymbol{x}}{r} \psi_0(\boldsymbol{x}, t)$$

$$\int d^3 k \, \mathrm{e}^{\mathrm{i}\boldsymbol{k} \cdot (\hat{\boldsymbol{k}}_0 r)} \psi(\boldsymbol{k}, t) \mathrm{i} k \cdot \hat{\boldsymbol{k}}_0 \approx \mathrm{i} k_0 \psi_0(\hat{\boldsymbol{k}}_0 r, t) \quad.$$

Here, we have $\boldsymbol{k}_0 \cdot \boldsymbol{x}/r \approx k_0$, since the transmitted packet $\psi_0(\boldsymbol{x}, t)$ is concentrated to the right of the scatterer for times after the scattering event. The total scattered current is by (18.44c)

$$\int d\Omega j_{r,\text{scat}} = \int d\Omega |f_{\boldsymbol{k}_0}(\vartheta, \varphi)|^2 \frac{\hbar k_0}{m} \int_{-\infty}^{\infty} dt |\psi_0(\hat{\boldsymbol{k}}_0 r, t)|^2 \quad . \tag{18.45}$$

For the total interference current, we need the following integral:

$$\int d\Omega \, j_{r,\text{int}} = \frac{\hbar k_0}{rm} 2 \operatorname{Re} \left[\psi_0(\hat{\boldsymbol{k}}_0 r) \int d\Omega \, f_{\boldsymbol{k}_0}(\vartheta, \varphi) \psi_0^*(\boldsymbol{x}, t) \right] \quad ,$$

$$\int d\Omega \, f_{\boldsymbol{k}_0}(\vartheta, \varphi) \psi_0^*(\boldsymbol{x}, t) \approx f_{\boldsymbol{k}_0}(0) \int d\Omega \, \psi_0^*(\boldsymbol{x}, t) = f_{\boldsymbol{k}_0}(0) \int d\Omega \int \frac{d^3 k}{(2\pi)^3} e^{-i\boldsymbol{k} \cdot \boldsymbol{x}} a_{\boldsymbol{k}}^*$$

$$= f_{\boldsymbol{k}_0}(0) \int \frac{d^3 k}{(2\pi)^3} a_{\boldsymbol{k}}^* 2\pi \frac{(e^{-ikr} - e^{ikr})}{-ikr} \approx f_{\boldsymbol{k}_0}(0) \int \frac{d^3 k}{(2\pi)^3} a_{\boldsymbol{k}}^* 2\pi \left(\frac{e^{-i\boldsymbol{k}\cdot\hat{\boldsymbol{k}}_0 r} - e^{i\boldsymbol{k}\cdot\hat{\boldsymbol{k}}_0 r}}{-ik_0 r} \right)$$

$$= f_{\boldsymbol{k}_0}(0) \frac{2\pi}{(-i)k_0 r} [\psi_0(\hat{\boldsymbol{k}}_0 r, t)^* - \psi_0(-\hat{\boldsymbol{k}}_0 r, t)^*] \quad .$$

Since $\psi_0(\hat{\boldsymbol{k}}_0 r, t)\psi_0(-\hat{\boldsymbol{k}}_0 r, t)^* = 0$, it follows that

$$\int d\Omega \, j_{r,\text{int}} = \frac{4\pi \hbar k_0}{k_0 r^2 m} \operatorname{Re} \left[i|\psi_0(\hat{\boldsymbol{k}}_0 r, t)|^2 f_{\boldsymbol{k}_0}(0) \right] \quad .$$

Hence, the total interference current becomes

$$\int dt \int d\Omega \, r^2 j_{r,\text{int}} = -\frac{4\pi \hbar}{m} \operatorname{Im} \left(f_{\boldsymbol{k}_0}(0) \int dt |\psi_0(\hat{\boldsymbol{k}}_0 r, t)|^2 \right) \quad . \tag{18.46}$$

The integrals of $j_{r,0}$ and j_{in} are equal from (18.44b) and (18.20). Thus, from the equality of the incident current with the sum of transmitted, scattered (18.45), and interference (18.46) currents, one has

$$\sigma = \frac{4\pi}{k_0} \operatorname{Im} f_{\boldsymbol{k}_0}(0) \quad . \tag{18.42'}$$

The interference of the scattered wave with the transmitted wave $\psi_0(\boldsymbol{x}, t)$ leads to a decrease of the current in the forward direction. This interference term is proportional to $\operatorname{Im} f_{\boldsymbol{k}_0}(0)$. Since this decrease just gives the total scattering cross section, the *optical theorem* again follows, valid also for non-spherical potentials.

18.5 The Born Approximation

Substituting (18.30), (17.46), and (5.30) into (18.10), one finds

$$f_k(\vartheta, \varphi) = -\frac{m}{2\pi\hbar^2} \int d^3x' e^{-i\mathbf{k}' \cdot \mathbf{x}'} v(r')\psi_k(\mathbf{x}')$$

$$= -\frac{m}{2\pi\hbar^2} \int d^3x' 4\pi \sum_{l=0}^{\infty} \sum_{m=-l}^{l} (-i)^l Y_{lm}(\Omega_{k'}) Y_{lm}^*(\Omega_{x'}) j_l(kr')v(r')$$

$$\times \sum_{l'=0}^{\infty} i^{l'}(2l'+1)P_{l'}(\cos\vartheta')R_{l'}(r')$$

$$= -\frac{2m}{\hbar^2} \sum_{l=0}^{\infty} (2l+1)P_l(\cos\vartheta) \int dr\, r^2 v(r)j_l(kr)R_l(r) \quad,$$

independent of φ, and thus one obtains for the partial wave scattering amplitudes

$$f_l = -\frac{2m}{\hbar^2} \int dr\, r^2 v(r)j_l(kr)R_l(r) \quad.$$

If the potential is weak and the effect on R_l is small, then $R_l \approx j_l(kr)$, and δ_l is small (see (18.36b)). It is then possible to expand (18.37) in δ_l, i.e., $\delta_l \approx kf_l$, and from the preceding equation one obtains

$$\delta_l \approx -\frac{2mk}{\hbar^2} \int_0^\infty dr\, r^2\, v(r)\, [j_l(kr)]^2 \quad. \tag{18.47}$$

This is the *Born approximation* for the partial wave l.

The Born approximation is good for large l. Classically, large l implies a large impact parameter $\hbar l = dp = d\hbar k$, or that the particle is incident far from the scatterer and hardly influenced. See remark (ii) at the end of Sect. 18.3.

If the influence of the potential on all partial waves is small, one can replace the wave function $\psi_k(\mathbf{x})$ in the integral representation for $f_k(\vartheta, \varphi)$ (18.10) directly by $\exp\{i\mathbf{k} \cdot \mathbf{x}\}$. An important form of the Born approximation is then

$$f_k(\vartheta, \varphi) = -\frac{m}{2\pi\hbar^2} \int d^3x' e^{i(\mathbf{k}-\mathbf{k}') \cdot \mathbf{x}'} v(\mathbf{x}') = -\frac{m}{2\pi\hbar^2} \tilde{v}(\mathbf{k}' - \mathbf{k}) \quad, \tag{18.48}$$

in which the scattering amplitude is proportional to the Fourier transform of the potential.

As an illustration, we consider the scattering from the *Yukawa potential*,

$$v(\mathbf{x}) = a\frac{e^{-\mu r}}{r} \quad, \tag{18.49}$$

whose Fourier transform takes the well known form

$$\tilde{v}(\boldsymbol{p}) = \frac{4\pi a}{p^2 + \mu^2} \quad . \tag{18.50}$$

Using

$$(\boldsymbol{k} - \boldsymbol{k}')^2 = k^2 + k'^2 - 2\boldsymbol{k} \cdot \boldsymbol{k}' = 2k^2(1 - \cos \vartheta) = \left(2k \sin \frac{\vartheta}{2}\right)^2 \quad ,$$

where, according to the definition of $\boldsymbol{k}'$ preceding (18.9), one has $|\boldsymbol{k}'| = |\boldsymbol{k}|$, we find for the scattering cross section

$$
\begin{aligned}
\frac{d\sigma}{d\Omega} &= \frac{m^2}{(2\pi\hbar^2)^2} \frac{(4\pi a)^2}{[(2k \sin \vartheta/2)^2 + \mu^2]^2} \\
&= \frac{a^2}{(4E_k \sin^2 \vartheta/2 + \hbar^2\mu^2/2m)^2} \quad .
\end{aligned}
\tag{18.51a}
$$

In the limit $\mu \to 0$ this reduces to scattering from a *Coulomb potential*. Although here the partial wave expansion must be modified, as we will sketch at the end of this chapter, and although the conditions for the Born approximation are not satisfied, remarkably (18.51a) still gives the exact scattering cross section. I.e., for $v(r) = \frac{Z_1 Z_2 e^2}{r}$ one obtains the Rutherford formula

$$\frac{d\sigma}{d\Omega} = \frac{Z_1 Z_2 e^2}{16 E_k^2 \sin^4 \vartheta/2} \quad . \tag{18.51b}$$

The classical computation also gives the exact result. This circumstance, which is related to the homogeneity of the Coulomb potential and the scattering cross section as a function of the momentum transfer, was quite conducive to the correct interpretation of experiments in the initial phase of atomic theory.

We now give an *estimate of the validity of the Born approximation:* For short-range potentials, the wave function $\psi_{\boldsymbol{k}}(\boldsymbol{x})$ only enters the scattered wave of (18.8) at small distances. We therefore calculate $\psi_{\boldsymbol{k}}(\boldsymbol{x})$ for small $\boldsymbol{x}$ in the Born approximation and set $\boldsymbol{x} = 0$ in the second term of (18.8):

$$\psi_{\boldsymbol{k}}(\boldsymbol{x}) \approx e^{i\boldsymbol{k} \cdot \boldsymbol{x}} - \frac{m}{2\pi\hbar^2} \int d^3x' \frac{e^{ik|\boldsymbol{x}'|}}{|\boldsymbol{x}'|} v(r') e^{i\boldsymbol{k} \cdot \boldsymbol{x}'} \quad .$$

Here the second term can be neglected in comparison to the plane wave term whenever

$$\left| \frac{2m}{\hbar^2} \int dr' \, r'^2 \frac{e^{ikr'}}{r'} v(r') \frac{\sin kr'}{kr'} \right| \ll 1$$

is satisfied. The Born approximation is thus valid for weak potentials and large incident energy because of the factor k^{-1}.

18.6 Inelastic Scattering

In (18.35), it followed from the conservation of particle flux that $|S_l(E)| = 1$. However, if absorption processes take place, in which the scatterer is excited or particles are annihilated and/or generated, then

$$S_l(E) = s_l(E)e^{2i\delta_l(E)} \quad \text{with} \quad 0 \leq s_l(E) \leq 1 \quad . \tag{18.52}$$

The amplitude of the lth partial wave is thus

$$f_l = \frac{S_l - 1}{2ik} = \frac{1}{2k}[s_l \sin 2\delta_l + i(1 - s_l \cos 2\delta_l)] \quad , \tag{18.53}$$

and the total elastic scattering cross section becomes

$$\sigma_{\text{el}} = 4\pi \sum_l (2l+1)|f_l|^2 = \frac{\pi}{k^2}\sum_l (2l+1)(1 + s_l^2 - 2s_l \cos 2\delta_l) \quad . \tag{18.54}$$

For the calculation of the total inelastic cross section (= reaction cross section), we require the total elastic flux

$$-\int d\Omega \, r^2 j_r = -\int d\Omega \, r^2 \sum_{l,l'} i^{-l+l'}(2l+1)(2l'+1)P_l P_{l'} \frac{\hbar}{m} \,\text{Im}\, R_l^* \frac{d}{dr} R_{l'}$$

$$= \text{Re}\, \frac{i\pi\hbar r^2}{m} \sum_{l=0}^{\infty}(2l+1)R_l^* \frac{d}{dr} R_l$$

$$= \text{Re}\, \frac{4\pi\hbar}{km} \sum_{l=0}^{\infty}(2l+1)[e^{ikr} - (-1)^l S_l^* e^{-ikr}][e^{-ikr} + (-1)^l S_l e^{ikr}]$$

$$= -\frac{\hbar k}{m}\frac{\pi}{k^2}\sum_l (2l+1)(-1 + |S_l|^2) \quad . \tag{18.55}$$

This is just the flux lost due to inelastic processes. The inelastic cross section, also called the reaction cross section, is defined as the ratio of this quantity to the incident flux $\hbar k/m$:

$$\sigma_{\text{inel}} = \frac{\pi}{k^2}\sum_l (2l+1)(1 - s_l^2) \quad . \tag{18.56}$$

The sum of (18.54) and (18.56) gives the total scattering cross section

$$\sigma_{\text{tot}} = \sigma_{\text{el}} + \sigma_{\text{inel}} = \frac{2\pi}{k^2}\sum_l (2l+1)(1 - s_l \cos 2\delta_l) \quad . \tag{18.57}$$

Comparing this with (18.53) and using $f(0) = \sum_l(2l+1)f_l$, we obtain the *optical theorem*

$$\text{Im}\, f(0) = \frac{k}{4\pi}\sigma_{\text{tot}} \quad . \tag{18.58}$$

The optical theorem thus holds even in the presence of inelastic scattering.

Remarks:

(i)   For $s_l = 1$, by (18.42), there is no inelastic scattering. For $s_l = 0$,

$$\sigma_{\text{inel}} = \frac{\pi}{k^2} \sum_l (2l + 1) = \pi a^2 \quad \text{and} \quad \sigma_{\text{el}} = \pi a^2 \quad ,$$

where a is the extension of the scattering object (see computation pre-ceding (18.100), where the sum is replaced by an integral). There is also elastic scattering! (Although $S_l = 0$, one also has $f_l = \text{i}/2k$.) The phys-ical reason is the shadow scattering , which we will later discuss in the context of high-energy elastic scattering.

(ii)   Examples of inelastic processes are inelastic scattering, absorption pro-cesses, particle capture, and particle decay processes. The latter reac-tions can only be treated in quantum field theory. The effect on the elastic scattering amplitude can be simulated phenomenologically in the Schrödinger equation by means of a complex potential. The potential $V(x) = V_1(x) + \text{i}V_2(x)$ leads to the absorption of particles. Here, the continuity equation takes the form

$$\frac{\partial \varrho(\boldsymbol{x}, t)}{\partial t} + \boldsymbol{\nabla} \cdot \boldsymbol{j}(\boldsymbol{x}, t) = -\frac{2V_2}{\hbar} \varrho(\boldsymbol{x}, t) \quad .$$

18.7 Scattering Phase Shifts

Next, we compute the phase shifts in order to derive important properties of the scattering cross section. We assume a short range potential which vanishes for $r > a$. The following conclusions are valid even if, at large distances, only the weaker assumption $v(r) < r^{-2}$ is satisfied.

For $r > a$, the partial wave takes the form

$$R_l^> (r) = \tfrac{1}{2}[h_l^*(kr) + e^{2\text{i}\delta_l} h_l(kr)] \quad , \quad h_l \equiv h_l^{(1)} \quad . \tag{18.59}$$

For $r < a$, we designate the partial wave by $R_l^<(r)$, which may be found analytically or determined numerically. In the following, only the logarithmic derivative at the position a matters,

$$\alpha_l \equiv \left. \frac{d \log R_l^<}{dr} \right|_{r=a} \quad . \tag{18.60}$$

This k-dependent quantity enters the continuity condition

$$\left. \frac{d/dr(h_l^*(kr) + e^{2\text{i}\delta_l} h_l(kr))}{h_l^*(kr) + e^{2\text{i}\delta_l} h_l(kr)} \right|_{r=a} = \alpha_l \quad , \tag{18.61}$$

whence

$$e^{2i\delta_l} - 1 = \frac{2(dj_l/dr - \alpha_l j_l)}{\alpha_l h_l - dh_l/dr}\bigg|_a \quad , \tag{18.62}$$

and

$$\cot \delta_l = \frac{dn_l/dr - \alpha_l n_l}{dj_l/dr - \alpha_l j_l}\bigg|_a \quad . \tag{18.62'}$$

As an example, we consider the scattering from a *hard sphere* of radius a, which is used as a model for low-energy nuclear scattering experiments. One then has $R_l(a) = 0$ and $\alpha_l = \infty$. Here, (18.62') takes the form

$$\cot \delta_l = \frac{n_l(ka)}{j_l(ka)} \quad . \tag{18.63}$$

From this and (17.12), the phase shift of the s-waves then becomes

$$\delta_0 = -ka \quad . \tag{18.64}$$

For a repulsive potential, the phase shift is negative.

Behavior of δ_l for small k (i.e., small energy)

The expansion of (18.62') for small k gives, with the aid of (17.13),

$$\tan \delta_l(k) = \frac{(2l+1)}{[(2l+1)!!]^2}(ka)^{2l+1}\frac{l - a\alpha_l}{l+1+a\alpha_l} \quad . \tag{18.65}$$

For $ka \to 0$ the wave number dependence of the scattering phase shift becomes

$$\delta_l \sim k^{2l+1} \quad . \tag{18.66}$$

For sufficiently small energy, the partial waves with $l \geq 1$ therefore do not contribute. One then has pure *s-wave scattering*:

$$\frac{d\sigma}{d\Omega} = \frac{\sin^2 \delta_0}{k^2} \quad \text{and} \quad \sigma = 4\pi\frac{\sin^2 \delta_0}{k^2} \quad . \tag{18.67}$$

For low-energy scattering from a hard sphere, (18.67) and (18.64) yield in the limit as k approaches zero

$$\sigma = 4\pi a^2 = 4 \times \text{(classical scattering cross section)} \quad . \tag{18.68}$$

For sufficiently small energy, the scattering is purely isotropic. Compare remark (ii) at the end of Sect. 18.3.

We now show that $\exp\{2i\delta_l\} - 1$ has poles at the energies of the bound states. For a bound state with angular momentum l, (17.21b) and $h_l \equiv h_l^{(1)}$, the wave function outside the potential is

$$R_l^>(r) = h_l(i\kappa r) \quad \text{with} \quad E_b = -\frac{\hbar^2 \kappa^2}{2m} \quad . \tag{18.69}$$

The logarithmic derivative is then given by

$$\alpha_l(E) = \frac{1}{h_l(i\kappa r)} \frac{d}{dr} h_l(i\kappa r) \bigg|_{r=a} \quad . \tag{18.70}$$

Substituting this expression into (18.61), one sees that $\exp\{2i\delta_l\} - 1$ has a simple pole at $k = i\kappa$.

18.8 Resonance Scattering from a Potential Well

For a spherical potential well of radius a and depth V_0 (Fig. 18.4), i.e.,

$$v(r) = -V_0\Theta(a - r) \quad , \tag{18.71}$$

the scattering solutions with energy E inside the well, by (17.33), are given by

$$R_l^<(r) = j_l(qr) \quad , \quad q = \frac{\sqrt{2m(V_0 + E)}}{\hbar} \quad ,$$

$$\alpha_l = q\frac{j_l'(qa)}{j_l(qa)} \quad . \tag{18.72}$$

We return now to (18.64) for the phase shift δ_l in the limit of low energy, i.e. small k. If the *resonance condition*

$$l + 1 + a\alpha_l(E) = 0 \tag{18.73}$$

is satisfied, then $\delta_l(k) = (n + 1/2)\pi$, and the partial-wave scattering cross section takes its maximum value

$$\sigma_l(k) = \frac{4\pi(2l + 1)}{k^2} \tag{18.74}$$

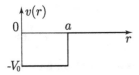

Fig. 18.4. The spherical potential well

(resonance scattering). Since for other k-values the phase shifts $\delta_l \sim (ka)^{2l+1}$, these resonances become very sharp for large l.

Now, we investigate resonance scattering in detail for small energies and a very deep potential well:

$$ka \ll l \ll qa \quad . \tag{18.75}$$

We can then substitute the asymptotic formula for $j_l(qa)$, Eq. (17.14), into (18.72) and the resonance condition (18.73) to obtain

$$\frac{l}{qa} = -\cot\left(qa - \frac{l\pi}{2}\right) \quad , \tag{18.76}$$

giving approximately

$$qa - \frac{l\pi}{2} \approx \left(n + \frac{1}{2}\right)\pi + \frac{l}{qa} \quad . \tag{18.76'}$$

This condition is equivalent to the equation determining bound states in the potential well, (17.27). We are dealing here with virtual levels, rather than bound states, since $E > 0$.

Aside from the determination of the position of the resonances, all of the considerations of this section also hold for other short-range potentials.

We now determine the energy dependence of the *phase shift and scattering cross section* near the resonance energy E_R. Expanding (18.65), we find for the phase shift

$$\tan \delta_l(k) = -\frac{\gamma(ka)^{2l+1}}{E - E_R} + O[(ka)^{2l+1}] \quad , \tag{18.77}$$

where we define $\gamma = -[(2l-1)!!]^{-2}[a\alpha_l'(E_R)]^{-1}$. At resonance, the phase shift takes the value $\pi/2$. One can show generally that $\gamma > 0$, see Problem 18.3. Furthermore, we introduce the abbreviation $\Gamma_k = 2(ka)^{2l+1}\gamma$. The partial scattering cross section then becomes

$$\sigma_l = \frac{4\pi(2l+1)}{k^2} \frac{(\Gamma_k/2)^2}{(E - E_R)^2 + (\Gamma_k/2)^2} \quad . \tag{18.78}$$

This is the *Breit–Wigner formula* for resonance scattering (Fig. 18.5).

We can find the scattering amplitude most simply by writing (18.37) in the form

$$f_l(k) = \frac{\tan \delta_l}{k(1 - i \tan \delta_l)} \quad . \tag{18.79}$$

Hence, using (18.77), one finds

$$f_l(k) = \frac{-\Gamma_k/2}{E - E_R + i\Gamma_k/2} \quad . \tag{18.80}$$

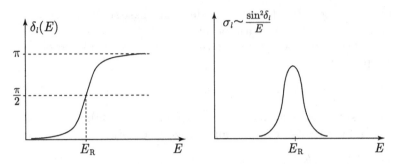

Fig. 18.5. The qualitative behavior of the phase shift and the scattering cross section in the region near a resonance

The scattering amplitude has a pole at

$$E = E_R - i\Gamma_k/2 \tag{18.81}$$

in the lower half–plane of the second Riemannian sheet ($k = \sqrt{E}$). Because of the k-dependence of the width, $\gamma(ka)^{2l+1}$, the low-energy resonances for $l \geq 1$ are very sharp. For s-waves, the width is proportional to k, and therefore resonances are, if present at all, not as sharp. Whether or not a maximum occurs in σ_0 depends on $\alpha_0'(E)$. The condition for the occurrence of a maximum in σ_0 is $|\partial\alpha_0/\partial k^2| > 1/\sqrt{2}k$ for $k = \sqrt{2mE_R}/\hbar$.

Finally, using (18.62) and (18.41), we give in Figs. 18.6 and 18.7 the wave number dependence of the phase shift of the lowest partial waves and the

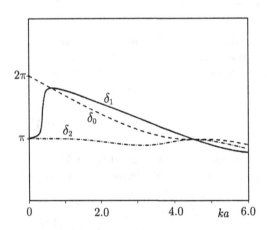

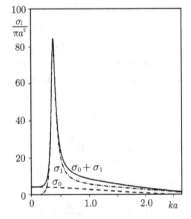

Fig. 18.6. The phase shift $\delta_l(k)$ for the spherically symmetric square well potential of strength $\zeta = 6.25$; (- - -): $l = 0$, (———): $l = 1$, ($-\cdot-\cdot-$): $l = 2$. For $ka \to \infty$, these phase shifts fall to zero, $\delta_l(\infty) = 0$

Fig. 18.7. The differential scattering cross section $\sigma_l(k)$ for the spherically symmetric square well potential of strength $\zeta = 6.25$; ($- - -$): σ_0, ($-\cdot-\cdot-$): σ_1, (———): $\sigma_0 + \sigma_1$

scattering cross section for the potential strength $\zeta = 6.25$, defined by

$$\zeta = \frac{\sqrt{2mV_0}\,a}{\hbar} \quad . \tag{18.82}$$

Remark: Very briefly, we introduce here the concept of the *Jost function*[3]. If instead of the scattering solutions one considers regular solutions of the radial Schrödinger equation for which

$$\lim_{r \to 0} r^{-l-1} u_l(k, r) = 1 \quad ,$$

then these have the asymptotic behavior

$$u_l(k, r) = \frac{1}{2ik} [f_l(k)e^{ikr} - f_l(-k)e^{-ikr}] \quad . \tag{18.83}$$

The functions $f_l(k) = f_l^*(-k^*)$ are known as Jost functions. Comparison with (18.31) shows that the S-matrix element of the lth partial wave (18.35) can be expressed by

$$S_l(k) = (-1)^l \frac{f_l(k)}{f_l(-k)} \quad . \tag{18.84}$$

The zeros and poles of $S_l(k)$ are thus determined by the zeros of $f_l(\pm k)$. As a function of energy, the Riemann surface of $S_l(E)$ always has a $\sqrt{E}$-branch cut along the positive real axis. Poles on the negative real axis in the first (physical) sheet correspond to bound states (see (18.70)). Furthermore, the poles in the second sheet occur in pairs located mirror symmetrically above and below the Re E-axis. At these positions, $S_l(E)$ has zeros in the first sheet. Poles in the lower half-plane of the second sheet lying near the positive real axis correspond to resonances (Fig. 18.8).

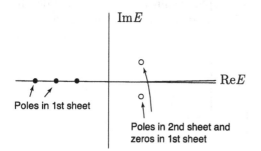

Poles in 1st sheet

Poles in 2nd sheet and zeros in 1st sheet

Fig. 18.8. Poles and zeros of $S_l(E)$ on the Riemannian sheets

[3] C.J. Joachain: *Quantum Collision Theory* (North-Holland, New York 1975);
R. Omnès, M. Froissart: *Mandelstam Theory and Regge Poles* (Benjamin, New York 1963);
H.M. Nussenzveig: *Causality and Dispersion Relations* (Academic Press, New York 1972)

18.9 Low Energy s-Wave Scattering; the Scattering Length

For s-waves within a spherical potential well,

$$u^<(r) = rR^<(r) = \sin(qr) \quad , \quad \text{with} \quad q = \frac{\sqrt{2m(E+V_0)}}{\hbar} \quad , \qquad (18.85)$$

and thus (18.60) yields for $l = 0$

$$\alpha_0 = q \cot(qa) - \frac{1}{a} \quad . \qquad (18.86)$$

Substituting now (17.12) and (18.86) into (18.62′), we obtain for the s-wave phase shift

$$\tan \delta_0(k) = \frac{(k/q)\tan(qa) - \tan(ka)}{1 + (k/q)\tan(qa)\tan(ka)} = \tan(pa - ka) \quad . \qquad (18.87)$$

In order to rewrite the original identity, we have defined the auxiliary variable p via $\tan(pa) = (k/q)\tan(qa)$, and we have used the addition theorem for the tangent. Hence, the phase shift becomes, as in (17.37),

$$\delta_0(k) = \arctan\left(\frac{k}{q}\tan(qa)\right) - ka \quad . \qquad (18.88)$$

For low energies, unless $\tan qa = \infty$ we obtain

$$\delta_0(k) \approx ka\left(\frac{\tan(qa)}{qa} - 1\right) \quad \text{mod } \pi \quad . \qquad (18.89)$$

We next study the dependence of the phase shift $\delta_0(k)$ on the well depth V_0 (Fig. 18.9). We recall equation (17.25), determining the binding energies

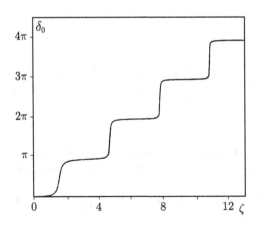

Fig. 18.9. The Levinson theorem (18.88) as a function of ζ for $ka = 0.2$

$\tan{(qa)} = -\hbar q/(2m|E|)^{1/2}$. As V_0 gets larger, for $\zeta \equiv \sqrt{2mV_0a^2}/\hbar \geq \pi/2$, $3\pi/2, \ldots$ the potential has one, two, $\ldots$ bound states.

For the values $\zeta = (2n+1)\pi/2$, exactly one of the bound states has energy $E = 0$. In this case, $\tan{(qa)} = \infty$, and it is evident from (18.88) that the phase is $\delta_0 = \pi/2 \mod \pi$.

Every time an additional bound state occurs, the phase passes through $\pi/2$ and increases by π. This is an example of the *Levinson* theorem:

$$\delta_0(0) = n_b\pi \quad , \tag{18.90}$$

where n_b is the number of bound states. For the sake of completeness we note that if for $l = 0$ the Jost function obeys $f_0(0) = 0$, then (18.90) is replaced by $\delta_0(0) = (n_b + 1/2)\pi$.

The general prof of Levinson's theorem[4] proceeds by evaluating a contour integral of $f'_l(z)/f_l(z)$ in the complex plane, where $f_l(z)$ are the Jost functions mentioned at the end of Sect. 18.8. The result is

$$\delta_l(0) = \begin{cases} n_b\pi & \text{for} \quad l \neq 0 \quad \text{or} \quad l = 0, \; f_0(0) \neq 0 \\ (n_b + \frac{1}{2})\pi & \text{for} \quad l = 0, \quad f_0(0) = 0. \end{cases} \tag{18.91}$$

Starting from (18.89), we can derive the *Ramsauer–Townsend effect* for the scattering of electrons from the noble gases Ar, Kr, Xe, which have closed shells and together with the nucleus provide an attractive short range potential. Substituting the low-energy phase shift (18.89) into the scattering cross section, we find

$$\sigma \approx 4\pi a^2 \left[\frac{\tan{(qa)}}{qa} - 1\right]^2 \quad . \tag{18.92}$$

When the energy of the incident electrons has the value $E \approx 0.7$ eV, the condition $\tan{(qa)}/qa = 1$ is fulfilled and σ vanishes, and at these small energies partial waves with $l \geq 1$ do not play any role. The phase shift is $\delta_0 = \pi$ for this energy.

We now return to (18.65) to investigate this relation further for very small energies and to introduce the concept of scattering length. For small energy, one can expand (18.65):

$$k \cot{\delta_0(k)} = -\frac{1}{a_0} + \frac{1}{2}r_0k^2 \quad . \tag{18.93}$$

In this expansion, a_0 is called the scattering length and r_0 the effective range of the potential. It can be proven for arbitrary short-range potentials that the second term in the expansion is $\propto k^2$. From the low-energy expansion discussed earlier it follows that

---

[4] See Joachain, p. 258, and Nussenzveig, p. 207, quoted in footnote 3

$$\frac{1}{a_0} = \frac{1}{a} \frac{1 + a\alpha_0(0)}{a\alpha_0(0)}$$

and $\sin^2 \delta_0 = (ka_0)^2 + O(k^4)$. The scattering cross section thus becomes

$$\sigma = 4\pi a_0^2 + O(k^2) \quad . \tag{18.94}$$

The scattering length thus determines the low-energy scattering cross section. We see that

$$e^{2i\delta_0} - 1 = \frac{2i}{\cot \delta_0 - i} = \frac{2ka_0}{i - ka_0} \tag{18.95}$$

has a pole for $k = i/a_0$. For positive $a_0 \gg a$, this pole corresponds to a bound state of energy $E_b = -\hbar^2 \kappa^2/2m$ and $\kappa = 1/a_0$. Since the bound state is of the form $\exp\{-\kappa r\}/r$, its extension is κ^{-1}. Thus, the scattering length is about as large as the extension of the bound state. The scattering amplitude becomes

$$f_0 = \frac{2ka_0}{i - ka_0} \frac{1}{2ik} \approx -a_0 \quad \text{for} \quad k \ll 1/a_0 \quad . \tag{18.96}$$

Finally, we note that, from (18.95), the energy dependence of the scattering cross section becomes

$$\sigma = \frac{2\pi\hbar^2/m}{-E_b + E} \tag{18.97}$$

and that its low-energy behavior is completely determined by the bound state lying (as we assumed) near the continuum.

We would now like to see how to read off the scattering length from the wave function. For the wave function outside the potential, (18.36b),

$$R_0^>(r) = e^{i\delta_0}[j_0(kr) \cos \delta_0 - n_0(kr) \sin \delta_0]$$

one has in the limit of small k

$$u^>(r) = rR^>(r) \propto \left(-\frac{1}{ka_0} \sin kr + \cos kr\right) \propto \left(1 - \frac{r}{a_0}\right) \quad . \tag{18.98}$$

In the last proportionality, we have used $kr \ll 1$. We see from (18.98) that the extrapolation of $u^>(r)$ intersects the r-axis at a_0.

In Fig. 18.10, one recognizes that the scattering length is negative for an attractive potential without a bound state, while the phase shift is positive. If the potential gets stronger, $a_0 \to -\infty$ and $\delta \to \pi/2$. As the potential increases further, one then has a bound state. As long as the binding energy remains small, the scattering length, which now is positive, is very large. The phase shift in this regime is negative or equivalently lies between $\pi/2$ and π.

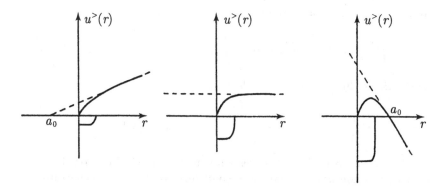

Fig. 18.10. The scattering length for a potential without a bound state, with a bound state at $E = 0$, and with a bound state at finite energy

For a repulsive potential, the phase shift is negative, while the scattering length is always positive and comparable to the width of the potential. In this case, poles do not occur at negative energies.

18.10 Scattering at High Energies

We consider the scattering from a *hard sphere* of radius a. At high energies, (18.63) together with (17.14) implies for the phase shift

$$\delta_l = -ka + \frac{l\pi}{2} \quad . \tag{18.99}$$

Inserting this into (18.40) and rearranging the terms pairwise, we find

$$
\begin{aligned}
\sigma &= \frac{4\pi}{k^2} \Bigg\{ \left[\sin^2 ka + \sin^2 \left(ka - \frac{\pi}{2} \right) \right] \\
&\quad + 2 \left[\sin^2 \left(ka - \frac{\pi}{2} \right) + \sin^2 (ka - \pi) \right] + \dots \Bigg\} \\
&= \frac{4\pi}{k^2} \sum_{l=0}^{ka} l = \frac{4\pi}{k^2} \frac{1}{2} ka(ka + 1) \quad .
\end{aligned}
$$

Hence, it follows for high energies, i.e., $ka \gg 1$, that

$$\sigma = 2\pi a^2 \quad , \tag{18.100}$$

and not, as perhaps expected, πa^2. The quantum mechanical scattering cross section at high energies is double the classical scattering cross section. The reason is the diffraction of the wave. This can be seen from the differential scattering cross section at high energies[5]:

$$\frac{d\sigma}{d\Omega} = \frac{1}{4}a^2 \left[1 + \cot^2 \frac{\vartheta}{2} \, J_1^2(ka \sin \vartheta) \right] \quad , \tag{18.101}$$

(J_1 is the Bessel function of first order) with

$$J_1(x) \xrightarrow[x \to 0]{} \frac{1}{2}x \quad , \quad J_1(x) \xrightarrow[x \to \infty]{} \left(\frac{2}{\pi x} \right)^{1/2} \cos\left(x - \frac{3}{4}\pi \right) \quad .$$

The first term is isotropic and corresponds to the classical differential scattering cross section $a^2/4$, while the second term is sharply concentrated in the forward direction and describes diffraction effects. Half of the the total scattering cross section is due to classical reflection scattering, the other half to diffraction.

This can also be understood as follows: The wave function $\psi = \psi_{\text{in}} + \psi_{\text{scat}}$ is the sum of the incident and the scattered wave. Immediately behind the sphere, in the shadow, ψ_{scat} must be just equal to ψ_{in} in magnitude and opposite in sign. The flux of ψ_{scat} is equal to that part of the flux of ψ_{in} obscured by the sphere (Fig. 18.11). The shadow results from the interference of the scattered wave in the forward direction with the incident wave. This interference must remove precisely as much intensity from the ray propagating in the forward direction as is reflected off into finite angles. For large distances, this additional contribution to the scattering amplitude is concentrated in the forward direction. Classically, one regards only the part reflected off the sphere as scattering.

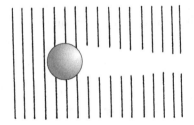

Fig. 18.11. Shadow scattering

[5] P. M. Morse, H. Feshbach: *Methods of Theoretical Physics* (New York, McGraw-Hill 1953), pp. 1485, 1551.

18.11 Additional Remarks

18.11.1 Transformation to the Laboratory Frame

Let us now establish the connection between the results of potential scattering and the scattering of two particles in the laboratory frame (L) as opposed to the center-of-mass frame (CM). In the laboratory frame, one of the particles is at rest prior to the scattering and is bombarded by the other particle. In the center-of-mass frame, the center of mass of the two particles is at rest. The former system usually corresponds to the experimental situation (e.g., scattering of electrons or photons from atoms). The latter system is immediately accessible to theory, because – to the extent that the interaction of the two particles can be represented by a potential – scattering in the center-of-mass frame is identical to potential scattering, provided the mass is replaced by the reduced mass of the two particles.

The relations between the differential scattering cross section $d\sigma/d\Omega$ in the center-of-mass frame and that of the laboratory frame $d\sigma/d\Omega_{\mathrm{L}}$ and between the deflection angles ϑ and ϑ_{L} are the kinematical relations familiar from classical mechanics (Fig. 18.12):

$$\frac{d\sigma}{d\Omega} = \frac{M_2^2(M_2 + M_1 \cos\vartheta)}{(M_1^2 + M_2^2 + 2M_1 M_2 \cos\vartheta)^{3/2}} \frac{d\sigma}{d\Omega_{\mathrm{L}}} \quad,$$

$$\cos\vartheta_{\mathrm{L}} = \frac{M_1 + M_2 \cos\vartheta}{\sqrt{M_1^2 + M_2^2 + 2M_1 M_2 \cos\vartheta}} \quad. \tag{18.102}$$

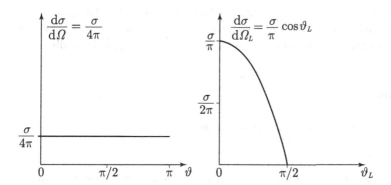

Fig. 18.12. Transformation from the center-of-mass system to the laboratory system: The differential scattering cross section for the scattering of hard spheres of equal mass ($M_2 = M_1$) is isotropic in the center-of-mass system. The angular dependence in the laboratory system follows from the kinematic relations, which for equal masses reduce to $d\sigma/d\Omega_{\mathrm{L}} = 4\cos\vartheta_{\mathrm{L}} d\sigma/d\Omega$ and $\cos\vartheta_{\mathrm{L}} = \cos\vartheta/2$

18.11.2 The Coulomb Potential

We found in the case of the Coulomb potential that the classical scattering cross section $d\sigma/d\Omega$ is identical to the scattering cross section in the Born approximation and, remarkably, also to the exact quantum mechanical result. One additional remark concerning the exact form of the Coulomb wave functions: For large separation r, we found for the bound states $(Z = 1)$

$$R_{nl} \sim r^{n-1}e^{-\kappa r}$$

instead of

$$\frac{e^{-\kappa r}}{r} \quad ,$$

valid for short-range potentials, where

$$E = -\frac{me_0^4}{2\hbar^2 n^2} = -\frac{\hbar^2\kappa^2}{2m} \quad .$$

Hence it follows that

$$n = \frac{me_0^2}{\hbar^2\kappa}$$

and

$$R_{nl} \sim \frac{1}{r}\exp\{-\kappa r + n\log r\} = \frac{1}{r}\exp\left\{-\kappa r + \frac{me_0^2}{\hbar^2\kappa}\log r\right\} \quad .$$

From this, replacing κ by ik, one obtains the scattering states. Because of the infinite range of the Coulomb potential, the scattering states are not spherical waves, but contain additionally a phase shift depending logarithmically on r.

Problems

18.1 In the case where the resonance condition

$$1 + a\alpha_0(E_R) = 0$$

(18.73) is satisfied, calculate the scattering cross section for s-wave scattering in the neighborhood of these resonances. Prove that a zero E_R of the s-wave resonance condition leads to a maximum of the scattering cross section σ_0 only if the inequality

$$\left|\frac{\partial\alpha_0}{\partial k^2}\right| \geq \frac{1}{\sqrt{2}k_R}$$

is satisfied for $k_R = \sqrt{2m\,E_R}/\hbar$. (See page 344.)

18.2 Investigate the scattering from the δ-shell potential of Problem 17.2

$$v(r) = -\lambda \frac{\hbar^2}{2m} \delta(r-a) \quad,$$

using the notation introduced there.

(a) Calculate the scattering phases $\delta_l(k)$.

(b) Give the scattering cross section for s-waves.

(c) Determine the condition for the maxima of the s-wave scattering cross section.

(d) From here on, assume $g \gg \pi$. Determine the maxima for $ka \ll g$.

(e) Show that there are sharp and broad resonances. Show that the Breit–Wigner formula holds near the sharp resonances.

(f) Determine the poles of $e^{2i\delta_l} - 1$ on the negative real E-axis and compare with Problem 17.2.

18.3 Prove that

$$\frac{\partial}{\partial E} \alpha_l(E) < 0 \quad,$$

where, according to (18.60),

$$\alpha_l(E) = \frac{\partial}{\partial r} \log R_l^<(r)\big|_{r=a} \quad.$$

Hint: Rewrite the Schrödinger equation for $u_l(r) = rR_l^<(r)$ in the form

$$-\frac{\partial}{\partial r}\left[u_l^2 \frac{\partial^2 \log u_l}{\partial(k^2)\partial r}\right] = u_l^2$$

and integrate this expression.

18.4 Calculate the total cross section for s-wave scattering on a completely impenetrable sphere, i.e., for

$$v(r) = \begin{cases} \infty, & r < a \\ 0, & r > a \end{cases} \quad.$$

18.5 Calculate the phase shift δ_0 from s-wave scattering states for an attractive and for a repulsive square-well potential.

18.6 Consider the potential of Problem 17.3, $V(\boldsymbol{x}, \boldsymbol{x}') = \lambda \varrho(|\boldsymbol{x}|) \varrho(|\boldsymbol{x}'|)$ and determine the scattering states. Show that this potential leads only to s-wave scattering.

18.7 To complete the discussion after (18.101), calculate the integral

$$\lim_{ka \to \infty} \int d\Omega \, \cot^2 \frac{\vartheta}{2} J_1^2 (ka \sin \vartheta) \quad.$$

19. Supersymmetric Quantum Theory

19.1 Generalized Ladder Operators

We recall the treatment of the harmonic oscillator by means of ladder operators a and $a^\dagger$ and pose the following question: Can one also represent other Hamiltonians as the "absolute square" of an operator and then construct their solutions algebraically?

Let the Hamiltonian H^0

$$H^0 = -\frac{1}{2}\frac{d^2}{dx^2} + V^0(x) \tag{19.1}$$

be given, where without loss of generality $V^0(x)$ is chosen in such a way that the ground state energy is zero. The ground state ψ_0 then satisfies

$$H^0\psi_0 \equiv \left[-\frac{1}{2}\frac{d^2}{dx^2} + V^0(x) \right]\psi_0 = 0 \quad, \tag{19.2}$$

whence

$$V^0(x) = \frac{1}{2}\frac{\psi_0''}{\psi_0} \tag{19.3}$$

and

$$H^0 = \frac{1}{2}\left[-\frac{d^2}{dx^2} + \frac{\psi_0''}{\psi_0} \right] \tag{19.4}$$

follow, and this suggests the introduction of the operators

$$Q^\pm = \frac{1}{\sqrt{2}}\left[\mp\frac{d}{dx} - \frac{\psi_0'}{\psi_0} \right] \quad. \tag{19.5}$$

One has

$$2Q^\pm Q^\mp = -\frac{d^2}{dx^2} \pm \left[\frac{d}{dx}, \frac{\psi_0'}{\psi_0} \right] + \left(\frac{\psi_0'}{\psi_0} \right)^2 = -\frac{d^2}{dx^2} \pm \frac{d}{dx}\frac{\psi_0'}{\psi_0} + \left(\frac{\psi_0'}{\psi_0} \right)^2$$

$$= -\frac{d^2}{dx^2} \pm \frac{\psi_0''}{\psi_0} + (1 \mp 1)\left(\frac{\psi_0'}{\psi_0} \right)^2 \quad. \tag{19.6}$$

Defining

$$V^1(x) = V^0(x) - \frac{d}{dx}\frac{\psi_0'}{\psi_0} \tag{19.7}$$

and

$$H^1 = -\frac{1}{2}\frac{d^2}{dx^2} + V^1(x) \quad , \tag{19.8}$$

it then follows that

$$H^0 = Q^+Q^- \quad , \tag{19.9a}$$

$$H^1 = Q^-Q^+ \quad . \tag{19.9b}$$

One refers to H^1 as the supersymmetric (SUSY) partner of H^0.

By virtue of

$$Q^+ = (Q^-)^\dagger \quad , \tag{19.10}$$

we have again represented the Hamiltonian H^0 as an absolute square. However, in the case of potentials which are not harmonic, the commutator of the operators $Q^\pm$ is a function of x:

$$[Q^-, Q^+] = -\frac{d}{dx}\frac{\psi_0'}{\psi_0} = H^1 - H^0 \quad . \tag{19.11}$$

We first demonstrate a few identities.

$$Q^-\psi_0 = 0 \quad . \tag{19.12}$$

Proof:

From (19.9a) and (19.2) it follows that $Q^+Q^-\psi_0 = 0$ and $\langle\psi_0|Q^+Q^-|\psi_0\rangle = 0$, and (19.10) implies the vanishing of the norm of $Q^-|\psi_0\rangle$ and thus of $Q^-\psi_0$ itself.

Multiplying (19.9a) and (19.9b) by Q^+ and Q^-, one further obtains the relations

$$Q^+H^1 - H^0Q^+ = 0 \quad , \quad Q^-H^0 - H^1Q^- = 0 \quad . \tag{19.13}$$

Let a state ψ_n^0 be given with eigenvalue E_n^0 of H^0:

$$Q^+Q^-\psi_n^0 = E_n^0\psi_n^0 \quad .$$

Multiplying by Q^-, one finds

$$Q^-Q^+(Q^-\psi_n^0) = E_n^0(Q^-\psi_n^0) \quad .$$

Thus, $Q^-\psi_n^0$ is an eigenstate of H^1 with eigenvalue E_n^0, except for the ground state ψ_0^0, by (19.12).

Let ψ_n^1 be an eigenstate of H^1 with eigenvalue E_n^1,

$$Q^-Q^+\psi_n^1 = E_n^1\psi_n^1 \quad .$$

Multiplication by Q^+ yields

$$Q^+Q^-(Q^+\psi_n^1) = E_n^1(Q^+\psi_n^1) \quad .$$

Hence, $Q^+\psi_n^1$ is an eigenfunction of H^0 with eigenvalue E_n^1.

The spectra of the two Hamiltonians can be derived from each other (see Fig. 19.1). This is especially useful if one of the problems is exactly solvable or can be treated more easily than the other by controlled approximations (that is, approximations whose accuracy can be verified by independent means).

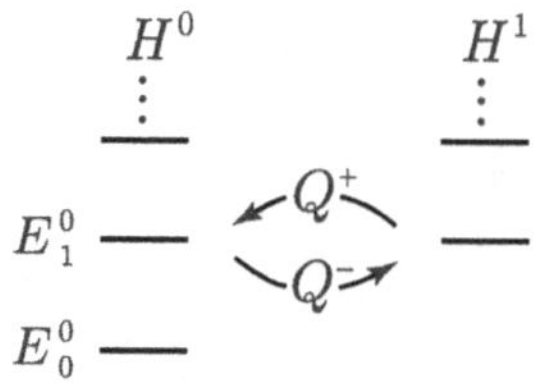

Fig. 19.1. Energy levels of H^0 and H^1

Let us now consider the normalization. From $\langle \psi_n^1 | Q^- Q^+ | \psi_n^1 \rangle = E_n^1 \langle \psi_n^1 | \psi_n^1 \rangle$ one sees the following: If $|\psi_n^1\rangle$ is normalized to unity, then

$$|\psi_n^0\rangle = \frac{1}{\sqrt{E_n^1}} Q^+ |\psi_n^1\rangle \tag{19.14a}$$

is also normalized to unity. An analogous relation holds for $|\psi_n^0\rangle$:

$$|\psi_n^1\rangle = \frac{1}{\sqrt{E_n^0}} Q^- |\psi_n^0\rangle \quad . \tag{19.14b}$$

With the modified notation for the ground state wave function ψ_0^0 (containing the upper suffix 0) of H^0, we define

$$\Phi = -\frac{\psi_0^{0\prime}}{\psi_0^0} \quad , \tag{19.15}$$

whereupon (19.5) yields

$$Q^{\pm} = \frac{1}{\sqrt{2}} \left[\mp \frac{d}{dx} + \Phi(x) \right] \quad , \tag{19.16}$$

and from (19.3) and (19.7) it follows that

$$V^0 = \tfrac{1}{2}(-\Phi' + \Phi^2) \quad , \quad V^1 = \tfrac{1}{2}(\Phi' + \Phi^2) \quad . \tag{19.17}$$

We can express the two Hamiltonians compactly by combining them into a matrix

$$\begin{pmatrix} H^1 & 0 \\ 0 & H^0 \end{pmatrix} = \frac{1}{2}p^2 + \frac{1}{2}\Phi^2 + \frac{1}{2}\sigma_z\Phi' \quad . \tag{19.18}$$

The part H^1 is known as the Fermi sector, and the part H^0 is called the Bose sector (Sect. 19.3).

Integration of (19.15) gives

$$\psi_0^0 = \mathcal{N} \exp \left\{ -\int dx\, \Phi \right\} \quad, \tag{19.19}$$

where $\mathcal{N}$ follows from the normalization of ψ_0^0.

We can invert the procedure by specifying a function Φ, seeing which Hamiltonians result, and then making use of the above results.

19.2 Examples

19.2.1 Reflection-Free Potentials

We first investigate

$$\Phi = \tanh x \quad . \tag{19.20}$$

For this case, according to (19.17), one has the two potentials

$$V^0 = \frac{1}{2}\left(1 - \frac{2}{\cosh^2 x}\right) \quad, \quad V^1 = \frac{1}{2} \quad . \tag{19.21}$$

The potential $-1/\cosh^2 x$ has the constant potential $1/2$ as a supersymmetric partner, corresponding to a free particle.

The normalized ground state wave function of H^0 follows from (19.19) together with $\int dx\, \tanh x = \log \cosh x$:

$$\psi_0^0(x) = \frac{1}{\cosh x}\frac{1}{\sqrt{2}} \quad . \tag{19.22}$$

The ground state energy is $E_0^0 = 0$.

The eigenstates ψ_k^1 of H^1 and the energy eigenvalues E_k^1 are

$$\psi_k^1 = e^{ikx} \quad \text{and} \quad E_k^1 = \tfrac{1}{2}(1 + k^2) \quad . \tag{19.23}$$

Thus, by means of the operators

$$Q_1^{\pm} = \frac{1}{\sqrt{2}}\left(\mp\frac{d}{dx} + \tanh x\right) \tag{19.24}$$

and Eqn. (19.14a), the remaining normalized eigenfunctions of H^0 are given by:

$$\psi_k^0 = \frac{1}{\sqrt{E_k^1}}Q_1^+\psi_k^1 = \frac{(-ik + \tanh x)}{\sqrt{1 + k^2}}e^{ikx} \quad . \tag{19.25}$$

The corresponding energy eigenvalues are also E_k^1. The continuum states (19.25) have the remarkable property of possessing no reflected waves. Potentials of the type (19.21) are therefore called *reflection-free potentials*. (See also Problem 3.6, where the eigenstates are determined by elementary methods.)

Remark: Why is there no state satisfying $Q^- Q^+ \psi = 0$?

For a normalizable wave function, this would lead to $Q^+ \psi = 0$. This implies $\psi = \cosh x$, a non-normalizable wave function. Since this contradicts the assumption, no state of the problem H^1 exists with energy $E = 0$. Moreover, for the examples investigated here, $V^1(x) \geq V^0(x)$, i.e., V^1 is more repulsive. This is in complete analogy to the nonexistence of eigenstates of $a^\dagger$ in the case of the harmonic oscillator (see Chap. 3).

For more general Φ,

$$\Phi = n \tanh x \quad ,$$

$$\Phi^2 = n^2 \tanh^2 x = n^2 \frac{\cosh^2 x - 1}{\cosh^2 x} \quad , \quad \Phi' = \frac{n}{\cosh^2 x} \quad , \tag{19.26}$$

we find

$$V^0 = \frac{1}{2}(\Phi^2 - \Phi') = \frac{1}{2}\left[n^2 - \frac{n(n+1)}{\cosh^2 x}\right] \quad ,$$

$$V^1 = \frac{1}{2}(\Phi^2 + \Phi') = \frac{1}{2}\left[n^2 - \frac{n(n-1)}{\cosh^2 x}\right] \quad . \tag{19.27}$$

The two SUSY partners are the reflection-free potentials $-1/\cosh^2 x$ with amplitudes $n(n-1)/2$ and $(n+1)n/2$, where the continuum begins at $n^2/2$. Thus, by successive application of SUSY quantum mechanics, one can obtain the eigenstates of the reflection-free potential $-n(n+1)/2 \cosh^2 x$ from the free motion of a particle. The number of bound states is n.

From (19.19) and $\int dx \Phi = n \log \cosh x$), one obtains for the ground state

$$\psi_0^0(x) \propto \frac{1}{\cosh^n x} \quad . \tag{19.28}$$

The ground state energy is 0. The remaining $(n-1)$ eigenvalues are identical to those of the problem V^1. The eigenfunctions may be obtained by applying $Q_n^+ = (-d/dx + n \tanh x)/\sqrt{2}$.

We note the close connection to classical nonlinear dynamics, where the solutions (19.22) and (19.25) occur as translational modes and as harmonic oscillations about a soliton. The case $n = 1$ is related to sine–Gordon theory, and $n = 2$ is related to ϕ^4-theory[1].

---

[1] J. Rubinstein: J. Math. Phys. **11**, 258 (1970); R. Rajaraman: Phys. Rep. **216**, 227 (1975)

Reflection-free potential $n = 2$

For $n = 2$, one has the two potentials

$$V^0 = 2 - \frac{3}{\cosh^2 x} \quad , \quad V^1 = 2 - \frac{1}{\cosh^2 x} \tag{19.29}$$

and the operators

$$Q_2^{\mp} = \frac{1}{\sqrt{2}} \left(\pm \frac{d}{dx} + 2 \tanh x \right) \quad . \tag{19.30}$$

We solved the problem V^1 for $n = 2$ earlier. Except for a shift of the zero of the energy by $3/2$, it is identical to V^0 for the $n = 1$ case. From the scattering states, we obtain

$$
\begin{aligned}
Q_2^+ \psi_k^1 &= \frac{1}{\sqrt{2}} \left(-\frac{d}{dx} + 2 \tanh x \right) \left(\frac{-ik + \tanh x}{\sqrt{1 + k^2}} \right) e^{ikx} \\
&= \frac{1}{\sqrt{2(1 + k^2)}} \left[-ik(-ik + \tanh x) - \frac{1}{\cosh^2 x} - 2ik \tanh x + 2 \tanh^2 x \right] e^{ikx} \\
&= \frac{1}{\sqrt{2(1 + k^2)}} (3 \tanh^2 x - 3ik \tanh x - 1 - k^2) e^{ikx} \quad . \tag{19.31}
\end{aligned}
$$

Now, since the eigenvalues are

$$E_k^2 = \tfrac{3}{2} + \tfrac{1}{2}(1 + k^2) = 2 + \tfrac{1}{2} k^2 \quad ,$$

the normalized scattering states become

$$\psi_k^0 = \frac{3 \tanh^2 x - 3ik \tanh x - 1 - k^2}{\sqrt{(4 + k^2)(1 + k^2)}} e^{ikx} \quad . \tag{19.31'}$$

In addition to the ground state of energy 0 found in (19.28),

$$\psi_0^0(x) = \frac{\sqrt{3}}{2} \frac{1}{\cosh^2 x} \quad , \quad E_0^0 = 0 \quad , \tag{19.32}$$

we find from (19.14a), (19.22), and (19.30) as a second normalized bound state

$$\psi_1^0 = \sqrt{\frac{2}{3}} \frac{1}{\sqrt{2}} \left(-\frac{d}{dx} + 2 \tanh x \right) \frac{1}{\sqrt{2} \cosh x} = \sqrt{\frac{3}{2}} \frac{\tanh x}{\cosh x} \quad , \tag{19.33}$$

with energy

$$E_1^0 = 2 - \tfrac{1}{2} = \tfrac{3}{2} \quad . \tag{19.34}$$

19.2.2 The δ-function

Choosing for Φ the step function

$$\Phi = \varepsilon(x) = \Theta(x) - \Theta(-x) \quad , \tag{19.35}$$

we find

$$\begin{aligned}
\Phi^2 &= 1 \quad , \quad \Phi' = 2\delta(x) \quad , \\
V^0 &= -\delta(x) + \tfrac{1}{2} \quad , \\
V^1 &= \delta(x) + \tfrac{1}{2} \quad .
\end{aligned} \tag{19.36}$$

The two supersymmetric partners are the attractive and the repulsive δ-potential.

19.2.3 The Harmonic Oscillator

From

$$\Phi = \omega x \quad , \tag{19.37}$$

one finds two harmonic oscillators

$$\begin{aligned}
H^0 &= -\frac{1}{2}\frac{d^2}{dx^2} + \frac{1}{2}\omega^2 x^2 - \frac{1}{2}\omega \quad , \\
H^1 &= -\frac{1}{2}\frac{d^2}{dx^2} + \frac{1}{2}\omega^2 x^2 + \frac{1}{2}\omega
\end{aligned} \tag{19.38}$$

with the zero point of energy shifted by ω. Using (19.12),

$$\sqrt{2}Q^-\psi_0^0 = \left(\frac{d}{dx} + \omega x\right)\psi_0^0 = 0 \quad ,$$

the ground state of H^0 becomes

$$\psi_0^0 = e^{-\omega x^2/2} \quad , \quad E_0^0 = 0 \quad . \tag{19.39}$$

Since H^1 differs from H^0 only by the shift ω, its lowest eigenstate is also

$$\psi_1^1 = e^{-\omega x^2/2} \quad , \quad E_1^1 = \omega \quad . \tag{19.40}$$

In order to obtain the first excited state of H^0, we need only to apply Q^+ to $\psi_1^1 \equiv \psi_0^0$:

$$Q^+ e^{-\omega x^2/2} \quad , \quad E_1^0 = \omega \quad . \tag{19.41}$$

This procedure is then iterated. Thus, we have again derived the algebraic solution of the harmonic oscillator given in Sect. 3.1.

19.2.4 The Coulomb Potential

As a further example for the treatment of problems solvable in an elementary way by means of SUSY quantum mechanics, we consider the bound states of the Coulomb potential. After introduction of the substitution $R_{nl} = u_{nl}/r$, the radial Schrödinger equation becomes

$$H_l\, u_{nl} = \left[-\frac{1}{2}\frac{d^2}{dr^2} + \frac{l(l+1)}{2r^2} - \frac{\gamma}{r} \right] u_{nl} = \frac{m}{\hbar^2} E_{nl}\, u_{nl} \quad , \qquad (19.42)$$

$$\gamma = \frac{Ze^2 m}{\hbar^2} \quad . \qquad (19.43)$$

We claim that (19.42) can be formulated as a problem of SUSY quantum mechanics with the operators

$$Q_l^{\pm} = \frac{1}{\sqrt{2}} \left[\mp \frac{d}{dr} - \frac{l+1}{r} + \kappa \right] \qquad (19.44)$$

and that the corresponding Φ is given by

$$\Phi = -\frac{l+1}{r} + \kappa \quad .$$

This expression yields

$$\Phi^2 = \frac{(l+1)^2}{r^2} - \frac{2\kappa(l+1)}{r} + \kappa^2 \quad , \quad \Phi' = \frac{l+1}{r^2} \quad ,$$

and from (19.17) one finds the two SUSY partners

$$V^0 = \frac{1}{2} \left[\frac{l(l+1)}{r^2} - \frac{2\kappa(l+1)}{r} + \kappa^2 \right] \qquad (19.45a)$$

and

$$V^1 = \frac{1}{2} \left[\frac{(l+1)(l+2)}{r^2} - \frac{2\kappa(l+1)}{r} + \kappa^2 \right] \quad . \qquad (19.45b)$$

Comparison of (19.45a) with (19.42) gives for the constant κ

$$\kappa = \frac{\gamma}{l+1} \quad . \qquad (19.46)$$

Introducing the abbreviation

$$\eta_l = \frac{1}{2} \left(\frac{\gamma}{l+1} \right)^2 \qquad (19.47)$$

we obtain from (19.45a,b), (19.42), and (19.9a,b)

$$Q_l^+ Q_l^- = H_l + \eta_l \tag{19.47a}$$

and

$$Q_l^- Q_l^+ = H_{l+1} + \eta_l \quad . \tag{19.47b}$$

Thus, we have succeeded in representing H_l and H_{l+1} as a supersymmetric pair. One can immediately read off the commutator of Q_l^- and Q_l^+ from (19.47a,b):

$$[Q_l^-, Q_l^+] = H_{l+1} - H_l = \frac{l+1}{r^2} \quad .$$

We label the states u_{nl} by the principal quantum number $n \geq l+1$. By (19.47a), the lowest stationary state with angular momentum quantum number l, that is $u_{l+1,l}$, is the eigenstate of $Q_l^+ Q_l^-$ of eigenvalue zero. This satisfies the linear differential equation

$$\sqrt{2} Q_l^- u_{l+1,l} = \left(\frac{d}{dr} - \frac{l+1}{r} + \kappa \right) u_{l+1,l} = 0 \quad . \tag{19.48}$$

Rewriting this as

$$\frac{d}{dr} \log u_{l+1,l} = \frac{l+1}{r} - \kappa$$

we find

$$u_{l+1,l} = N r^{l+1} e^{-\kappa r} \quad . \tag{19.49a}$$

Here, N is a normalization constant. The energy eigenvalue follows from (19.42) and (19.47a)

$$E_{l+1,l} = -\eta_l \frac{\hbar^2}{m} = -\frac{Z^2 e^4 m}{2(l+1)^2 \hbar^2} \quad . \tag{19.49b}$$

In (19.49a), we have found for each l the lowest state (see Fig. 19.2). The states $u_{n,l}$ with $n > l+1$ are now to be determined.

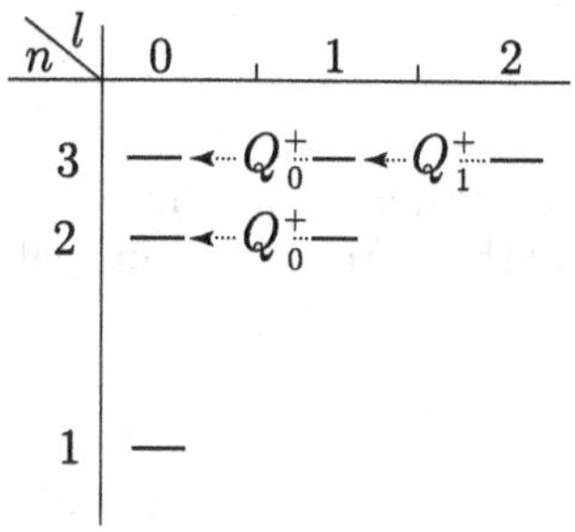

Fig. 19.2. The action of the ladder operators on the bound states of the Coulomb potential

For this, we need only examine the action of the ladder operators $Q_l^{\pm}$. In relations (19.47a,b), the same constant η_l enters, and therefore all the algebraic results for the operators H^0 and H^1 of the preceding section can be applied to H_l and H_{l+1}. In particular,

$$H_l\, u_{nl} = \left[-\frac{1}{2}\frac{d^2}{dr^2} + \frac{l(l+1)}{2r^2} - \frac{\gamma}{r} \right] u_{nl} = \frac{m}{\hbar^2} E_{nl} u_{nl} \tag{19.42}$$

yields the Schrödinger equation

$$H_{l+1}\left(Q_l^{-} u_{nl}\right) = \frac{m}{\hbar^2} E_{nl}\left(Q_l^{-} u_{nl}\right) \quad . \tag{19.50}$$

Hence, $Q_l^{-} u_{nl}$ is an eigenfunction of H_{l+1} with eigenvalue $mE_{nl}/\hbar^2$. One can immediately convince oneself of (19.50) directly by replacing H_l by (19.47a) in (19.42), multiplying by Q_l^{-}, and substituting (19.47b). Similarly,

$$H_{l+1}\, u_{n,l+1} = \frac{m}{\hbar^2} E_{n,l+1}\, u_{n,l+1} \tag{19.51}$$

yields the eigenvalue equation

$$H_l\, Q_l^{+} u_{n,l+1} = \frac{m}{\hbar^2} E_{n,l+1}\, Q_l^{+} u_{n,l+1} \quad . \tag{19.52}$$

Hence, $Q_l^{+} u_{n,l+1}$ is an eigenfunction of H_l.

We must now determine the value of the principal quantum numbers of $Q_l^{-} u_{n,l}$ and $Q_l^{+} u_{n,l+1}$. We will see that this is also n. Beginning with $u_{l+1,l}$, $l \geq 1$, we can construct $Q_{l-1}^{+} u_{l+1,l}$, with energy eigenvalue $E_{l+1,l} = -\eta_l \hbar^2/m$. The ground state for $l-1$, $u_{l,l-1}$, has energy $E_{l,l-1} = -\eta_{l-1}\hbar^2/m$. We claim that no further state lies between these two, i.e., $Q_{l-1}^{+} u_{l+1,l} \propto u_{l+1,l-1}$. Suppose that there were another state lying between these two. One could then apply the operator Q_{l-1}^{-} to it and would then obtain an eigenstate of H_l with eigenvalue below $E_{l+1,l}$, which is impossible. Continuing this line of argument, one obtains

$$Q_l^{+} u_{n,l+1} \propto u_{n,l} \quad . \tag{19.53}$$

Now that all the bound states and eigenvalues, beginning with $u_{l+1,l}$, can be determined (see Fig. 19.2), it then also follows that

$$Q_l^{-} u_{n,l} \propto u_{n,l+1} \quad (n > l+1) \quad .$$

We now express these results in terms of the principal quantum number n. Starting with (19.49), the state $u_{n,n-1}$, we obtain all the states with principal quantum number n:

$$u_{n,n-1}\,,\ Q_{n-2}^{+} u_{n,n-1}\,,\ Q_{n-3}^{+} Q_{n-2}^{+} u_{n,n-1}\,,\ \cdots$$
$$\cdots\,,\ Q_0^{+} \cdots Q_{n-2}^{+} u_{n,n-1} \quad . \tag{19.54}$$

These all have energy

$$E_{n,l} = -\eta_{l=n-1}\frac{\hbar^2}{m} = -\frac{Z^2 e^4 m}{2n^2 \hbar^2} \quad .$$

(19.55)

Hence, without a great deal of effort, we have recovered the bound states of the Coulomb potential and their energies.

19.3 Additional Remarks

Supersymmetric quantum mechanics, which has a close relationship to supersymmetric field theory, is the study of quantum mechanical systems whose Hamiltonian H is constructed from anticommuting charges Q which are the square root of H:

$$2H = \{Q, Q^\dagger\} = QQ^\dagger + Q^\dagger Q \quad ,$$

(19.56)

$$0 = \{Q, Q\} \quad .$$

(19.57)

This implies

$$[Q, H] = 0 \quad ;$$

(19.58)

i.e., the charge is conserved. This type of Hamiltonian contains coordinates which are quantized by commutators and anticommutators. These are mixed by supersymmetry transformations. For a particle with spin, the position and spin orientation form a pair of such coordinates. The explicit realization of Q and $Q^\dagger$ is then

$$Q = (p + i\Phi(x))\hat{\psi}^\dagger \quad , \quad Q^\dagger = (p - i\Phi(x))\hat{\psi} \quad ,$$

(19.59)

in which x and p are Bose degrees of freedom, whereas $\hat{\psi}$ and $\hat{\psi}^\dagger$ are Fermi degrees of freedom, with the corresponding commutation (anticommutation) relations ($\hbar = 1$)

$$[x, p] = i \quad ,$$
$$\{\hat{\psi}^\dagger, \hat{\psi}\} = 1 \quad , \quad \{\hat{\psi}, \hat{\psi}\} = \{\hat{\psi}^\dagger, \hat{\psi}^\dagger\} = 0 \quad .$$

(19.60)

This yields $\{Q^\dagger, Q^\dagger\} = \{Q, Q\} = 0$ and

$$H = \tfrac{1}{2}p^2 + \tfrac{1}{2}\Phi^2(x) - \tfrac{1}{2}[\hat{\psi}^\dagger, \hat{\psi}]\Phi'(x) \quad .$$

(19.61)

Using the (2×2) representation

$$\hat{\psi}^\dagger = \sigma_- = \begin{pmatrix} 0 & 0 \\ 1 & 0 \end{pmatrix} \quad , \quad \hat{\psi} = \sigma_+ = \begin{pmatrix} 0 & 1 \\ 0 & 0 \end{pmatrix} \quad ,$$
$$[\hat{\psi}^\dagger, \hat{\psi}] = -\sigma_z \quad ,$$

(19.62)

we find from (19.61)

$$H = \tfrac{1}{2}(p^2 + \Phi^2(x)) + \tfrac{1}{2}\sigma_z \Phi'(x) \quad . \tag{19.63}$$

Equation (19.63) is identical to the previous matrix representation of the two Hamiltonians H^0 and H^1. The sectors H^0 and H^1 are known as Bose and Fermi sectors, respectively. These two sectors have the same energy levels. The only exception is the case in which the ground state of the Bose sector has zero energy and is thus nondegenerate.

Remark: Finally, it deserves to be mentioned that the supersymmetry transformations are represented by the unitary operator[2]

$$U = \exp\{\varepsilon Q + \varepsilon^\dagger Q^\dagger\} \tag{19.64}$$

with anticommuting c-numbers ε and $\varepsilon^\dagger$ (Grassmann algebra).

Supersymmetric one-particle quantum mechanics serves as a model for the investigation of spontaneous breaking of supersymmetry, which is supposed to occur in supersymmetric field theories.[2] The ground state $|0\rangle$ is invariant with respect to supersymmetry transformations provided $U|0\rangle = |0\rangle$. This is satisfied if and only if $Q|0\rangle = Q^\dagger|0\rangle = 0$, i.e., if the ground state energy is zero. If the ground state energy is greater than zero, supersymmetry is spontaneously broken. An example of spontaneously broken supersymmetry is

$$\Phi = g(x^2 - a^2) \quad ,$$
$$H = \frac{p^2}{2} + \frac{g^2}{2}(x^2 - a^2)^2 + gx\sigma_z \quad . \tag{19.65}$$

The two potentials V^0 and V^1 satisfy

$$V^0(-x) = V^1(x) \quad . \tag{19.66}$$

There is no normalizable state in this case with $E_0^0 = 0$, since

$$\int dx\, \Phi(x) = g(\tfrac{1}{3}x^3 - a^2 x) \quad .$$

The ground state energy then is positive $(QQ^\dagger!)$ and degenerate.

The following literature is recommended for further study:

F. Cooper, B. Freedman: Annals of Physics (N.Y.) **156**, 262 (1983)

P.A. Deift: Duke Math. Journ. **45**, 267 (1978)

L. Infeld, T.E. Hull: Rev. Mod. Phys. **23**, 21 (1951)

A. Joseph: Rev. Mod. Phys. **37**, 829 (1967)

G. Junker: *Supersymmetric Methods in Quantum and Statistical Physics* (Springer, Berlin, New York 1995)

---

[2] See, for example, D. Lancaster: Nuov. Cim. **79**, 28 (1984)

D. Lancaster: Nuov. Cim. **79**, 28 (1984)

E. Schrödinger: Proc. Roy. Irish Acad. A**46**, 9 (1940), 183 (1941)

J. Wess, B. Zumino: Nucl. Phys. B **70**, 39 (1974)

E. Witten: Nucl. Phys. B **185**, 513 (1981); **202**, 253 (1982)

Problems

19.1 Solve the spherical harmonic oscillator using supersymmetric quantum mechanics.

19.2 We are given the potential

$$V(x) = \begin{cases} \left(\dfrac{\pi}{a}\right)^2 \left(\dfrac{1}{\sin^2 \frac{\pi x}{a}} - \dfrac{1}{2}\right) & \text{for } 0 < x < a \\ \infty & \text{otherwise} \end{cases} .$$

Calculate the energy spectrum with the aid of supersymmetric quantum theory. Hint: Set $\hbar = m = 1$.

19.3 Find the stationary states for the potential

$$V^0(x) = \frac{9}{2} - \frac{6}{\cosh^2 x}$$

by means of supersymmetric quantum mechanics.

19.4 Calculate the hydrogen states $u_{2,l}$ by means of supersymmetric quantum mechanics.

20. State and Measurement in Quantum Mechanics

20.1 The Quantum Mechanical State, Causality, and Determinism

> "There was a time when newspapers said that only twelve men understood the theory of relativity. I do not believe that there ever was such a time. ... On the other hand, I think it is safe to say that no one understands quantum mechanics."
>
> R. P. Feynman
> *The Character of Physical Law* (1967) p. 129[1]

Here, "to understand" does not mean just the mastery of the mathematical formalism, but rather an understanding within the framework of our conceptual ideas acquired on the basis of classical and nonrelativistic phenomena. Indeed, one can understand (in this sense of the word) such consequences of special relativity as the Lorentz contraction or time dilation as soon as one has a clear notion of the relativity of simultaneity in coordinate systems which are in motion with respect to one another. Although the Newtonian equations are indeed modified in relativity theory, so that the resulting equations are covariant with respect to Lorentz transformations, the concept of a state – specification of position and velocity – is not altered.

In contrast, the conceptual changes in quantum theory are considerably more incisive. For one thing, the state is given by a vector in a linear, infinite-dimensional space, and the observables are represented by in general non-commuting operators, which ultimately leads to decisive consequences for the measurement process. Let us recall the axioms of quantum theory:

I.   The state is described by a vector $|\psi\rangle$ in a linear space.
II. The observables are represented by hermitian operators, and f(observable) is represented by f(operator).
III. The expectation value of an observable is $\langle\psi|\text{operator}|\psi\rangle$.
IV. The dynamics are given by the Schrödinger equation

$$i\hbar\partial|\psi\rangle/\partial t = H|\psi\rangle.$$

V.   In a measurement of the observable A with the result a, the original state changes into $|a\rangle$.

---

[1] Cited according to J.G. Cramer: Rev. Mod. Phys. **589**, 647 (1986)

From axioms II and III it follows that a measurement of A gives the eigenvalues a with probability $|c_a|^2$, where

$$|\psi\rangle = \underset{a}{S}\, c_a |a\rangle \quad \text{and} \quad A|a\rangle = a|a\rangle \quad .$$

By classical determinism one means the concept that position and momentum, which of course define the state in classical mechanics, can be specified with arbitrary accuracy. In quantum theory, position and momentum do not simultaneously take precise values, but rather, depending on the wave function, one has a statistical distribution. This situation is nondeterministic. On the other hand, classical physics and quantum mechanics are causal in the following sense: For initial values x, p, the values $x(t), p(t)$ at a later time can be computed from the Newtonian equations; if the initial wave function $\psi(x)$ is given, the wave function $\psi(x,t)$ follows from the Schrödinger equation. The change of the state in a measurement, axiom V, is occasionally regarded as noncausality. What enters here, however, is not so much a lack of causality, but rather a manifestation of the nondeterministic quantum mechanical state.

Here lies one of the root causes of the conceptual difficulties. One must also be aware that the uncertainty relation, which results from the noncommutivity of the observables, cannot be circumvented, no matter how clever the experimental setup. Furthermore, in an experiment the state is changed, the change depending on the type of experiment. For example, let us consider the Heisenberg thought experiment for the determination of position by means of a microscope with resolving power $\Delta x = \lambda/\sin\varphi$ (Fig. 20.1). The uncertainty of the recoil is given by

$$\Delta p_x = \left(\frac{2\pi\hbar}{\lambda}\right)\sin\varphi \;\to\; \Delta x \Delta p_x \sim 2\pi\hbar \quad .$$

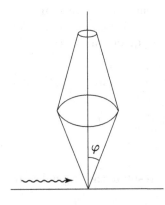

Fig. 20.1. The Heisenberg thought experiment

Can the impact of the photon onto the lens be used for a measurement of p_x? For this purpose, one would have to know the momentum of the lens more precisely than $(2\pi/\lambda)\hbar \sin \varphi$. The position of the lens would become correspondingly imprecise (the mass of the lens does not enter). Since then the principal maximum and the secondary maxima of the image would be blurred, the position of the observed particle would then become uncertain. Here, we also see that, depending on the type of measurement, the state will be changed.

In the course of this chapter, the following problems will be addressed: Clarification of the concept of a state, investigation of the question of whether quantum theory can be extended by hidden variables, and to what extent axiom V can be derived by a quantum mechanical treatment of the whole system. The chapter is subdivided into three main parts: first, consideration of the density matrix, a mathematical scheme for describing the questions posed; second, analysis of an idealized Stern–Gerlach experiment; and third, a look at the Bell inequality and experiments proving the impossibility of local hidden variables.

We remark at the outset that the conceptual difficulties which are inherent in the quantum mechanical state do not imply a weakness of quantum theory, whose validity has never encountered limitations, but only a weakness of our imagination.

20.2 The Density Matrix

20.2.1 The Density Matrix for Pure and Mixed Ensembles

The density matrix is of prime significance for the formulation of quantum statistics. In this field, the terms "statistical operator" and "density operator" are used interchangeably for the density matrix.

Let the system be in the state $|\psi\rangle$. In this state, the observable A has the expectation value

$$\langle A \rangle = \langle \psi | A | \psi \rangle \quad . \tag{20.1}$$

The structure of the expectation value suggests that we define the density matrix

$$\varrho = |\psi\rangle\langle\psi| \quad . \tag{20.2}$$

One has

$$\langle A \rangle = \mathrm{tr}\,(\varrho A) \quad , \tag{20.3a}$$

$$\mathrm{tr}\,\varrho = 1 \quad , \tag{20.3b}$$

$$\varrho^2 = \varrho \tag{20.3c}$$

and

$$\varrho^\dagger = \varrho \quad . \tag{20.3d}$$

Here, the definition of the trace is

$$\text{tr } X = \sum_n \langle n|X|n\rangle \quad , \tag{20.4}$$

where $\{|n\rangle\}$ is an arbitrary complete orthonormal system.

Remark: Proofs of (20.3a–c):

$$\text{tr } \varrho A = \sum_n \langle n|\psi\rangle\langle\psi|A|n\rangle = \sum_n \langle\psi|A|n\rangle\langle n|\psi\rangle = \langle\psi|A|\psi\rangle \quad ,$$

$$\text{tr } \varrho = \text{tr } \varrho 1 = \langle\psi|1|\psi\rangle = 1 \quad ,$$

$$\varrho^2 = |\psi\rangle\langle\psi|\psi\rangle\langle\psi| = |\psi\rangle\langle\psi| = \varrho \quad .$$

Let $\{|n\rangle\}$ and $\{|m\rangle\}$ be two different basis systems. One then has

$$\text{tr } X = \sum_n \langle n|X|n\rangle = \sum_n \sum_m \langle n|m\rangle\langle m|X|n\rangle = \sum_m \sum_n \langle m|X|n\rangle\langle n|m\rangle$$

$$= \sum_m \langle m|X|m\rangle \quad ,$$

and thus the trace is independent of the basis.

If the systems or objects under investigation are all in one and the same state $|\psi\rangle$, one speaks of a *pure ensemble,* or one says that the system is in a *pure state.*

In order to verify the probability predictions contained in the wave function $|\psi\rangle$ experimentally, one must in fact investigate an ensemble of identically prepared objects. If for example

$$|\psi\rangle = \sum_n c_n|n\rangle \quad , \tag{20.5}$$

the eigenvalue a_n will result in an ensemble of N such objects N_n times. The larger N, the more precisely N_n/N approaches the probability $|c_n|^2$, i.e.,

$$|c_n|^2 = \lim_{N\to\infty} \frac{N_n}{N} \quad , \tag{20.6}$$

and the expectation value correspondingly becomes

$$\langle A\rangle = \sum_n |c_n|^2 a_n = \lim_{N\to\infty} \frac{1}{N} \sum_n N_n a_n \quad . \tag{20.7}$$

In addition to this inherent statistical character residing in the states themselves, an ensemble can also contain a statistical distribution of states. If an ensemble with distinct states is present, one refers to this as a *mixed ensemble,* a *mixture,* or one speaks of a *mixed state.* Suppose that, of the N representatives of the ensemble, $\mathcal{N}_1$ are in the state $|\psi_1\rangle, \ldots, \mathcal{N}_i$ in the state $|\psi_i\rangle, \ldots$; then the probability that an arbitrarily chosen element of the ensemble is in the state $|\psi_i\rangle$ is given by $p_i = \mathcal{N}_i/N$, with

$$\sum_i p_i = 1 \quad .$$

The expectation value of A is then

$$\langle A \rangle = \sum_i p_i \langle \psi_i | A | \psi_i \rangle \quad . \tag{20.8}$$

This expectation value can also be represented by the density matrix, which we now define as

$$\varrho = \sum_i p_i |\psi_i\rangle\langle\psi_i| \quad . \tag{20.9}$$

One has

$$\langle A \rangle = \operatorname{tr} \varrho A \quad , \tag{20.9a}$$

$$\operatorname{tr} \varrho = 1 \quad , \tag{20.9b}$$

$$\varrho^2 \neq \varrho \quad \text{and} \quad \operatorname{tr} \varrho^2 < 1 \quad , \quad \text{if } p_i \neq 0 \text{ for more than one } i \quad , \tag{20.9c}$$

and

$$\varrho^\dagger = \varrho \quad . \tag{20.10}$$

Remark: *Proofs* of (20.9a–c):

$$\operatorname{tr} \varrho A = \sum_n \sum_i p_i \langle \psi_i | A | n \rangle \langle n | \psi_i \rangle = \sum_i p_i \langle \psi_i | A | \psi_i \rangle = \langle A \rangle \quad .$$

This additionally yields, with $A = 1$, (20.9b).

$$\varrho^2 = \sum_i \sum_j p_i p_j |\psi_i\rangle\langle\psi_i|\psi_j\rangle\langle\psi_j| \neq \varrho \quad .$$

For each $|\psi\rangle$, the expectation value of ϱ,

$$\langle\psi|\varrho|\psi\rangle = \sum_i p_i |\langle\psi|\psi_i\rangle|^2 \geq 0 \quad ,$$

is positive semidefinite. Since ϱ is Hermitian, the eigenvalues P_m of ϱ are positive real:

$$\varrho|m\rangle = P_m|m\rangle \quad,$$

$$\varrho = \sum_{m=1}^{\infty} P_m|m\rangle\langle m| \quad,$$

$$P_m \geq 0 \quad, \quad \sum_{m=1}^{\infty} P_m = 1 \quad, \quad \langle m|m'\rangle = \delta_{mm'} \quad. \tag{20.11}$$

In this basis, $\varrho^2 = \sum_m P_m^2|m\rangle\langle m|$, and evidently $\operatorname{tr}\varrho = \sum_m P_m^2 < 1$, if more than one state is present.

One can also show (20.9c) directly from (20.9), provided at least two distinct, but not necessarily orthogonal, states occur in (20.8):

$$\begin{aligned}
\operatorname{tr}\varrho^2 &= \sum_n \sum_{i,j} p_i p_j \langle\psi_i|\psi_j\rangle\langle\psi_j|n\rangle\langle n|\psi_i\rangle \\
&= \sum_{i,j} p_i p_j |\langle\psi_i|\psi_j\rangle|^2 < \sum_i p_i \sum_j p_j = 1 \quad.
\end{aligned}$$

The criterion for a pure or a mixed state is $\operatorname{tr}\varrho^2 = 1$ or $\operatorname{tr}\varrho^2 < 1$, respectively.

For the density matrix (20.9), the expectation value of a *projection operator* $|n\rangle\langle n|$ is

$$\operatorname{tr}(|n\rangle\langle n|\varrho) = \sum_i p_i|\langle n|\psi_i\rangle|^2 = \sum_i p_i|c_n^{(i)}|^2 \quad, \tag{20.12a}$$

that is, it is equal to the probability of obtaining the state n as the result of a measurement. For projection operators with a continuous spectrum, as in the case of projection onto position eigenfunctions, one has

$$\operatorname{tr}(|x\rangle\langle x|\varrho) = \sum_i p_i|\langle x|\psi_i\rangle|^2 = \sum_i p_i|\psi_i(x)|^2 \quad. \tag{20.12b}$$

Let the system consist of *two subsystems* 1 and 2 with orthonormal states $\{|1n\rangle\}$ and $\{|2m\rangle\}$. A general pure state in the direct product space is then

$$|\psi\rangle = \sum_{n,m} c_{nm}|1n\rangle|2m\rangle \quad, \quad \sum_{n,m}|c_{nm}|^2 = 1 \tag{20.13}$$

and the corresponding density matrix is

$$\varrho = |\psi\rangle\langle\psi| = \sum_{n,m} \sum_{n',m'} c_{nm}c_{n'm'}^*|1n\rangle|2m\rangle\langle 1n'|\langle 2m'| \quad. \tag{20.14}$$

If we carry out measurements concerning only subsystem 1, that is, if the operators corresponding to the observables A under investigation only act on the states $|1n\rangle$, then

$$\langle A\rangle = \operatorname{tr}_1 \operatorname{tr}_2 \varrho A = \operatorname{tr}_1[(\operatorname{tr}_2 \varrho)A] \quad. \tag{20.15}$$

Here, tr_i signifies taking the trace over subsystem i. Thus, the pertinent object for discussing these questions is the density matrix

$$\hat{\varrho} = \mathrm{tr}_2\ \varrho = \sum_n \sum_{n'} \sum_m c_{nm} c^*_{n'm} |1n\rangle\langle 1n'|\quad, \tag{20.16}$$

i.e., ϱ averaged over subsystem 2. Its square is

$$\hat{\varrho}^2 = \sum_n \sum_{n'} \sum_m c_{nm} c^*_{n'm} \sum_{n_1} \sum_{n'_1} \sum_{m_1} c_{n_1 m_1} c^*_{n'_1 m_1} |1n\rangle\langle 1n'|1n_1\rangle\langle 1n'_1|$$

$$= \sum_{n,n',n'_1} \left(\sum_m c_{nm} c^*_{n'm} \right) \left(\sum_{m_1} c_{n'm_1} c^*_{n'_1 m_1} \right) |1n\rangle\langle 1n'_1|\quad.$$

In general, $\hat{\varrho}^2 \neq \hat{\varrho}$. Only if the c_{nm} take the form

$$c_{nm} = b_n d_m \text{ with } \sum_n |b_n|^2 = 1 \text{ and } \sum_m |d_m|^2 = 1$$

is $\hat{\varrho}^2 = \hat{\varrho}$. Under this assumption,

$$|\psi\rangle = \left(\sum_n b_n |1, n\rangle \right) \left(\sum_m d_m |2, m\rangle \right)$$

is thus the direct product of two pure states of the subspaces 1 and 2. One also sees from

$$\mathrm{tr}_1\ \hat{\varrho}^2 = \sum_{n,n'} \left(\sum_m c_{nm} c^*_{n'm} \right) \left(\sum_{m_1} c_{n'm_1} c^*_{nm_1} \right)$$

that $\mathrm{tr}_1\ \hat{\varrho}^2 = 1$ holds only under the above assumption, whereas otherwise $\hat{\varrho}$ represents the density matrix of a mixed ensemble. If the information of a subspace is ignored, the pure state becomes a mixed state. Although the total system is in the pure state (20.13), the density matrix $\hat{\varrho}$, which yields all expectation values pertaining only to subsystem 1, in general represents a mixed ensemble.

Projection operators

The projection operator in subspace 2, $|2m\rangle\langle 2m|$, projects onto the state $|2m\rangle$. Application to the state (20.13) yields

$$|\psi\rangle \rightarrow |2m\rangle\langle 2m|\psi\rangle = \sum_n c_{nm} |1n\rangle |2m\rangle\quad. \tag{20.17a}$$

Under projection onto the state $|2m\rangle$, the density matrix is changed as follows:

$$\varrho \rightarrow \frac{|2m\rangle\langle 2m|\varrho|2m\rangle\langle 2m|}{\mathrm{tr}_1\langle 2m|\varrho|2m\rangle}\quad. \tag{20.17b}$$

If ϱ represents a pure ensemble, then this is also the case for the projected density matrix.

Filters can be represented mathematically by the application of projection operators onto the density matrix.

20.2.2 The von Neumann Equation

We now derive the equation of motion for an arbitrary density matrix (20.9). The Schrödinger equation

$$i\hbar\frac{\partial}{\partial t}|\psi_i\rangle = H|\psi_i\rangle \quad , \quad -i\hbar\frac{\partial}{\partial t}\langle\psi_i| = \langle\psi_i|H$$

yields

$$\frac{\partial}{\partial t}\varrho = \sum_i p_i(|\dot\psi_i\rangle\langle\psi_i| + |\psi_i\rangle\langle\dot\psi_i|) = \frac{1}{i\hbar}\sum_i p_i(H|\psi_i\rangle\langle\psi_i| - |\psi_i\rangle\langle\psi_i|H) \quad ,$$

and thus finally

$$\frac{\partial}{\partial t}\varrho = -\frac{i}{\hbar}[H, \varrho] \quad . \tag{20.18}$$

This is the *von Neumann equation*, which also holds for time dependent Hamiltonians. It describes the time evolution of the density matrix in the Schrödinger picture. Note that the equation of motion for ϱ should under no circumstances be confused with the equation of motion for operators in the Heisenberg representation, see Sect. 8.5.2. The von Neumann equation is the quantum mechanical analog of the Liouville equation of classical statistical mechanics.

From (20.18), for an arbitrary density matrix at initial time t_0, the formal solution (16.4) of the Schrödinger equation

$$|\psi(t)\rangle = U(t, t_0)|\psi(t_0)\rangle \tag{20.19}$$

yields, together with (16.5), the solution of the initial-value problem

$$\varrho(t) = U(t, t_0)\varrho(t_0)U(t, t_0)^\dagger \quad . \tag{20.20}$$

Theorem. The quantity tr ϱ^2 is time independent. Hence, a pure (mixed) state remains pure (mixed).

Proof:

$$\text{tr } \varrho^2(t) = \text{tr } U\varrho(t_0)U^\dagger U\varrho(t_0)U^\dagger = \text{tr } \varrho^2(t_0) \quad ,$$

by the cyclic invariance of the trace.

The density matrix in the Heisenberg representation

$$\varrho(t_0) = U(t, t_0)^\dagger \varrho(t)U(t, t_0) \tag{20.21}$$

is time independent.

The expectation value of the observable A is given in these two representations by

$$\langle A \rangle = \operatorname{tr}\left(\varrho(t)A\right) = \operatorname{tr}\left(\varrho(t_0)A_{\mathrm{H}}(t)\right) \quad , \tag{20.22}$$

where $A_{\mathrm{H}}(t) = U^\dagger(t,t_0)AU(t,t_0)$ is the Heisenberg operator. The time dependence of the expectation value derives, in the Schrödinger picture, from the density matrix, and, in the Heisenberg picture, from the operator.

20.2.3 Spin-1/2 Systems

The situation is especially transparent in state spaces of low dimension. In particular, we consider spin-1/2 systems. The states $|\uparrow\rangle$, i.e.,

$$\chi_+ = \begin{pmatrix} 1 \\ 0 \end{pmatrix} \quad ,$$

and $|\downarrow\rangle$, i.e.,

$$\chi_- = \begin{pmatrix} 0 \\ 1 \end{pmatrix} \quad ,$$

are eigenstates of σ_z; $(\sigma_z\chi_\pm = \pm\chi_\pm)$.

We will use both Dirac notation and the spinor representation, depending on what is convenient, and we recall the Pauli spin matrices (9.14).

Let us first discuss rotations in spin space. A rotation through an angle ϑ about the axis $\hat{n}$ is represented by the unitary transformation

$$U = \exp\left\{\frac{\mathrm{i}}{2}\vartheta\hat{n}\cdot\boldsymbol{\sigma}\right\} = \mathbb{1}\cos\frac{\vartheta}{2} + \mathrm{i}\hat{n}\cdot\boldsymbol{\sigma}\sin\frac{\vartheta}{2} \quad . \tag{20.23}$$

Indeed, with the help of (9.18b), one sees that

$$U\boldsymbol{\sigma}U^\dagger = \hat{n}(\hat{n}\cdot\boldsymbol{\sigma}) - \hat{n}\times(\hat{n}\times\boldsymbol{\sigma})\cos\vartheta + \hat{n}\times\boldsymbol{\sigma}\sin\vartheta \quad . \tag{20.24}$$

For rotations about the x-axis, (20.23) and (20.24) simplify to

$$U_x = \mathbb{1}\cos\frac{\vartheta}{2} + \mathrm{i}\begin{pmatrix} 0 & 1 \\ 1 & 0 \end{pmatrix}\sin\frac{\vartheta}{2} \tag{20.23'}$$

and

$$U_x\boldsymbol{\sigma}U_x^\dagger = \begin{pmatrix} 1 & 0 & 0 \\ 0 & \cos\vartheta & -\sin\vartheta \\ 0 & \sin\vartheta & \cos\vartheta \end{pmatrix}\begin{pmatrix} \sigma_x \\ \sigma_y \\ \sigma_z \end{pmatrix} \quad . \tag{20.24'}$$

With the help of the rotation operation (20.23′), we can now determine the eigenstates of $\hat{t}\cdot\boldsymbol{\sigma}$, i.e., the eigenstates in the direction $\hat{t} = (0, -\sin\vartheta, \cos\vartheta)$.

From (20.24′) it follows that

$$U_x\hat{t}\cdot\boldsymbol{\sigma}U_x^\dagger = (0, -\sin\vartheta, \cos\vartheta)\begin{pmatrix} \sigma_x \\ \cos\vartheta\,\sigma_y - \sin\vartheta\,\sigma_z \\ \sin\vartheta\,\sigma_y + \cos\vartheta\,\sigma_z \end{pmatrix} = \sigma_z \quad . \tag{20.25}$$

Applying (20.25) to $\chi_\pm$, we find

$$U_x(\hat{t} \cdot \boldsymbol{\sigma})U_x^\dagger \chi_\pm = \sigma_z \chi_\pm = \pm \chi_\pm$$

and hence $(\hat{t} \cdot \boldsymbol{\sigma})(U_x^\dagger \chi_\pm) = \pm (U_x^\dagger \chi_\pm)$.

The two eigenfunctions of $\hat{t} \cdot \boldsymbol{\sigma}$ are therefore given by

$$U_x^\dagger \chi_+ = \begin{pmatrix} \cos \vartheta/2 \\ -i\sin \vartheta/2 \end{pmatrix} \quad, \quad U_x^\dagger \chi_- = \begin{pmatrix} -i\sin \vartheta/2 \\ \cos \vartheta/2 \end{pmatrix} . \tag{20.26}$$

We now present a few special cases.

For $\vartheta = -\pi/2$, one obtains the eigenfunctions of σ_y:

$$\chi_+^y = \frac{1}{\sqrt{2}}\begin{pmatrix} 1 \\ i \end{pmatrix} = \frac{1}{\sqrt{2}}(\chi_+ + i\chi_-) \quad,$$

$$\chi_-^y = \frac{1}{\sqrt{2}}\begin{pmatrix} i \\ 1 \end{pmatrix} = \frac{1}{\sqrt{2}}(i\chi_+ + \chi_-) \quad. \tag{20.27a}$$

Analogously, the eigenfunctions of σ_x are

$$\chi_+^x = \frac{1}{\sqrt{2}}\begin{pmatrix} 1 \\ 1 \end{pmatrix} \quad, \quad \chi_-^x = \frac{1}{\sqrt{2}}\begin{pmatrix} 1 \\ -1 \end{pmatrix} . \tag{20.27b}$$

A rotation through $\vartheta = 2\pi$ gives $U = -\mathbb{1}$ and $\chi \to -\chi$. In a rotation through $360°$, the spinor changes its sign. It takes a rotation through $\vartheta = 4\pi$ to produce $U = +\mathbb{1}$ and $\chi \to +\chi$. This illustrates the spinor properties of $\chi_\pm$, whereas (20.24') expresses the vector character of $\boldsymbol{\sigma}$.

We now return to the density matrix and discuss its spin part. For definiteness, we can imagine that we are dealing with electron beams. An electron beam of spin $\uparrow$ has the density matrix $\varrho_\uparrow = |\uparrow\rangle\langle\uparrow|$, while an electron beam of spin $\downarrow$ has the density matrix $\varrho_\downarrow = |\downarrow\rangle\langle\downarrow|$. If one mixes the two beams in the ratio 50:50, the density matrix is

$$\varrho_M = \tfrac{1}{2}(|\uparrow\rangle\langle\uparrow| + |\downarrow\rangle\langle\downarrow|) \quad. \tag{20.28}$$

This is indeed a mixed state, since

$$\varrho_M^2 = \tfrac{1}{2}\varrho_M \quad. \tag{20.29}$$

Let us compare this to the pure state

$$|\rangle = \frac{1}{\sqrt{2}}(|\uparrow\rangle + e^{i\alpha}|\downarrow\rangle) \quad, \tag{20.30}$$

consisting of the linear superposition of $|\uparrow\rangle$ and $|\downarrow\rangle$. The corresponding density matrix is

$$\varrho_\alpha = \tfrac{1}{2}(|\uparrow\rangle\langle\uparrow| + |\downarrow\rangle\langle\downarrow| + e^{-i\alpha}|\uparrow\rangle\langle\downarrow| + e^{i\alpha}|\downarrow\rangle\langle\uparrow|) \quad. \tag{20.31}$$

In contrast to ϱ_M, here, interference terms occur. Let us also give the matrix representation

$$\varrho_{nm} = \langle n|\varrho|m \rangle \tag{20.32}$$

with $n, m = \uparrow, \downarrow$ for the two density matrices (20.28) and (20.31):

$$\varrho_{\mathrm{M}} = \frac{1}{2} \begin{pmatrix} 1 & 0 \\ 0 & 1 \end{pmatrix} \quad , \quad \varrho_{\alpha} = \frac{1}{2} \begin{pmatrix} 1 & e^{-i\alpha} \\ e^{i\alpha} & 1 \end{pmatrix} \quad . \tag{20.33a,b}$$

The difference between the two density matrices also makes itself felt in the expectation values and thus has measurable consequences. For the pure state,

$$\langle A \rangle_{\alpha} = \tfrac{1}{2} (\langle \uparrow | A | \uparrow \rangle + \langle \downarrow | A | \downarrow \rangle + 2 \operatorname{Re} e^{i\alpha} \langle \uparrow | A | \downarrow \rangle) \quad , \tag{20.34}$$

whereas for the mixture

$$\langle A \rangle_{\mathrm{M}} = \tfrac{1}{2} (\langle \uparrow | A | \uparrow \rangle + \langle \downarrow | A | \downarrow \rangle) \tag{20.35}$$

holds. In both cases, the expectation value of σ_z is zero; in contrast, the expectation value of σ_x for the pure state is $\operatorname{Re} \exp\{i\alpha\} = \cos\alpha$ and thus is not equal to zero for $\alpha \neq \pi/2$, whereas it vanishes for the mixture.

Interference terms are present in the density matrix and in the expectation values of the pure state. Comparison of the two density matrices shows that the mixture arises by superposition of the ϱ_{α} or averaging over the phases:

$$\varrho_{\mathrm{M}} = \frac{1}{2\pi} \int_0^{2\pi} d\alpha \, \varrho_{\alpha} \quad . \tag{20.36}$$

Polarization of spin-1/2 particles

The most general density matrix in spin space is

$$\varrho = \tfrac{1}{2} (\mathbb{1} + \boldsymbol{b} \cdot \boldsymbol{\sigma}) \quad , \tag{20.37}$$

since every 2×2-matrix can be represented as a linear combination of the unit matrix and the Pauli matrices, and furthermore $\operatorname{tr} \mathbb{1} = 2$ and $\operatorname{tr} \sigma_i = 0$. Choosing $\boldsymbol{b}$ to point in the z-direction, we find

$$\varrho = \frac{1}{2} \begin{pmatrix} 1 + b & 0 \\ 0 & 1 - b \end{pmatrix} \quad .$$

Since $(1 \pm b)/2$ are probabilities, $|b| \leq 1$ must hold. When does ϱ characterize a pure case? For this we compute ϱ^2 using (9.18b):

$$\varrho^2 = \frac{1}{4} (\mathbb{1} + 2\boldsymbol{\sigma} \cdot \boldsymbol{b} + (\boldsymbol{\sigma} \cdot \boldsymbol{b})(\boldsymbol{\sigma} \cdot \boldsymbol{b})) = \frac{1}{2} \left(\frac{1 + b^2}{2} \mathbb{1} + \boldsymbol{\sigma} \cdot \boldsymbol{b} \right) \quad .$$

Thus $|\boldsymbol{b}| = 1$ is a pure case. Finally let us also compute the expectation value of $\boldsymbol{\sigma}$ using $\operatorname{tr} \sigma_i \sigma_j = 2\delta_{ij}$:

$$\langle \sigma_i \rangle = \operatorname{tr} \varrho \sigma_i = \operatorname{tr} [\tfrac{1}{2} (\mathbb{1} + b_j \sigma_j) \sigma_i] = b_i \quad . \tag{20.38}$$

The expectation value of $\boldsymbol{\sigma}$ is just $\boldsymbol{b}$, and the degree of polarization is characterized by $b = |\boldsymbol{b}|$. The case $b = 0$ represents an unpolarized beam, whereas the pure state with $b = 1$ represents a completely polarized beam.

20.3 The Measurement Process

20.3.1 The Stern–Gerlach Experiment

Let a beam of atoms or electrons move through the inhomogeneous field of a magnet. Because of the force $m_z(\partial B_z/\partial z)$, (1.9), the beam is split depending on the values of the magnetic moment m_z. The experimental setup is shown schematically in Fig. 20.2.

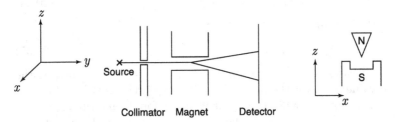

Fig. 20.2. The Stern–Gerlach experiment; the cross section of the magnet in the beam direction

In the following, we assume that the total angular momentum, later designated for brevity as the spin, is $1/2$. The motion in the y-direction is force-free and can be separated off. The Hamiltonian which remains is

$$H = \frac{p_z^2}{2m} + B(z)\mu_B\sigma_z \approx \frac{p_z^2}{2m} + (B + B'z + \ldots)\mu_B\sigma_z \quad , \tag{20.39}$$

describing the motion in the z-direction and the spin in the strong and inhomogeneous magnetic field oriented along the z-direction. Here, we have expanded $B(z)$ in a Taylor series. The motion is then determined by the Pauli equation

$$i\hbar\frac{\partial}{\partial t}\Psi = H\Psi \quad , \quad \Psi = \begin{pmatrix} u_+ \\ u_- \end{pmatrix} \quad . \tag{20.40}$$

With respect to the z-dependence, the problem is equivalent to free fall in a gravitational field,

$$i\hbar\frac{\partial}{\partial t}u_\pm = \left(\frac{p_z^2}{2m} \pm B\mu_B \pm B'\mu_B z\right)u_\pm \quad . \tag{20.41}$$

Let the spatial wave function be a wave packet $f(z)$ concentrated about the z-axis before entering the magnetic field at the time $t = 0$. At the time t, it is then approximately

$$u_\pm(z,t) = f(z \pm Ct^2)e^{\mp i\alpha t} \quad . \tag{20.42}$$

The particles of spin ↑ and ↓ are deflected downwards and upwards, respectively. The constant

$$C = \frac{B' \mu_B}{2m} \tag{20.43}$$

corresponds, in the case of the earth's gravitational field, to $g/2$, one-half the acceleration of gravity, and here

$$\alpha = B \mu_B / \hbar \quad . \tag{20.44}$$

20.3.2 The Quasiclassical Solution

Although the precise derivation of (20.42) is not important for what follows, we briefly explain how this classical motion follows from the Schrödinger equation. We start with the WKB solution for a general potential $V(z)$. The stationary states corresponding to the energy E are

$$\exp \left\{ \pm \frac{i}{\hbar} \int dz \sqrt{2m(E - V(z))} - iEt/\hbar \right\} \quad . \tag{20.45}$$

We construct a wave packet from states of different energy. The phase is stationary for (see Sect. 2.10.1)

$$\begin{aligned} 0 &= \frac{\partial}{\partial E} \left(\pm \int dz \sqrt{2m(E - V(z))} - Et \right) \\ &= \pm \int dz \sqrt{\frac{m}{2(E - V(z))}} - t. \end{aligned} \tag{20.46}$$

One must replace E by its value where the wave packet has a maximum. Thus the center of the wave packet follows the classical orbit, because one knows from classical mechanics that

$$\frac{m\dot{x}^2}{2} + V(x) = E \quad , \tag{20.47}$$

$$\dot{x} = \pm \sqrt{(-V(x) + E)\frac{2}{m}} \quad ,$$

$$\int dt = \pm \int \frac{dx}{\sqrt{2(E - V(x))/m}} \quad . \tag{20.48}$$

The heavier the particle and the larger the energy, the better the quasiclassical approximation.

20.3.3 The Stern–Gerlach Experiment as an Idealized Measurement

We now utilize the Stern–Gerlach experiment as a model for the process of measurement in quantum mechanics. Here, the object being measured is the spin of the particle and the measuring apparatus is the position of the particle after traversing the field. We know from (20.42) that

for $\uparrow$, $z < 0$, and for $\downarrow$, $z > 0$.

This apparatus, whose readout (position of the pointer) is the z-coordinate of the particle, is thus appropriate for distinguishing $\uparrow$ from $\downarrow$ (and − by means of the amount of deflection − for determining the size of the moment). The requirement that the z-coordinate of the particle serve as the pointer of a measuring instrument implies that the deflections must be macroscopically distinguishable. Formally, this implies that the overlap of the two wave packets $f(z + Ct^2)$ and $f(z - Ct^2)$ in (20.42) must be negligible. After having calibrated the apparatus in this way, we can now investigate general states.

Let us consider the initial state

$$\Psi(z,0) = \frac{1}{\sqrt{2}}(\chi_+ + \chi_-)f(z) \quad, \tag{20.49}$$

which after traversing the field becomes

$$\Psi(z,t) = \frac{1}{\sqrt{2}}\left[\chi_+ f(z + Ct^2)e^{-i\alpha t} + \chi_- f(z - Ct^2)e^{i\alpha t}\right] \quad. \tag{20.50}$$

The polarization and the pointer (the z-coordinate) are coupled. There is a unique correlation between the state of the spin and the state of the pointer. Neither the spin nor the pointer is in an eigenstate. In the basis of the states $|z\rangle|\pm\rangle$, the density matrix

$$\varrho_{zz'} = \langle\pm|\langle z|\psi(t)\rangle\langle\psi(t)|z'\rangle|\pm'\rangle \tag{20.51}$$

of the state (20.50) is given by

$$\varrho_{zz'} = \frac{1}{2}\begin{pmatrix} f(z+Ct^2)f(z'+Ct^2)^* & f(z+Ct^2)f(z'-Ct^2)^*e^{-i2\alpha t} \\ f(z-Ct^2)f(z'+Ct^2)^*e^{i2\alpha t} & f(z-Ct^2)f(z'-Ct^2)^* \end{pmatrix}. \tag{20.52}$$

Measurement of spin observables: After the spin-1/2 particle has passed through the Stern–Gerlach apparatus, suppose that its spin is measured. The measurement can take place (i) ignoring the pointer position z or (ii) for a particular pointer position z.

(i)   If one ignores the pointer position, then ϱ is equivalent to

$$\hat{\varrho} \equiv \int dz\langle z|\varrho|z\rangle = \frac{1}{2}\begin{pmatrix} 1 & 0 \\ 0 & 1 \end{pmatrix} \quad, \tag{20.53}$$

where the second equality follows from the nonoverlapping of the pointer wave functions $f(z+Ct^2)$ and $f(z-Ct^2)$. For an observable $F(\boldsymbol{\sigma})$, depending only on spin operators,

$$\mathrm{tr}_{z,\sigma}\left(\varrho F(\boldsymbol{\sigma})\right) = \mathrm{tr}_\sigma\left(\hat{\varrho}F(\boldsymbol{\sigma})\right) \quad. \tag{20.54}$$

The density matrix $\hat{\varrho}$ corresponds to a mixed ensemble (see (20.33a)). The pure ensemble (ϱ) is replaced by the mixed ensemble $(\hat{\varrho})$.

(ii)   Filtering out a "pointer" position: We now consider only the particles with pointer position "z-positive", i.e., we construct

$$\mathrm{tr}_{z>0}\, \varrho = \int_0^\infty dz\, \varrho_{zz} = \frac{1}{2} \begin{pmatrix} 0 & 0 \\ 0 & 1 \end{pmatrix} \; .$$

Because of the normalization, the density matrix is then

$$\begin{pmatrix} 0 & 0 \\ 0 & 1 \end{pmatrix} \equiv \varrho_\downarrow \; . \tag{20.55}$$

The particles deflected up have the spin wave function $|\downarrow\rangle$.

For a measurement with the result z-positive (spin negative) the state goes over to $|\downarrow\rangle$. This is consistent with *axiom V*, which was postulated earlier from the requirement that repeated experiments give the same result. For this simple experiment the time dependent Schrödinger equation for object and apparatus can be solved explicitly, and axiom V follows from axioms I–IV of quantum mechanics.

The fact that the particles which have been filtered off at a particular pointer position are in that eigenstate corresponding to the eigenvalue measured is known as *reduction* of the wave function. Going over from ϱ to $\hat{\varrho}$ with respect to all observables related to the spin can also be regarded as a reduction of the wave function. The density matrix $\hat{\varrho}$ (20.53) describes an ensemble composed of 50% spin-up and 50% spin-down states. If N particles are subjected to this Stern–Gerlach experiment, then as far as their spin properties are concerned they are completely equivalent to $N/2$ particles in the state $|\uparrow\rangle$ and $N/2$ particles in the state $|\downarrow\rangle$.

20.3.4 A General Experiment and Coupling to the Environment

We now investigate a general experiment. Let O be the object and A the apparatus including its readout. At the time $t = 0$, let the state of the whole system consisting of object and apparatus be

$$|\psi(0)\rangle = \sum_n c_n |O, n\rangle |A\rangle \; . \tag{20.56}$$

The $|O, n\rangle$ are object states, and $|A\rangle$ is the (metastable) initial state of the apparatus. At a later time t, after the interaction of the object with the measuring apparatus, the state

$$|\psi(t)\rangle = \sum_n c_n |O, n\rangle |A(n)\rangle \tag{20.57}$$

is realized, where the final states of the apparatus $|A(n)\rangle$, $n = 1, 2, \ldots$, must be macroscopically distinguishable. The density matrix for the (pure) state (20.57) is

$$\varrho(t) = |\psi(t)\rangle\langle\psi(t)| \quad . \tag{20.58}$$

If we do not read off the result of the measurement, the density matrix for observables relevant to the object O is

$$\hat{\varrho} = \text{tr}_{A(n)}\ \varrho(t) = \sum_n |c_n|^2 |O, n\rangle\langle O, n| \quad . \tag{20.59}$$

The fact that the final states of the apparatus are macroscopically distinguishable, and thus do not overlap, enters here: $\langle A(n)|A(m)\rangle = \delta_{nm}$. If we do not read off the result of the measurement, a mixture thus occurs with respect to O.

If on the other hand we read off a particular value, e.g., $A(m)$, the density matrix which then applies is

$$|O, m\rangle\langle O, m| \quad . \tag{20.60}$$

The probability of measuring the value $A(m)$ on the apparatus is, by (20.57), $|c_m|^2$.

The fact that in a measurement with the result $A(m)$ the density matrix changes from (20.58) to (20.60) is known as reduction of the wave packet (state).

We now take into account the fact that the object and the apparatus are *never completely isolated* from the *environment*, and we take E to be an additional variable representing all further macroscopic consequences which couple to the state A of the apparatus. The initial state is then

$$|\psi(0)\rangle = \sum_n c_n |O, n\rangle |A\rangle |E\rangle \quad , \tag{20.56'}$$

and after passage through the apparatus this evolves into

$$|\psi(t)\rangle = \sum_n c_n |O, n\rangle |A(n)\rangle |E(n)\rangle \quad . \tag{20.57'}$$

A correlation of the object and the apparatus with the environment develops. If we do "not read off E", which always happens in practice, since we cannot keep track of all the macroscopic consequences, the pure density matrix $|\psi(t)\rangle\langle\psi(t)|$ has to be traced over the environmental degrees of freedom E. Assuming $\langle E(n)|E(n')\rangle = \delta_{nn'}$, because of the different influence of the states n and n' on the environment the density matrix of the object and the apparatus becomes the mixture

$$\hat{\varrho} = \sum_n |c_n|^2 |A(n)\rangle |O, n\rangle\langle O, n|\langle A(n)| \quad . \tag{20.61}$$

The subsystem object + apparatus is thus in a mixed state. N such subsystems behave like $N|c_1|^2$ subsystems in the state $|O,1\rangle|A(1)\rangle$, $N|c_2|^2$ subsystems in the state $|O,2\rangle|A(2)\rangle,\ldots,N|c_n|^2$ subsystems in the state $|O,n\rangle|A(n)\rangle,\ldots$. We emphasize the difference between the density matrix (20.61) resulting from tracing over the environmental degrees of freedom E, which no longer contains any offdiagonal elements, and the pure density (20.58) without coupling to the environment.

(i)   If we do not read off $A(n)$ either, then $\hat{\varrho}$ is equivalent to

$$\hat{\varrho} = \sum_n |c_n|^2 |O,n\rangle\langle O,n| \quad .$$

(ii) If we read off $A(n)$, then the probability of obtaining the particular reading $A(m)$ is

$$\mathrm{tr}_{O,A}\left(|A(m)\rangle\langle A(m)|\,\hat{\varrho}\right) = |c_m|^2 \quad .$$

And in this case, the density matrix which then applies is

$$|A(m)\rangle|O,m\rangle\langle O,m|\langle A(m)| \quad .$$

From then on, it does not matter if we disregard A. Taking the trace over $A(n)$ yields for the observable O the density matrix $|O,m\rangle\langle O,m|$.

The key problem in the theory of measurement is the reduction of the wave function and in particular the question of when it takes place. This problem is illustrated quite drastically by means of *"Schrödinger's cat"*: Suppose that a cat within a closed box is killed by a $|\uparrow\rangle$-particle but not by a $|\downarrow\rangle$-particle. Now consider the effect of the state $(|\uparrow\rangle + |\downarrow\rangle)$, which for instance can be produced by a Stern–Gerlach apparatus. Suppose that a particle in the state $(|\uparrow\rangle + |\downarrow\rangle)$ hits the cat. The state of the spin and the cat makes a transition to $|\uparrow\rangle|$dead cat$\rangle + |\downarrow\rangle|$living cat$\rangle$, a pure state. When is it decided whether the cat is dead or alive? Just when the observer opens the cat's box? – An objective statement independent of the conscious mind of the observer would be impossible. – What is the consequence of including the observer himself in the quantum mechanical description?

According to the point of view presented above in connection with (20.56′) to (20.61), the cat (together with the mechanism for killing it which was not mentioned above) is linked to other macroscopic objects. These are influenced differently in the two final states so that their respective wave functions do not overlap. For everything that follows, these macroscopic consequences are not recorded; they are traced over. The final state of the cat is described by a mixture of states, Eq. (20.61),

$$\hat{\rho} = \frac{1}{2}\left(|\uparrow\rangle\,|\mathrm{dead}\rangle\,\langle\mathrm{dead}|\,\langle\uparrow| + |\downarrow\rangle\,|\mathrm{living}\rangle\,\langle\mathrm{living}|\,\langle\downarrow|\right) \quad ,$$

containing the states of a dead cat and a living cat. The cat is thus *either* dead *or* alive and not in a pure state |dead cat⟩ + |living cat⟩, which would include both possibilities.

This concludes the *essential part* of our consideration of *measurements*; the two following sections (20.3.5, 20.3.6) just give a few illustrations.

Remark: Generalizing the above considerations to arbitrary macroscopic systems the following picture emerges. Suppose we prepared a macroscopic system in a linear superposition of orthogonal states which would lead to distinguishable consequences in the environment. The interaction with the environment would then cause the offdiagonal elements of the density matrix of the system to go to zero. The system considered by itself would finally find itself in one of these states with a probability which follows both from the initial superposition as well as the density matrix.

The interaction and the entanglement of the object with the environment leads to mixed density matrices for the subsystems (often paraphrased as a local disappearance of phase relations). For this phenomenon the term "decoherence" is now generally used.

Linear superpositions of quantum states of *macroscopic* systems would be affected by decoherence within extremely short times, so that such superpositions would never be observed in practice. Hence, the interaction with the environment in effect imposes superselection rules onto the possible quantum states of macroscopic systems. The kind of quantum states which arise because of decoherence are determined by the nature and the form of the interactions the object experiences. For example, the scattering processes by molecules of a gas or by photons lead to localisation of an object. The coherence between macroscopically different positions is rapidly destroyed by scattering processes[2]. Classical properties emerge in a practically irreversible manner through the unavoidable and ubiquitous interaction with the environment.

The interaction with the "environment" leads to the emergence of a preferred set of states[3] of the system amongst which no phase relations (i.e., no linear superpositions) or nondiagonal terms in the density matrix survive. These states belong to the basis of operators which commute with the interaction Hamiltonian of the system and the environment and they can be distinguished through their effect on the environment. The continuous interaction with the environment leads to a rapid destruction of superpositions of these states. This again, leads us to the environment-induced superselection rules and the ensuing classical properties.

The term superselection rule used above has been introduced by Wick, Wigner and Wightman to denote fundamental restrictions on the linear superposition of states. For example, the superposition of states with different electric charge or integer and half-odd-integer spin is excluded. In the case of decoherence the superselection rules are generated dynamically by the interaction with the environment, and possess a range of approximate validity dependent on the specific situation.

Returning again to (20.56′) to (20.61), we may interpret the transition from the pure state (20.56′) (designated in this context as "coherent") to the mixture (20.61) as follows: Due to the entanglement of the object-apparatus states with those of

---

[2] E. Joos and D. Zeh, Z. Phys. B, Cond. Mat., **59**, 223 (1985)

[3] J. P. Paz, J. Habib and W. H. Zurek, Phys. Rev. D **47**, 488 (1993)

the environment states, the coherence of the wave function of the object-apparatus disappears locally. Nevertheless, the extended system (object, apparatus and environment) is still in a pure state and there has been no reduction (collapse) of the overall wave function. Yet, for the object and the apparatus considered by itself, the consequences of the coupling to the environment are, for all practical purposes, as though the environment had acted as an "observer" without revealing the outcome.

20.3.5 Influence of an Observation on the Time Evolution

In order to further analyze the measurement process and its impact, let us return to the Stern–Gerlach experiment and consider the following additional setup. After the atomic beams have traversed the Stern–Gerlach apparatus, we recombine them by means of a complicated field configuration in such a way that all of the deformation and spreading of the wave function is carefully undone, i.e., the state

$$f(z)(c_1 e^{i\varphi_+}\chi_+ + c_2 e^{i\varphi_-}\chi_-) \tag{20.62}$$

is formed. This again is essentially the initial wave function – up to the free motion in the y-direction, which is suppressed here. The phases $\varphi_\pm$ which have been inserted characterize path length differences.

Now, in the region where the beams $+$ and $-$ are macroscopically separated, we can set up a real measuring device whose pointer Z reacts to z by an interaction $U(z - Z)$, so that positive (negative) z leads to positive (negative) Z. We then have for the initial state

$$|\psi_a\rangle = f(z)(c_1\chi_+ + c_2\chi_-)|Z = 0\rangle \quad , \tag{20.63}$$

for the intermediate state

$$|\psi_c\rangle = c_1\chi_+ f(z + Ct^2)|Z = -1\rangle + c_2\chi_- f(z - Ct^2)|Z = +1\rangle \quad , \tag{20.64}$$

and for the state after traversing the entire setup

$$|\psi_e\rangle = f(z)(c_1\chi_+ e^{i\varphi_+}|Z = -1\rangle + c_2\chi_- e^{i\varphi_-}|Z = +1\rangle) \quad . \tag{20.65}$$

The pointer positions are described by $|Z = 0\rangle$ and $|Z = \pm 1\rangle$.

We can now compare the following two situations:

I.   We turn on the coupling to the measuring device Z and obtain the final state (20.65).

II. We turn off the coupling to the measuring device Z and obtain the final state (20.62), multiplied by $|Z = 0\rangle$.

The resulting density matrices are quite different. Although the state (20.65) is a pure state, it is equivalent to a mixture, as far as statements relating only to the particle are concerned. This is due to the fact that the macroscopic states $|Z = \pm 1\rangle$ do not overlap. In situation II, both the total

state and the state of the atom are pure states. In situation I, the final state
of the total system atom + pointer (characterized by spin, z, and Z) is a
pure state, whereas the density matrix of the subsystem (spin, z) is mixed,
unless c_1 or c_2 vanish.

Evidently the physical situation differs according to whether the interaction $U(z - Z)$ between system and pointer is turned on or not. Even if we do
not read off the result, U still influences the atomic system.

In order to further illustrate the back-reaction of the experiment on
the object, we consider the following thought experiment.[4] We connect two
Stern–Gerlach devices in series (Fig. 20.3). Let the initial state of the atoms
in the atomic beam be $|a\rangle$, let $\{|e\rangle\}$ be a basis for the final states, and let
$\{|c\rangle\}$ be a basis for the intermediate states. We now determine the probability of a transition to the final state $|e\rangle$, by representing this in terms of the
transition amplitudes $U_{ac}^{(1)}$ and $U_{ce}^{(2)}$ of $|a\rangle$ to $|c\rangle$ and $|c\rangle$ to $|e\rangle$.

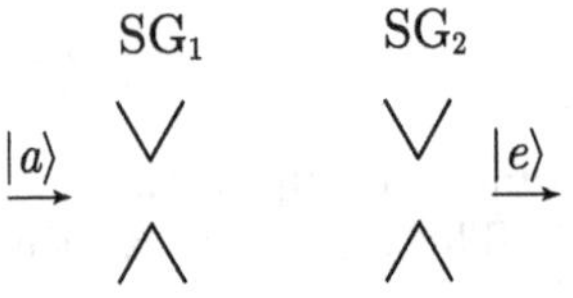

Fig. 20.3. Two Stern-Gerlach apparatuses SG_1 and SG_2 in series

For an isolated system, the transition probability is

$$P_{a \to e}^{I} = \left| \sum_{c} U_{ac}^{(1)} U_{ce}^{(2)} \right|^2 \quad , \tag{20.66}$$

because $\sum_c U_{ac} U_{ce} = U_{ae}$ holds in an isolated system.

On the other hand, one could also say that the transition probability is
the product of the probabilities $|U_{ac}^{(1)}|^2 |U_{ce}^{(2)}|^2$, summed over all intermediate
states c:

$$P_{a \to e}^{II} = \sum_{c} |U_{ac}^{(1)}|^2 |U_{ce}^{(2)}|^2 \quad . \tag{20.67}$$

The probabilities (20.66) and (20.67) correspond to different experiments.

$P_{a \to e}^{II}$: Here, there is a measurement in the intermediate region, and this
introduces unknown phase factors $\exp\{i\varphi_c\}$ which have to be averaged over.

Experiment 1. Between SG_1 and SG_2, the atoms remain unperturbed.
There is no coupling to the external world, and the transition probability
is $P_{a \to e}^{I}$ from (20.66).

---

[4] W. Heisenberg: *The Physical Principles of the Quantum Theory* (Dover, New
York 1950)

Experiment 2. Between SG_1 and SG_2, there is an influence on the atoms, making possible a determination of the stationary state. However, the result of the measurement is not recorded, and a mixture is formed. The transition probability is $P^{II}_{a \to e}$ from (20.67).

Experiment 3. Between SG_1 and SG_2 an influence on the atoms occurs making possible a determination of the stationary state. Let us assume that c is found. The probability for state e behind SG_2 is then given by $|U_{ce}^{(2)}|^2$.

20.3.6 Phase Relations in the Stern–Gerlach Experiment

It was pointed out earlier that the change of the density matrix of a pure ensemble into that of a mixed one corresponds to averaging over the phases. Let us make an order-of-magnitude analysis of the Stern–Gerlach experiment in order to understand qualitatively why the phase relations between spin-up and spin-down components are lost. We consider a beam with velocity v and width b which is split in the field of the Stern–Gerlach setup (see Fig. 20.4).

The deflection angle of the upper atomic beam is classically

$$\varphi = \frac{\mu_B B' t^2}{2mvt} = \frac{\mu_B B' t}{2p} \quad .$$

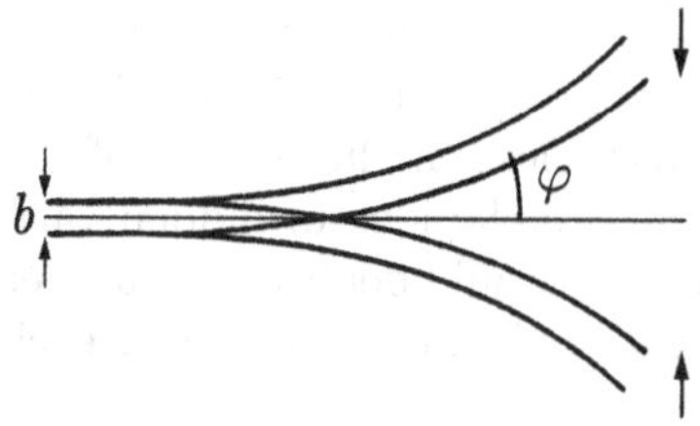

Fig. 20.4. The bending angle in the Stern–Gerlach experiment

The quantum mechanical spreading of the wave packet in the z-direction after a time t is, by (2.12) and (2.16), $\Delta z \approx \hbar t/bm$. Therefore, the uncertainties of the beam angles are

$$\Delta\varphi = \frac{\Delta z}{vt} \approx \frac{\hbar}{bp} \quad , \tag{20.68}$$

corresponding to a spread of $\Delta\varphi \approx \lambda/b$.

The condition that the two atomic beams be separable is $\varphi > \Delta\varphi$. Hence, the inequality

$$\frac{\mu_B B' t b}{2\hbar} > 1 \tag{20.69}$$

follows. The position dependent part of the Larmor energy is given by $\mu_B B' z$. Substituting the beam width b for z and multiplying by $t/\hbar$, we obtain the phase uncertainty of the $\exp\{iEt/\hbar\}$-dependence of the wave function:

$$\Delta\alpha = \frac{\mu_B B' b t}{\hbar} \quad .$$

Together with the preceding inequality, one finds

$$\Delta\alpha > 2 \quad ; \tag{20.70}$$

the phase relations are completely smeared out.[5]

20.4 The EPR Argument, Hidden Variables, the Bell Inequality

20.4.1 The EPR (Einstein–Podolsky–Rosen) Argument

The nondeterministic character of quantum mechanics is unfamiliar to the imagination, which is trained in classical phenomena[6]. Hence, there were repeated attempts to replace quantum theory by a statistical theory. According to these, there exist hidden variables, whose values prescribe the values of all observables for any particular object, except that the hidden variables are unknown to the experimenter, thus yielding the probabilistic character of the theory. The probabilistic character of quantum mechanics would then be quite analogous to that of classical statistical mechanics, where one can imagine that the motion of all particles is in principle known. For example, let us consider a particle of spin $1/2$ in an eigenstate of S_x with eigenvalue $\hbar/2$. According to quantum mechanics, the z-component is not fixed. If one measures it for a very large number of such particles, one finds 50 % of the time $\hbar/2$ and 50 % of the time $-\hbar/2$. According to the idea of hidden variables, for each particle, parameters unknown to us would determine whether $+\hbar/2$ or $-\hbar/2$ results. These hidden variables would prescribe $\pm\hbar/2$ each 50 % of the time.

---

[5] In order to visualize the consequence of this, we imagine that we recombine the two partial beams along the y-axis by means of an additional device – without attempting to remove the spreading and the associated variation of the phase, so that the result will not be the state (20.62). Then the density matrix would still be equivalent to that of a mixture for spin measurements. Indeed, this happens not as in (20.53) because of the nonoverlapping of the two beams, but rather because there is an effective averaging over all phases in the z-integration (compare (20.36)).

[6] Einstein expressed his rejection of quantum theoretical indeterminism by the remark "God does not play dice". Another often cited remark, which reflects Einstein's rejection of the fact that the value of a nondiagonal observable is fixed only when an experiment is performed, is the question: "Is the moon there when nobody looks?"

Even Schrödinger, one of the founders of quantum theory, who sought to construct a classical continuum theory of the microworld, was dissatisfied with the probability interpretation: "If we have to keep these damned quantum jumps, I regret that I ever had anything to do with quantum theory."

By means of a number of thought experiments, Einstein attempted to demonstrate the incompleteness of the quantum mechanical description and to get around the indeterminism and the uncertainty relation. Each of these arguments was refuted in turn by Bohr.

An argument – sometimes referred to as a paradox – due to Einstein, Podolsky, and Rosen (EPR), played a pivotal role in the discussion of indeterminism and the existence of hidden variables; we consider this argument as reformulated by D. Bohm.

Let two spin-1/2 particles in a singlet state

$$|0,0\rangle = \frac{1}{\sqrt{2}}(|\uparrow\rangle|\downarrow\rangle - |\downarrow\rangle|\uparrow\rangle) \tag{20.71}$$

be emitted from a source and move apart. Even if the two particles are separated by an arbitrarily large distance and can no longer communicate with one another, one finds in the state (20.71) the following correlations in a measurement of the one-particle spin states: If one measures the z-component of the spin and finds for particle 1 spin up, particle 2 then has spin down. If one finds for particle 1 spin down, particle 2 has spin up. If instead one measures S_x, then $+\hbar/2$ for particle 1 implies the value $-\hbar/2$ for particle 2, etc.

This expresses the nonlocality of quantum theory. The experiment on particle 1 influences the result of the experiment on particle 2, although they are widely separated. The nonlocality is a consequence of the existence of correlated many-particle states such as the direct product $|\uparrow\rangle|\downarrow\rangle$ and the fact that one can linearly superimpose such states. The nonlocality of quantum mechanics does not lead to contradictions with relativity theory. Although a measurement of a spin component of particle 1 immediately reveals the value of that component for particle 2, no information can be transmitted in this way. Since particle 1 takes values $\pm\hbar/2$ 50 % of the time, this remains true for particle 2, even after the measurement of particle 1. Only by subsequent comparison of the results is it possible to verify the correlation.

Einstein, Podolsky, and Rosen[7] gave the following argument in favor of hidden parameters in conjunction with the EPR thought experiment:

---

[7] The original argumentation of EPR does not refer to singlet states, but to two identical particles, say decay products, moving apart in opposite directions with equal speeds. If the position of particle 1 is measured and the value x is found, then particle 2 has position $-x$. If the momentum of particle 1 is measured with value p, then particle 2 has the momentum $-p$. Here as well, the position and the momentum are fixed neither for particle 1 nor for particle 2, but there is a nonlocal *entanglement* (Verschränkung)[8] between the values for particles 1 and 2. The significance of the reformulated variant of the EPR thought experiment consists in the experimental realization and the possibility of testing the existence of hidden variables.

[8] States such as (20.71), (20.57) and (20.57′) are examples of superpositions of nonoverlapping product states of two or more different degrees of freedom. These cannot be factorized into a single direct product. Following Schrödinger (see

Through the measurement of S_z or S_x of particle 1, the values of S_z or S_x of particle 2 are known. Because of the separation of the particles, there was no influence on particle 2, and therefore the values of S_z, S_x etc. must have been fixed before the experiment. Thus, there must be a more complete theory with hidden variables. In the EPR argument, the consequences of the quantum mechanical state (20.71) are used, but the inherent nonlocality of quantum theory is denied.

There is no room here to go into the various stages of hidden-variable theories and their connection to the von Neumann counterargument and Bohm's new formulation of the guide wave interpretation, implying classical equations of motion with an additional quantum potential. We refer the interested reader to the book by Baumann and Sexl[9].

In what follows, we will consider local hidden variables. These would predetermine which value each of the components of S of particle 1 has and likewise for particle 2. Each of the particles would carry this information independently of the other.

20.4.2 The Bell Inequality

We now show that such local hidden variables lead to results different from those of quantum mechanics. We then compare with experiment.

To this end, we consider a correlation experiment in which a particle of total spin zero decays into two particles each with spin $1/2$. At a sufficiently large distance from the source, a rotatable polarizer and a detector are set up for each particle (see Fig. 20.5), so that the particles can be registered and investigated for a correlation of the spin orientation. Polarizer 1 with angular setting α only lets particle 1 through if its spin in the direction $\hat{n}_\alpha$ takes the value $+\hbar/2$, and polarizer 2 with angular setting β lets particle 2 through only if its spin in the direction $\hat{n}_\beta$ takes the value $+\hbar/2$. Two detectors 1 and 2 register the particles. If they respond, the spin is positive, otherwise negative.

---

literature at the end of this chapter) they are called entangled (verschränkt) for the following reason. For any of the degrees of freedom the outcome of an experiment is uncertain and only characterized by a probability amplitude. Yet, if one of the degrees of freedom is measured, the other is known too. The investigation of entangled states has been fundamental in the study of the measurement process and other basic questions of quantum theory. Recently, the possibility of using entangled states for applications in quantum information processing has been focused on: quantum cryptography, quantum teleportation and quantum computing. In fact the appearance of decoherence represents one of the basic difficulties to the experimental realization of quantum computers.

[9] K. Baumann, R.U. Sexl: *Die Deutungen der Quantentheorie* (Vieweg, Wiesbaden 1984). See also F. Selleri (ed.): *Quantum Mechanics versus local Realism* (Plenum Press, New York 1988)

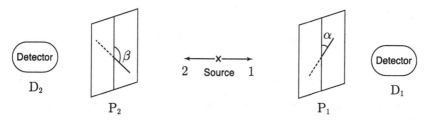

Fig. 20.5. The EPR correlation experiment, with source, polarizers P_1 and P_2 oriented perpendicular to the beams, and detectors D_1 and D_2

We consider the correlation between various angular settings of the polarization experiment. A measure of the correlation is $N(\alpha; \beta)$, defined by the relative number of experiments resulting in particle 1 at angle α positive and particle 2 at angle β positive.

Quantum mechanics gives

$$
\begin{aligned}
N(\alpha; \beta) &\equiv \langle 0, 0 | \delta_{\boldsymbol{\sigma}_1 \cdot \hat{\boldsymbol{n}}_\alpha, 1} \delta_{\boldsymbol{\sigma}_2 \cdot \hat{\boldsymbol{n}}_\beta, 1} | 0, 0 \rangle \\
&= \langle 0, 0 | \tfrac{1}{2}(1 + \boldsymbol{\sigma}_1 \cdot \hat{\boldsymbol{n}}_\alpha) \tfrac{1}{2}(1 + \boldsymbol{\sigma}_2 \cdot \hat{\boldsymbol{n}}_\beta) | 0, 0 \rangle \\
&= \langle 0, 0 | \tfrac{1}{2}(1 + \boldsymbol{\sigma}_1 \cdot \hat{\boldsymbol{n}}_\alpha) \tfrac{1}{2}(1 - \boldsymbol{\sigma}_1 \cdot \hat{\boldsymbol{n}}_\beta) | 0, 0 \rangle \\
&= \tfrac{1}{4}(1 - \hat{\boldsymbol{n}}_\alpha \cdot \hat{\boldsymbol{n}}_\beta) \quad,
\end{aligned}
\tag{20.72}
$$

since $\langle 0, 0 | \boldsymbol{\sigma}_1 | 0, 0 \rangle = 0$ in the singlet state $|0, 0\rangle$. For coplanar detectors (2.72) reduces to

$$
N(\alpha; \beta) = \frac{1}{2} \sin^2 \frac{\beta - \alpha}{2} \quad.
\tag{20.72'}
$$

If hidden variables were really present in nature, we could represent $N(\alpha; \beta)$ by the following sum:

$$
N(\alpha; \beta) = N(\alpha\gamma; \beta) + N(\alpha; \gamma\beta) \quad.
\tag{20.73}
$$

Here, $N(\alpha\gamma; \beta)$ is the relative number of particle pairs in which particle 1 has positive spin at the angles α and γ and negative spin at β, while $N(\alpha; \beta\gamma)$ is the relative number of particle pairs in which instead particle 1 has negative spin at γ. In theories with hidden variables, all of these quantities are available. Now, one has $N(\alpha\gamma; \beta) \le N(\gamma; \beta)$, since $N(\gamma; \beta) = N(\alpha\gamma; \beta) + N(\gamma; \beta\alpha)$ and since both quantities on the right-hand side are nonnegative; similarly, $N(\alpha; \gamma\beta) \le N(\alpha; \gamma)$. Hence, (20.73) implies

$$
N(\alpha; \beta) \le N(\alpha; \gamma) + N(\gamma; \beta) \quad.
\tag{20.74}
$$

This is a simple version of the *Bell inequality*.

Remarks:

(i)   In experiments one often works with the correlation defined by

$$P(\alpha;\beta) = \langle 0,0|\boldsymbol{\sigma}_1 \cdot \hat{\boldsymbol{n}}_\alpha\, \boldsymbol{\sigma}_2 \cdot \hat{\boldsymbol{n}}_\beta|0,0\rangle$$
$$= 4N(\alpha;\beta) - 1 \quad , \tag{20.75}$$

instead of $N(\alpha;\beta)$ itself. Quantum mechanics (20.72') yields for coplanar polarizers

$$P(\alpha - \beta) \equiv P(\alpha;\beta) = -\cos(\alpha - \beta) \quad . \tag{20.72''}$$

In terms of these quantities the Bell inequality reads

$$P(\alpha;\beta) - 1 \leq P(\alpha;\gamma) + P(\gamma;\beta) \quad .$$

(ii)   The limit prescribed by the Bell inequality can be determined as follows. In (20.74) one substitutes for α, β, γ the values $0, \pi, \pi/2$, respectively:

$$N(0;\pi) \leq N\left(0;\frac{\pi}{2}\right) + N\left(\frac{\pi}{2};\pi\right) \quad .$$

In the singlet state, $N(0;\pi) = \frac{1}{2}$ and $N(0;\frac{\pi}{2}) = N(\frac{\pi}{2};\pi)$; hence $N(0;\frac{\pi}{2}) \geq \frac{1}{4}$. Additional values are obtained by successive combination of angles.

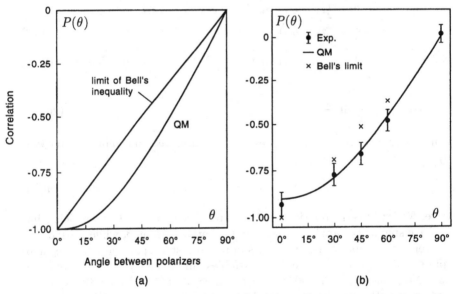

Fig. 20.6. (a) The Correlation $P(\theta) \equiv P(\theta;0)$ according to quantum mechanics (20.72'') and the Bell inequality. (b) Experimental results on the spin correlation of the proton pairs in experiments carried out by Lamehi-Rachti and Mittig in comparison with quantum mechanics (QM) and the limits of Bell's inequality ($\times$) [10]

---

[10] M. Lamehi-Rachti, W. Mittig, Phys. Rev. D **14**, 2543 (1976)

Finally, we contrast the consequences of the Bell inequality with quantum mechanics and compare with experiments. To this end, we compute $N(\alpha; \beta)$, $N(\alpha; \gamma)$, and $N(\gamma; \beta)$ for the three angles $\alpha = 0$, $\gamma = 45°$, and $\beta = 90°$ according to the quantum mechanical formula (20.72'): $(1/2)\sin^2 45°$, $(1/2)\sin^2 22.5°$, $(1/2)\sin^2 22.5°$. Substituting into the Bell inequality (20.74), one would obtain for these probability densities the inequality

$$\sin^2 45° \leq 2\sin^2 22.5°$$

or $0.5 \leq 0.29$, which is evidently violated. Hence, quantum mechanics and local hidden variables are incompatible. The comparison of quantum mechanics and the Bell inequality is illustrated in Fig. 20.6(a). The violation of the Bell inequality was demonstrated experimentally by Lamehi-Rachti and Mittig with pairs of protons and by Aspect et al. with photons; see Fig. 20.6(b). *Experiment provides decisive evidence in favor of quantum mechanics and against local hidden variables.*

Further literature concerning this chapter can be found in the following references:

A. Aspect, P. Grangier, G. Roger: Phys. Rev. Lett. **47**, 460 (1981); Phys. Rev. Lett. **49**, 91 (1982)

A. Aspect, J. Dalibard, G. Roger: Phys. Rev. Lett. **49**, 1804 (1982)

K. Baumann, R. U. Sexl: *Die Deutungen der Quantentheorie* (Vieweg, Wiesbaden 1984)

J. S. Bell: "On the Einstein Podolsky Rosen Experiment", Physics **1**, 195 (1964)

D. Bohm: "A Suggested Interpretation of the Quantum Theory in Terms of 'Hidden' Variables", Phys. Rev. **85**, 166 (1952)

J. G. Cramer: "The Transactional Interpretation of Quantum Mechanics", Rev. Mod. Phys. **589**, 647 (1986)

B. d'Espagnat: Scientific American, Nov. 1979, p. 128

K. Gottfried: *Quantum Mechanics* (Benjamin, New York, Amsterdam 1966) Chap. IV

D. Giulini, E. Joos, C. Kiefer, J. Kupsch, I.-O. Stamatescu, H. D. Zeh, *Decoherence and the appearance of a classical world in quantum theory* (Springer, Berlin Heidelberg 1996)

W. Heisenberg: *The Physical Principles of the Quantum Theory* (Dover, New York 1950)

C. Kiefer and E. Joos: "Decoherence: Concepts and Examples", in *Quantum Future*, Lecture Notes in Physics 517, p. 105, Eds. Ph. Blanchard, A. Jadczyk (Springer, Berlin Heidelberg 1997)

M. Lamehi-Rachti, W. Mittig: Phys. Rev. D**14**, 2543 (1976)

N. D. Mermin: "Is the Moon There when Nobody Looks? Reality and the Quantum Theory", Physics Today, April 1985, 38

N. F. Mott: "The Wave Mechanics of α-Ray Tracks", Roy. Soc. Proc. A**124**, 375 (1929)

J. von Neumann: *Mathematical Foundations of Quantum Mechanics* (Princeton University Press, New Jersey 1964)

W. Pauli: "Die allgemeinen Prinzipien der Wellenmechanik", in *Handbuch der Physik, Encyclopedia of Physics* V/1, *Principles of Quantum Theory I* (Springer, Berlin, Heidelberg 1958)

L. I. Schiff: *Quantum Mechanics*, 3rd ed. (McGraw-Hill, New York 1968) p. 335

E. Schrödinger: "Die gegenwärtige Situation der Quantenmechanik", Die Naturwissenschaften **23**, 807, 823, 844 (1935)

F. Selleri: *Die Debatte um die Quantentheorie* (Vieweg, Wiesbaden 1983)

F. Selleri (ed.): *Quantum Mechanics versus Local Realism* (Plenum, New York 1988)

R. F. Streater and A. S. Wightman, *TCP, Spin and Statistics and all That* (Addison-Wesley, Reading 1964)

E. Wigner: "The Problem of Measurement", Am. J. Phys. **31**, 6 (1963)

E. Wigner: "On Hidden Variables and Quantum Mechanical Probabilities", Am. J. Physics **38**, 1005 (1970)

W. H. Zurek: "Decoherence and the transition from Quantum to Classical", Phys. Today **44**, No. 10, 36 (1991)

Problems

20.1 Show that the singlet state (20.71) takes the form

$$|0,0\rangle = \frac{1}{\sqrt{2}} \left(|e,+\rangle|e,-\rangle - |e,-\rangle|e,+\rangle \right)$$

in an arbitrary basis (Recall (9.5)).

20.2 (a) Show the property

$$\text{tr } AB = \text{tr } BA \quad .$$

(b) Show that: $\text{tr } P_a = \langle a|a\rangle$, where $P_a = |a\rangle\langle a|$.

20.3 Investigate the following EPR experiment. Let the two analyzers be oriented at the angles $0°$, $120°$, or $240°$ and suppose that these orientations are independently set completely stochastically.

(a) What does the Bell inequality say?

(b) What does quantum theory give?

(c) Let N_{equ} (N_{opp}) be the average number of measurements in which the two analyzers give the same (opposite) values for the spin components. Calculate $N_{\text{equ}} - N_{\text{opp}}$:

(α) according to quantum theory $N_{\text{equ}} - N_{\text{opp}} = \sum_i \sum_j \langle 0,0|(\hat{\boldsymbol{a}}_i \cdot \boldsymbol{S}_1)(\hat{\boldsymbol{a}}_j \cdot \boldsymbol{S}_2)|0,0\rangle$ and

(β) under the assumption of hidden variables. Here, $\hat{\boldsymbol{a}}_i, i = 1,2,3$, are unit vectors with the orientations $0°$, $120°$, $240°$.

20.4 Consider a system with Hamiltonian H_0 acted upon by an external, time dependent field $F(t)$, so that the total Hamiltonian is given by

$$H = H_0 + BF(t) \quad .$$

B is an operator, F a classical field. Solve the von Neumann equation with the aid of time dependent perturbation theory under the assumption that $F(t \to -\infty)$ vanishes. For $t = -\infty$, ϱ thus has the usual equilibrium form. What is the expectation value for an operator A at the time t in first order in F?

20.5 Consider a general entangled state, (20.13). Show that two orthogonal basis systems exist, such that

$$|\psi\rangle = \sum_k c_k |1k\rangle |2k\rangle$$

is just given by a single sum. Find the solutions which determine $\{|1k\rangle\}$ and $\{|2k\rangle\}$. These two bases are termed a biorthogonal system or a Schmidt basis.

E. Schrödinger, Proc. Cambridge Phil. Soc. **31**, 555 (1935); E. Schmidt, Math. Annalen **63**, 433 and **64**, 161 (1907)

Appendix

A. Mathematical Tools for the Solution of Linear Differential Equations

A.1 The Fourier Transform

Let $f(t)$ be continuous with at most finitely many discontinuities of the first kind (i.e., $f(t+0)$ and $f(t-0)$ exist) and

$$\int_{-\infty}^{+\infty} dt\, |f(t)| < \infty \quad .$$

Then the Fourier transform

$$\tilde{f}(\omega) = \int_{-\infty}^{+\infty} dt\, e^{i\omega t} f(t) \tag{A.1}$$

exists, and the inverse transform gives

$$\int_{-\infty}^{+\infty} \frac{d\omega}{2\pi} e^{-i\omega t} \tilde{f}(\omega) = \begin{cases} f(t) & \text{at continuous points} \\ \frac{1}{2}\left(f(t+0) + f(t-0)\right) & \\ & \text{at the discontinuities.} \end{cases} \tag{A.2}$$

A.2 The Delta Function and Distributions

This section is intended to give a heuristic understanding of the δ-function and other related distributions as well as a feeling for the essential elements of the underlying mathematical theory.

Definition of a "test function" $F(x), G(x), \dots$: All derivatives exist and vanish at infinity faster than any power of $1/|x|$, e.g., $\exp\{-x^2\}$. In order to introduce the δ-function heuristically, we start with (for arbitrary $F(x)$)

$$F(x) = \int_{-\infty}^{+\infty} \frac{d\omega}{2\pi} e^{-i\omega x} \int_{-\infty}^{+\infty} dx'\, e^{i\omega x'} F(x') \quad,$$

and exchange – without investigating the admissibility of these operations – the order of the integrations:

$$F(x) = \int_{-\infty}^{+\infty} dx'\, F(x') \int_{-\infty}^{+\infty} \frac{d\omega}{2\pi} e^{i\omega(x'-x)} = \int_{-\infty}^{+\infty} dx'\, F(x') \delta(x'-x) \quad .$$

From this, we read off

$$\int_{-\infty}^{+\infty} \frac{d\omega}{2\pi} e^{i\omega(x'-x)} = \delta(x'-x) = \begin{cases} 0 & \text{for } x' \neq x \\ \infty & \text{for } x' = x \end{cases} \qquad (A.3)$$

This "function" of x' thus has the property of vanishing for all $x' \neq x$ and taking the value infinity for $x' = x$, as illustrated schematically in Fig. A.1.

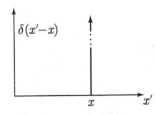

Fig. A.1. The δ-function, schematic representation of (A.3)

It is thus the analogue for integrals of the Kronecker-δ for sums,

$$\sum_{n'} K_{n'} \delta_{n,n'} = K_n \quad .$$

The Dirac δ-function is not a function in the usual sense. In order to give it a precise meaning, we consider in place of the above integral (A.3) one that exists. We can either allow the limits of integration to extend only to some finite value or else introduce a weighting function falling off at infinity. Accordingly, we define the following sequence of functions parameterized by n,

$$\delta_n(x) = \int_{-\infty}^{+\infty} \frac{d\omega}{2\pi} \exp\left\{ i\omega x - \frac{1}{n}|\omega| \right\} = \frac{1}{\pi} \frac{1/n}{x^2 + (1/n)^2} \qquad (A.4a)$$

with the following properties:

I. $$\lim_{n \to \infty} \delta_n(x) = \begin{cases} \infty & \text{for } x = 0 \\ 0 & \text{for } x \neq 0 \end{cases} , \qquad (A.4b)$$

II. $$\lim_{n \to \infty} \int_{-a}^{b} dx\, \delta_n(x) G(x) = G(0) \quad . \qquad (A.4c)$$

Proof of II:

$$\lim_{n \to \infty} \int_{-an}^{bn} dy\, \frac{1/\pi}{y^2 + 1} G\left(\frac{y}{n}\right) = G(0) \int_{-\infty}^{+\infty} dy\, \frac{1/\pi}{y^2 + 1} = G(0) \quad .$$

We thus define the δ-function (distribution) by

$$\int_{-a}^{b} dx\, \delta(x) G(x) = \lim_{n \to \infty} \int_{-a}^{b} dx\, \delta_n(x) G(x) \quad . \qquad (A.5)$$

This definition suggests the following generalization.

Let a sequence of functions $d_n(x)$ be given whose limit as $n \to \infty$ does not necessarily yield a function in the usual sense. Let

$$\lim_{n \to \infty} \int dx \, d_n(x) G(x)$$

exist for each G. One then defines the distribution $d(x)$ via

$$\int dx \, d(x) G(x) = \lim_{n \to \infty} \int dx \, d_n(x) G(x) \quad . \tag{A.6}$$

The generalization (A.6) allows one to introduce additional definitions of importance for distributions.

(i)   Definition of the equality of two distributions: Two distributions are equal,

$$a(x) = b(x) \quad , \tag{A.7a}$$

if $\int dx \, a(x) G(x) = \int dx \, b(x) G(x)$ for every $G(x)$.

(ii) Definition of the sum of two distributions:

$$c(x) = a(x) + b(x) \quad ; \tag{A.7b}$$

$c(x)$ is defined by $c_n(x) = a_n(x) + b_n(x)$.

(iii) Definition of the multiplication of a distribution by a function $F(x)$:

$$d(x)F(x) \quad \text{is defined by} \quad d_n(x)F(x) \quad . \tag{A.7c}$$

(iv) Definition of an affine transformation:

$$d(\alpha x + \beta) \quad \text{is defined by} \quad d_n(\alpha x + \beta) \quad . \tag{A.7d}$$

(v)   Definition of the derivative of a distribution:

$$d'(x) \quad \text{is defined by} \quad d'_n(x) \quad . \tag{A.7e}$$

From these definitions, one has that the same linear operations can be performed for distributions as for ordinary functions. It is not possible to define the product of two arbitrary distributions in a natural way.

Properties of the δ-function:

$$\int_{-\infty}^{+\infty} dx \, \delta(x - x_0) F(x) = F(x_0) \quad , \tag{A.8}$$

$$\int_{-\infty}^{+\infty} dx \, \delta'(x) F(x) = -F'(0) \quad , \tag{A.9}$$

$$\delta(x)F(x) = \delta(x)F(0) \quad , \tag{A.10}$$

$$\delta(xa) = \frac{1}{|a|}\delta(x) \quad . \tag{A.11}$$

Remark: *Proof of (A.11):*

$$\int_{-\infty}^{+\infty} dx\, \delta(xa) F(x) = \lim_{n\to\infty} \int_{-\infty}^{+\infty} dx\, \delta_n(xa) F(x) = \lim_{n\to\infty} \int_{-\infty}^{+\infty} dx\, \delta_n(x|a|) F(x)$$

$$= \lim_{n\to\infty} \frac{1}{|a|} \int_{-\infty}^{+\infty} dy\, \delta_n(y) F\left(\frac{y}{|a|}\right) = \frac{1}{|a|} F(0) \quad .$$

$$\delta(f(x)) = \sum_i \frac{1}{|f'(x_i)|} \delta(x - x_i) \quad , \quad x_i \text{ simple zeros of } f \quad . \tag{A.12}$$

From (A.10) and (A.11), it follows that

$$x\delta(x) = x^2\delta(x) = \ldots = 0 \quad , \tag{A.13}$$

$$\delta(-x) = \delta(x) \quad . \tag{A.14}$$

Fourier transform of the δ-function:

$$\int_{-\infty}^{+\infty} dx\, e^{-i\omega x} \delta(x) = 1 \quad . \tag{A.15}$$

Three-dimensional δ-function:

$$\delta^{(3)}(\boldsymbol{x} - \boldsymbol{x}') = \delta(x_1 - x_1')\delta(x_2 - x_2')\delta(x_3 - x_3') \quad . \tag{A.16a}$$

In spherical coordinates:

$$\delta^{(3)}(\boldsymbol{x} - \boldsymbol{x}') = \frac{1}{r^2}\delta(r - r')\delta(\cos\vartheta - \cos\vartheta')\delta(\varphi - \varphi') \quad . \tag{A.16b}$$

Step function:

$$\Theta_n(x) = \frac{1}{2} + \frac{1}{\pi} \arctan nx \quad ,$$
$$\Theta_n'(x) = \delta_n(x) \quad ,$$
$$\to \Theta'(x) = \delta(x) \quad . \tag{A.17}$$

Other sequences which also represent the δ-function:

$$\delta_n(x) = \frac{1}{\pi x} \sin nx = \int_{-n}^{n} \frac{dk}{2\pi} e^{ikx} \quad , \tag{A.18}$$

$$\delta_n(x) = \sqrt{\frac{n^2}{\pi}} e^{-n^2 x} \quad . \tag{A.19}$$

If a sequence $d_n(x)$ defines a distribution $d(x)$, one then writes symbolically

$$d(x) = \lim_{n\to\infty} d_n(x) \quad .$$

Integral representations

We conclude this section by giving a few integral representations for $\delta(x)$ and related distributions:

$$\delta(x) = \frac{1}{2\pi} \int_{-\infty}^{+\infty} dk \, e^{ikx} , \tag{A.20}$$

$$\Theta(x) = \lim_{\varepsilon \to 0} \frac{1}{2\pi i} \int_{-\infty}^{+\infty} dk \, \frac{e^{ikx}}{k - i\varepsilon} . \tag{A.21}$$

We also define the distributions

$$\delta_+(x) = \frac{1}{2\pi} \int_0^\infty dk \, e^{ikx} , \tag{A.22a}$$

$$\delta_-(x) = \frac{1}{2\pi} \int_{-\infty}^0 dk \, e^{ikx} . \tag{A.22b}$$

These can also be represented in the form

$$\delta_\pm(x) = \mp \frac{1}{2\pi i} \lim_{\varepsilon \to 0} \frac{1}{x \pm i\varepsilon} . \tag{A.23}$$

Further, one has

$$\lim_{\varepsilon \to 0} \frac{1}{x \pm i\varepsilon} = P\frac{1}{x} \mp i\pi\delta(x) , \tag{A.24}$$

where P designates the Cauchy principal value,

$$P \int dx \frac{1}{x} G(x) = \lim_{\varepsilon \to 0} \left(\int_{-\infty}^{-\varepsilon} + \int_\varepsilon^\infty \right) dx \frac{1}{x} G(x) . \tag{A.25}$$

The distributions $\delta_\pm$ have the properties

$$\delta_\pm(-x) = \delta_\mp(x) , \tag{A.26}$$

$$x\delta_\pm(x) = \mp \frac{1}{2\pi i} , \tag{A.27}$$

$$\delta_+(x) + \delta_-(x) = \delta(x) , \tag{A.28}$$

$$\delta_+(x) - \delta_-(x) = \frac{i}{\pi} P\frac{1}{x} . \tag{A.29}$$

Further literature concerning sections A.1 and A.2:

M.J. Lighthill: *Introduction to Fourier Analysis and Generalised Functions* (Cambridge University Press, Cambridge 1958)
I.M. Gel'fand, G.E. Shilov: *Generalized Functions*, Vol. 1–5 (Academic Press, New York 1968)

A.3 Green's Functions

Starting from a linear differential operator D and a function $f(x)$, we study the linear inhomogeneous differential equation

$$D\psi(x) = f(x) \tag{A.30}$$

for $\psi(x)$.

Replacing the inhomogeneity by a δ-distribution located at x', one finds

$$DG(x, x') = \delta(x - x') \quad . \tag{A.31}$$

The quantity $G(x, x')$ is called the Green's function of the differential operator D. For translationally invariant D, $G(x, x') = G(x - x')$.

Using the Green's function, one finds for the general solution of (A.30)

$$\psi(x) = \psi_0(x) + \int dx' \, G(x, x') f(x') \quad , \tag{A.32}$$

where $\psi_0(x)$ is the general solution of the homogeneous differential equation

$$D\psi_0(x) = 0 \quad . \tag{A.33}$$

Equation (A.32) contains a particular solution of the inhomogeneous differential equation (A.30), given by the second term, which is not restricted to any special form of the inhomogeneity $f(x)$. A great advantage of the Green's function is that, once it has been determined from (A.31), it enables one to compute a particular solution for arbitrary inhomogeneities.

In scattering theory, we require the Green's function for the wave equation

$$(\boldsymbol{\nabla}^2 + k^2)G(\boldsymbol{x} - \boldsymbol{x}') = \delta^{(3)}(\boldsymbol{x} - \boldsymbol{x}') \quad . \tag{A.34}$$

The Fourier transform of $G(\boldsymbol{x} - \boldsymbol{x}')$

$$\tilde{G}(\boldsymbol{q}) = \int d^3y \, e^{-i\boldsymbol{q} \cdot \boldsymbol{y}} G(\boldsymbol{y}) \tag{A.35}$$

becomes, with (A.34),

$$(-q^2 + k^2)\tilde{G}(\boldsymbol{q}) = 1 \quad . \tag{A.36}$$

Inverting (A.35) and using (A.36), one first obtains for the Green's function

$$G(\boldsymbol{y}) = \int \frac{d^3q}{(2\pi)^3} e^{i\boldsymbol{q} \cdot \boldsymbol{y}} \frac{1}{-q^2 + k^2} \quad . \tag{A.37}$$

However, because of the poles at $\pm k$, the integral in (A.37) does not exist ($k > 0$). In order to obtain a well defined integral, we must displace the poles by an infinitesimal amount from the real axis:

$$G_{\pm}(\boldsymbol{x}) = -\lim_{\varepsilon \to 0} \int \frac{d^3q}{(2\pi)^3} \frac{e^{i\boldsymbol{q} \cdot \boldsymbol{x}}}{q^2 - k^2 \mp i\varepsilon} \quad . \tag{A.38}$$

In the integrand of G_+, the poles are at the locations $q = \pm (k + i\varepsilon/2k)$, and in the integrand of G_-, they are at $q = \pm (k - i\varepsilon/2k)$. From this one sees that the shift of the poles of G_+ in the limit $\varepsilon \to 0$ is equivalent to deforming the path of integration along the real axis as shown in Fig. 18.2. After carrying out the angular integration, one finds

$$G_\pm(x) = -\frac{1}{4\pi^2 i r} \int_{-\infty}^{+\infty} dq \, \frac{q e^{iqr}}{q^2 - k^2 \mp i\varepsilon} \quad . \tag{A.39}$$

Since $r = |x| > 0$, the path of integration can be closed by an infinite semicircle in the upper half-plane, so that the residue theorem then yields

$$G_\pm(x) = -\frac{e^{\pm ikr}}{4\pi r} \quad . \tag{A.40}$$

The quantity G_+ is called the *retarded Green's function*. The solution (A.32) is composed of a free solution of the wave equation and an outgoing spherical wave.

The quantity G_- is called the *advanced Green's function*. The solution (A.32) then consists of a free solution of the wave equation and an incoming spherical wave.

B. Canonical and Kinetic Momentum

In this appendix, we collect some formulae from the classical mechanics of charged particles moving in an electromagnetic field.

We first recall that the Hamiltonian

$$H = \frac{1}{2m} \left(p - \frac{e}{c} A(x,t) \right)^2 + e\Phi(x,t) \tag{B.1}$$

leads to the classical equations of motion (B.3). For this, we compute (note the summation convention)

$$\dot{x}_i = \frac{\partial H}{\partial p_i} = \frac{1}{m} \left(p_i - \frac{e}{c} A_i(x,t) \right) \quad , \tag{B.2a}$$

$$\begin{aligned} \dot{p}_i = -\frac{\partial H}{\partial x_i} &= -\frac{1}{m} \left(p_j - \frac{e}{c} A_j(x,t) \right) \left(-\frac{e}{c} A_{j,i} \right) - e\Phi_{,i} \\ &= \dot{x}_j \frac{e}{c} A_{j,i} - e\Phi_{,i} \end{aligned} \tag{B.2b}$$

with $f_{,i} \equiv \partial f / \partial x_i$. From (B.2a,b), the Newtonian equation of motion

$$m\ddot{x}_i = \dot{p}_i - \frac{e}{c} A_{i,j} \dot{x}_j - \frac{e}{c} \dot{A}_i = \dot{x}_j \frac{e}{c} A_{j,i} - e\Phi_{,i} - \frac{e}{c} A_{i,j} \dot{x}_j - \frac{e}{c} \dot{A}_i$$

follows, i.e.,

$$m\ddot{x}_i = \left(\frac{e}{c} \dot{x} \times B + e E \right)_i \quad . \tag{B.3}$$

Here, we have also used

$$(\dot{\boldsymbol{x}} \times \boldsymbol{B})_i = \varepsilon_{ijk}\dot{x}_j\varepsilon_{krs}A_{s,r} = \dot{x}_j(A_{j,i} - A_{i,j})$$

and

$$(\operatorname{curl} \boldsymbol{A})_k = B_k, \quad \boldsymbol{E} = -\operatorname{grad} \Phi - \frac{1}{c}\frac{\partial \boldsymbol{A}}{\partial t} \quad .$$

One refers to $\boldsymbol{p}$ as the canonical momentum and $m\dot{\boldsymbol{x}}$ from (B.2a) as the kinetic momentum.

From (B.1) and (B.2), we obtain the Lagrangian

$$L = \boldsymbol{p}\dot{\boldsymbol{x}} - H = m\dot{\boldsymbol{x}}^2 + \frac{e}{c}\boldsymbol{A}\dot{\boldsymbol{x}} - \frac{m}{2}\dot{\boldsymbol{x}}^2 - e\Phi \quad ,$$

$$L = \frac{m}{2}\dot{\boldsymbol{x}}^2 + \frac{e}{c}\boldsymbol{A}\dot{\boldsymbol{x}} - e\Phi \quad . \tag{B.4}$$

The Lagrange equations of motion

$$\frac{d}{dt}\frac{\partial L}{\partial \dot{\boldsymbol{x}}} = \frac{\partial L}{\partial \boldsymbol{x}}$$

with

$$\frac{\partial L}{\partial \dot{\boldsymbol{x}}} = m\dot{\boldsymbol{x}} + \frac{e}{c}\boldsymbol{A} \quad , \quad \left(\frac{\partial L}{\partial \boldsymbol{x}}\right)_i = \frac{e}{c}\dot{x}_j A_{j,i} - e\Phi_{,i} \quad ,$$

and

$$\frac{d}{dt}\left(\frac{\partial L}{\partial \dot{\boldsymbol{x}}}\right)_i = m\ddot{x}_i + \frac{e}{c}A_{i,j}\dot{x}_j + \frac{e}{c}\dot{A}_i$$

lead again to Newton's second law with the Lorentz force:

$$m\ddot{\boldsymbol{x}} = e\boldsymbol{E} + \frac{e}{c}\dot{\boldsymbol{x}} \times \boldsymbol{B} \quad .$$

C. Algebraic Determination
of the Orbital Angular Momentum Eigenfunctions

We now determine the eigenfunctions of orbital angular momentum algebraically. For this we define

$$x_{\pm} = x \pm iy \quad . \tag{C.1}$$

The following commutation relations hold:

$$[L_z, x_{\pm}] = \pm \hbar x_{\pm}, \quad [L_{\pm}, x_{\pm}] = 0 \quad , \quad [L_{\pm}, x_{\mp}] = \pm 2\hbar z \quad ,$$

$$[\boldsymbol{L}^2, x_+] = L_z\hbar x_+ + \hbar x_+ L_z + \hbar^2 x_+ - 2\hbar z L_+$$
$$= 2\hbar x_+ L_z + 2\hbar^2 x_+ - 2\hbar z L_+ \quad , \tag{C.2a}$$

where

$$\boldsymbol{L}^2 = L_z^2 + \hbar L_z + L_- L_+ \tag{C.2b}$$

has been used.

From (C.2a), (C.2b) and relations (5.14) and (5.15), which were also shown algebraically, it follows that

$$L_z x_+ |l, l\rangle = x_+ L_z |l, l\rangle + \hbar x_+ |l, l\rangle = \hbar(l + 1)x_+ |l, l\rangle \tag{C.3a}$$

and

$$\boldsymbol{L}^2 x_+ |l, l\rangle = \hbar^2 l(l + 1)x_+ |l, l\rangle + 2\hbar^2 (l + 1)x_+ |l, l\rangle$$
$$= \hbar^2 (l + 1)(l + 2)x_+ |l, l\rangle \quad . \tag{C.3b}$$

The quantity x_+ is thus the ladder operator for the states $|l, l\rangle$,

$$x_+ |l, l\rangle = N|l + 1, l + 1\rangle \quad . \tag{C.4}$$

Hence, the eigenstates of angular momentum can be represented as follows:

$$|l, m\rangle = N' L_-^{l-m} (x_+)^l |0, 0\rangle \quad . \tag{C.5}$$

N and N' in (C.4) and (C.5) are constants. Since $\boldsymbol{L}|0, 0\rangle = 0$, it follows that (compare (5.4))

$$\langle \boldsymbol{x}|U_{\delta\varphi}|0, 0\rangle = \langle U_{\delta\varphi}^{-1}\boldsymbol{x}|0, 0\rangle = \langle \boldsymbol{x}|0, 0\rangle \quad ,$$

and thus

$$\psi_{00}(\boldsymbol{x}) = \langle \boldsymbol{x}|0, 0\rangle \tag{C.6}$$

does not depend on the polar angles ϑ, φ. The norm of $|0, 0\rangle$

$$\langle 0, 0|0, 0\rangle = \int d\Omega \langle 0, 0|\boldsymbol{x}\rangle\langle \boldsymbol{x}|0, 0\rangle$$

is unity for

$$\psi_{00}(\boldsymbol{x}) = \frac{1}{\sqrt{4\pi}} \quad . \tag{C.7}$$

The norm of the state $|l, l\rangle \propto (x_+/r)^l |0, 0\rangle$, whose coordinate representation is

$$\langle x|\left(\frac{x_+}{r}\right)^l |0, 0\rangle = \frac{1}{\sqrt{4\pi}} \sin^l \vartheta\, e^{il\varphi} \quad ,$$

becomes

$$\langle 0,0| \left(\frac{x_-}{r}\right)^l \left(\frac{x_+}{r}\right)^l |0,0\rangle = \langle 0,0| \left(\frac{x^2+y^2}{r^2}\right)^l |0,0\rangle = \langle 0,0| \left(1-\frac{z^2}{r^2}\right)^l |0,0\rangle$$

$$= \langle 0,0| \sin^{2l} \vartheta |0,0\rangle$$

$$= \int_0^{2\pi} d\varphi \int_0^\pi d\vartheta \, \sin\vartheta \, \frac{1}{4\pi} \sin^{2l}\vartheta$$

$$= \frac{1}{2} \int_{-1}^1 d(\cos\vartheta) \, \sin^{2l}\vartheta = I_l \quad,$$

$$I_l = \int_0^1 d\eta(1-\eta^2)^l = \eta(1-\eta^2)^l \Big|_0^1 + 2l \int_0^1 d\eta \, \eta(1-\eta^2)^{l-1}\eta$$

$$= -2lI_l + 2lI_{l-1} \quad,$$

$$I_l = \frac{2l}{2l+1}I_{l-1} = \frac{2l}{2l+1}\frac{2(l-1)}{2(l-1)+1} \cdots \frac{2\times 1}{2+1}I_0$$

$$= \frac{2l(2l-2)\ldots 2}{(2l+1)(2l-1)\ldots 3} = \frac{2^{2l}(l!)^2}{(2l+1)!} \quad,$$

$$I_0 = 1 \quad.$$

One thus has

$$\psi_{ll}(\boldsymbol{x}) = \frac{1}{\sqrt{4\pi I_l}} \sin^l\vartheta \, e^{il\varphi} \tag{C.8}$$

and the definition of the spherical harmonics

$$Y_{ll}(\vartheta,\varphi) = (-1)^l \sqrt{\frac{(2l+1)!}{4\pi}} \frac{1}{2^l l!} \sin^l\vartheta \, e^{il\varphi} \quad. \tag{C.9}$$

$Y_{ll}(\vartheta,\varphi)$ can also be found directly from the equations

$$L_z Y_{ll} = \hbar l \, Y_{ll} \quad\text{and}\quad L_+ Y_{ll} = 0 = e^{i\varphi} \left(\frac{\partial}{\partial\vartheta} + i\cot\vartheta \frac{\partial}{\partial\varphi}\right) e^{il\varphi} f(\vartheta) \quad.$$

The first implies

$$Y_{ll} = e^{il\varphi} f(\vartheta) \quad,$$

and the second implies

$$\frac{\partial}{\partial\vartheta} f(\vartheta) = l \cot\vartheta f(\vartheta) \quad,$$

$$\frac{df}{f} = l \cot\vartheta d\vartheta \quad,$$

$$\log|f| = l \log\sin\vartheta + A \quad,$$

$$f = \alpha \sin^l\vartheta \quad \text{q.e.d} \quad.$$

The remaining eigenfunctions are obtained by application of L_-:

$$(L_-)^{l-m}|l,l\rangle = N'|l,m\rangle \quad . \tag{C.10}$$

In order to determine N', we start from

$$L_-|l,m\rangle = \hbar\sqrt{(l+m)(l-m+1)}|l,m-1\rangle \quad ;$$

hence,

$$(L_-)^{l-m}|l,l\rangle = [2l \times 1 \times (2l-1) \times 2 \dots (l+m+1)(l-m)]^{1/2}\hbar^{l-m}|l,m\rangle$$

$$= \left(\frac{(2l)!(l-m)!}{(l+m)!}\right)^{1/2}\hbar^{l-m}|l,m\rangle \quad ,$$

and

$$Y_{lm}(\vartheta,\varphi) = \sqrt{\frac{(l+m)!}{(2l)!(l-m)!}}(L_-/\hbar)^{l-m}Y_{ll}(\vartheta,\varphi) . \tag{C.11}$$

We now apply the operator L_-:

$$(L_-/\hbar)f(\vartheta)\mathrm{e}^{\mathrm{i}m\varphi} = \mathrm{e}^{-\mathrm{i}\varphi}\left(-\frac{\partial}{\partial\vartheta} + \mathrm{i}\cot\vartheta\frac{\partial}{\partial\varphi}\right)f(\vartheta)\mathrm{e}^{\mathrm{i}m\varphi}$$

$$= \mathrm{e}^{\mathrm{i}(m-1)\varphi}(-1)(f'(\vartheta) + m\cot\vartheta f) \quad .$$

Comparing this with

$$\frac{d}{d\cos\vartheta}(f\sin^m\vartheta) = -(f' + mf\cot\vartheta)\sin^{m-1}\vartheta \quad ,$$

we see that

$$(L_-/\hbar)f(\vartheta)\mathrm{e}^{\mathrm{i}m\varphi} = \mathrm{e}^{\mathrm{i}(m-1)\varphi}\sin^{1-m}\vartheta\frac{d(f\sin^m\vartheta)}{d\cos\vartheta} \quad .$$

Applying L_- $(l-m)$ times yields

$$(L_-/\hbar)^{l-m}\mathrm{e}^{\mathrm{i}l\varphi}\sin^l\vartheta = \mathrm{e}^{\mathrm{i}m\varphi}\sin^{-m}\vartheta\frac{d^{l-m}\sin^{2l}\vartheta}{(d\cos\vartheta)^{l-m}} \tag{C.12}$$

and

$$Y_{lm}(\vartheta,\varphi) = (-1)^l\sqrt{\frac{(l+m)!(2l+1)}{(l-m)!4\pi}}\frac{1}{2^l l!}\mathrm{e}^{\mathrm{i}m\varphi}\sin^{-m}\vartheta\frac{d^{l-m}\sin^{2l}\vartheta}{(d\cos\vartheta)^{l-m}} \tag{C.13}$$

$$= (-1)^{l+m}\frac{1}{2^l l!}\sqrt{\frac{(l-m)!(2l+1)}{(l+m)!4\pi}}\mathrm{e}^{\mathrm{i}m\varphi}\sin^m\vartheta\frac{d^{l+m}\sin^{2l}\vartheta}{(d\cos\vartheta)^{l+m}} \quad . \tag{C.13'}$$

This is in accord with (5.22), and the spherical harmonics obey

$$Y_{l,m}(\vartheta,\varphi) = (-1)^m Y^*_{l,-m}(\vartheta,\varphi) \quad . \tag{C.14}$$

This concludes the algebraic derivation of the angular momentum eigenfunctions.

Remark: In going from (C.13) over to the conventional representation (C.13′), we have used the fact that the associated Legendre function

$$P_l^m(\eta) = \frac{1}{2^l l!}(1-\eta^2)^{m/2}\frac{d^{l+m}}{d\eta^{l+m}}(\eta^2-1)^l \tag{C.15}$$

satisfies the identity

$$P_l^{-m} = (-1)^m\frac{(l-m)!}{(l+m)!}P_l^m \quad . \tag{C.16}$$

For the derivation of this identity, we note that both P_l^m and P_l^{-m} are lth-order polynomials in η for even m; for odd m, they are polynomials of order $(l-1)$, multiplied by $\sqrt{1-\eta^2}$. Further, the differential equation for P_l^m contains the coefficient m only quadradically, and therefore P_l^{-m} is also a solution and must be proportional to the regular solution P_l^m which we began with. In order to determine the coefficient of proportionality, we compare the highest powers of η in the expressions for P_l^{-m} and P_l^m, multiplied by $(1-\eta^2)^{m/2}$:

$$(1-\eta^2)^{m/2}P_l^{-m} = \frac{1}{2^l l!}\frac{d^{l-m}(\eta^2-1)^l}{d\eta^{l-m}} = \frac{(2l)!}{2^l l!(l+m)!}\eta^{l+m} + \cdots$$

and

$$(1-\eta^2)^{m/2}P_l^m = \frac{(1-\eta^2)^m}{2^l l!}\frac{d^{l+m}(\eta^2-1)^l}{d\eta^{l+m}} = \frac{(2l)!(-1)^m}{2^l l!(l-m)!}\eta^{l+m} + \cdots \quad ,$$

which yields (C.16).

We now prove algebraically that for the angular momentum operator the quantum number l is a nonnegative integer. (A shorter derivation is studied in Problem 5.7.) To this end, we construct a "ladder operator", which lowers the quantum number l by 1; for half-integral l-values, it would then take us out of the region $l \geq 0$. We introduce the definition

$$\boldsymbol{a}^{(l)} = \mathrm{i}\hat{\boldsymbol{x}} \times \boldsymbol{L} - \hbar l\hat{\boldsymbol{x}} = \mathrm{i}\begin{cases} \hat{x}_y L_z - \hat{x}_z L_y \\ \hat{x}_z L_x - \hat{x}_x L_z \\ \hat{x}_x L_y - \hat{x}_y L_x \end{cases} - \hbar l\hat{\boldsymbol{x}} \quad , \tag{C.17}$$

where $\hat{\boldsymbol{x}} = \boldsymbol{x}/|\boldsymbol{x}|$ is the radial unit vector. It turns out to be useful to introduce the decomposition

$$a_\pm^{(l)} = a_x^{(l)} \pm \mathrm{i}a_y^{(l)} = \mp\hat{x}_z L_\pm \pm \hat{x}_\pm(L_z \mp \hbar l)$$
$$a_z^{(l)} = \hat{x}_- L_+ + \hat{x}_z(L_z - \hbar l) - \hat{\boldsymbol{x}} \cdot \boldsymbol{L} = \hat{x}_- L_+ + \hat{x}_z(L_z - \hbar l) \quad , \tag{C.18}$$

where we have used $\hat{\boldsymbol{x}} \cdot \boldsymbol{L} = 0$, a property which is valid specifically for the orbital angular momentum, and where we have defined $\hat{x}_\pm = \hat{x}_x \pm \mathrm{i}\hat{x}_y$. The commutation relations read

$$[a_+^{(l)}, a_-^{(l)}] = 2\hbar\hat{\boldsymbol{x}}^2 L_z = 2\hbar L_z \quad , \tag{C.19a}$$

$$[L_+, a_-^{(l)}] = 2\hbar a_z^{(l)} \quad , \tag{C.19b}$$

$$[L_z, a_-^{(l)}] = -\hbar a_-^{(l)} \quad . \tag{C.19c}$$

Equations (C.18) and (5.15) then imply

$$a_+^{(l)}|l, l\rangle = 0 \tag{C.20a}$$

and

$$a_z^{(l)}|l, l\rangle = 0 \quad . \tag{C.20b}$$

Together with the commutator (C.19a), this yields

$$a_+^{(l)} a_-^{(l)}|l, l\rangle = 2\hbar^2 l \hat{x}^2|l, l\rangle \quad . \tag{C.21}$$

Multiplication of (C.21) by $\langle l, l|$ thus yields $a_-^{(l)}|l, l\rangle \neq 0$ for all $l \neq 0$. For the state $|0, 0\rangle$, both (C.18) and (C.21) imply

$$a_-^{(0)}|0, 0\rangle = (\hat{x}_z L_- - \hat{x}_-(L_z + 0))|0, 0\rangle = 0 \quad .$$

We now determine the eigenvalues of the state $a_-^{(l)}|l, l\rangle$: Using (C.19b) and (C.20b), one finds

$$L_+ a_-^{(l)}|l, l\rangle = a_-^{(l)} L_+|l, l\rangle + 2\hbar a_z^{(l)}|l, l\rangle = 0 \tag{C.22}$$

and, from (C.19c),

$$L_z a_-^{(l)}|l, l\rangle = \hbar(l - 1) a_-^{(l)}|l, l\rangle \quad . \tag{C.23}$$

With $\boldsymbol{L}^2 = L_- L_+ + \hbar L_z + L_z^2$, we obtain from (C.23) and (C.22)

$$\boldsymbol{L}^2 a_-^{(l)}|l, l\rangle = \hbar^2((l - 1) + (l - 1)^2) a_-^{(l)}|l, l\rangle = \hbar^2 l(l - 1) a_-^{(l)}|l, l\rangle \quad . \tag{C.24}$$

In summary, (C.23) and (C.24) imply

$$a_-^{(l)}|l, l\rangle \propto |l - 1, l - 1\rangle \quad . \tag{C.25}$$

In Sect. 5.2, it was already shown that the algebra of angular momentum operators inevitably leads to half-integral or integral l. If half-integral l were to occur, then starting from $|l, l\rangle$ with $a_-^{(l)}|l, l\rangle \propto |l - 1, l - 1\rangle, \ldots a_-^{(l-1)} a_-^{(l)}|l, l\rangle \propto |l - 2, l - 2\rangle$, and so on, one would eventually encounter negative half-integral l. This contradicts the inequality $l \geq 0$ derived in Sect. 5.2! Together with (5.16), this implies that the *orbital angular momentum eigenvalues* l are given by the nonnegative integers $0, 1, 2, \ldots$[1]

---

[1] Further literature concerning this appendix can be found in C.C. Noack: Phys. Bl. **41**, 283 (1985). A different algebraic proof is presented in F. Schwabl, *Quantenmechanik*, 7. Auflage (Springer, Berlin Heidelberg 2007).

D. The Periodic Table
and Important Physical Quantities

Conversion factors:

$1\,\text{eV} = 1.60219 \times 10^{-19}\,\text{J}$

$1\,\text{N} = 10^5\,\text{dyn}$

$1\,\text{J} = 1 \times 10^7\,\text{erg}$

$1\,\text{C} = 2.997925 \times 10^9\,\text{esu} = 2.997925 \times 10^9\,\sqrt{\text{dyn cm}^2}$

$1\,\text{K} \cong 0.86171 \times 10^{-4}\,\text{eV}$

$1\,\text{eV} \cong 2.4180 \times 10^{14}\,\text{Hz} \cong 1.2399 \times 10^{-4}\,\text{cm}$

$1\,\text{T} = 10^4\,\text{gauss (G)}$

$1\,\text{Å} = 10^{-8}\,\text{cm}$

The Periodic Table

Legend: Element Z / Ground state configuration / Ground state $^{2S+1}L_J$

Column group headers:
- I — Alkali metals
- II — Alkali earth metals
- (Transition block) — Noble metals
- Chalcogens
- Halogens
- Noble gases

I	II	III	IV	V	VI	VII	VIII			(Noble metals)		III	IV	Chalco-gens	Halo-gens	Noble gases
H^1 $1s$ $^2S_{1/2}$																He^2 $1s^2$ $^1S_0$
Li^3 $1s^2 2s$ $^2S_{1/2}$	Be^4 $1s^2 2s$ $^1S_0$											B^5 $2s^2 2p$ $^2P_{1/2}$	C^6 $2s^2 2p^2$ $^3P_0$	N^7 $2p^3$ $^4S_{3/2}$ / O^8 $2p^4$ $^3P_2$	F^9 $2p^5$ $^2P_{3/2}$	Ne^{10} $2p^6$ $^1S_0$
Na^{11} $3s$ $^2S_{1/2}$	Mg^{12} $3s^2$ $^1S_0$											Al^{13} $3s^2 3p$ $^2P_{1/2}$	Si^{14} $3s^2 3p^2$ $^3P_0$	P^{15} $3p^3$ $^4S_{3/2}$ / S^{16} $3p^4$ $^3P_2$	Cl^{17} $3p^5$ $^2P_{3/2}$	Ar^{18} $3s^2 3p^6$ $^1S_0$
K^{19} $4s$ $^2S_{1/2}$	Ca^{20} $4s^2$ $^1S_0$	Sc^{21} $4s^2 3d$ $^2D_{3/2}$	Ti^{22} $4s^2 3d^2$ $^3F_2$	V^{23} $4s^2 3d^3$ $^4F_{3/2}$	Cr^{24} $4s 3d^5$ $^7S_3$	Mn^{25} $4s^2 3d^5$ $^6S_{5/2}$	Fe^{26} $4s^2 3d^6$ $^5D_4$	Co^{27} $4s^2 3d^7$ $^4F_{9/2}$	Ni^{28} $4s^2 3d^8$ $^3F_4$	Cu^{29} $4s 3d^{10}$ $^2S_{1/2}$	Zn^{30} $4s^2 3d^{10}$ $^1S_0$	Ga^{31} $4s^2 3d^{10} 4p$ $^2P_{1/2}$	Ge^{32} $3d^{10} 4p^2$ $^3P_0$	As^{33} $3d^{10} 4p^3$ $^4S_{3/2}$ / Se^{34} $3d^{10} 4p^4$ $^3P_2$	Br^{35} $3d^{10} 4p^5$ $^2P_{3/2}$	Kr^{36} $4s^2 4p^6$ $^1S_0$
Rb^{37} $5s$ $^2S_{1/2}$	Sr^{38} $5s^2$ $^1S_0$	Y^{39} $5s^2 4d$ $^2D_{3/2}$	Zr^{40} $5s^2 4d^2$ $^3F_2$	Nb^{41} $5s 4d^4$ $^6D_{1/2}$	Mo^{42} $5s 4d^5$ $^7S_3$	Tc^{43} $5s^2 4d^5$ $^6S_{5/2}$	Ru^{44} $5s 4d^7$ $^5F_5$	Rh^{45} $5s 4d^8$ $^4F_{9/2}$	Pd^{46} $4d^{10}$ $^1S_0$	Ag^{47} $5s 4d^{10}$ $^2S_{1/2}$	Cd^{48} $5s^2 4d^{10}$ $^1S_0$	In^{49} $5s^2 4d^{10} 5p$ $^2P_{1/2}$	Sn^{50} $4d^{10} 5p^2$ $^3P_0$	Sb^{51} $4d^{10} 5p^3$ $^4S_{3/2}$ / Te^{52} $4d^{10} 5p^4$ $^3P_2$	I^{53} $4d^{10} 5p^5$ $^2P_{3/2}$	Xe^{54} $5s^2 5p^6$ $^1S_0$
Cs^{55} $6s$ $^2S_{1/2}$	Ba^{56} $6s^2$ $^1S_0$	La^{57} $6s^2 5d$ $^2D_{3/2}$	Hf^{72} $6s^2 5d^2$ $^3F_2$	Ta^{73} $6s^2 5d^3$ $^4F_{3/2}$	W^{74} $6s^2 5d^4$ $^5D_0$	Re^{75} $6s^2 5d^5$ $^6S_{5/2}$	Os^{76} $6s^2 5d^6$ $^5D_4$	Ir^{77} $6s^2 5d^7$ $^4F_{9/2}$	Pt^{78} $6s 5d^9$ $^3D_3$	Au^{79} $6s 5d^{10}$ $^2S_{1/2}$	Hg^{80} $6s^2 5d^{10}$ $^1S_0$	Tl^{81} $6s^2 5d^{10} 6p$ $^2P_{1/2}$	Pb^{82} $6p^2$ $^3P_0$	Bi^{83} $6p^3$ $^4S_{3/2}$ / Po^{84} $6p^4$ $^3P_2$	At^{85} $6p^5$ $^2P_{3/2}$	Rn^{86} $6p^6$ $^1S_0$
Fr^{87} $7s$ $^2S_{1/2}$	Ra^{88} $7s^2$ $^1S_0$	Ac^{89} $7s^2 6d$ $^2D_{3/2}$														

Lanthanides (rare earths):

La^{57} $6s^2 5d$ $^2D_{3/2}$	Ce^{58} $6s^2 5d\,4f$ $^3H_5$	Pr^{59} $6s^2 4f^3$ $^4I_{9/2}$	Nd^{60} $6s^2 4f^4$ $^5I_4$	Pm^{61} $6s^2 4f^5$ $^6H_{5/2}$	Sm^{62} $6s^2 4f^6$ $^7F_0$	Eu^{63} $6s^2 4f^7$ $^8S_{7/2}$	Gd^{64} $6s^2 5d\,4f^7$ $^9D_2$	Tb^{65} $6s^2 5d\,4f^8$	Dy^{66} $6s^2 4f^{10}$	Ho^{67} $6s^2 4f^{11}$	Er^{68} $6s^2 4f^{12}$	Tm^{69} $6s^2 4f^{13}$ $^2F_{7/2}$	Yb^{70} $6s^2 4f^{14}$ $^1S_0$	Lu^{71} $6s^2 5d\,4f^{14}$ $^2D_{3/2}$

Actinides:

Ac^{89} $7s^2 6d$ $^2D_{3/2}$	Th^{90} $7s^2 6d^2$	Pa^{91} $6d\,5f^2$	U^{92} $6d\,5f^3$ $^5L_6$	Np^{93} $5f^5$	Pu^{94} $5f^6$	Am^{95} $5f^7$ $^8S_{7/2}$	Cm^{96} $6d\,5f^7$	Bk^{97} $5f^9$	Cf^{98} $5f^{10}$	Es^{99} $5f^{11}$	Fm^{100} $5f^{12}$	Md^{101} $5f^{13}$	No^{102} $5f^{14}$	Lr^{103} $6d\,5f^{14}$

Important Constants

Quantity	Symbol or formular in cgs Representation	Symbol or formular in SI Representation	Numerical value and units in cgs	System	SI
Planck's constant	h	$\hbar = h/2\pi$	6.6262×10^{-27} erg s 1.0546×10^{-27} erg s		6.6262×10^{-34} J s $\cong 4.1357 \times 10^{-15}$ eV s 1.0546×10^{-34} J s $\cong 6.5822 \times 10^{-16}$ eV s
Elementary charge		e_0	4.80324×10^{-10} esu		1.60219×10^{-19} C
Speed of light in vacuum		c	2.997925×10^{10} cm s^{-1}		2.997925×10^8 m s^{-1}
Atomic mass unit		$\frac{1}{12} m_{C^{12}}$	1.66053×10^{-24} g		1.66053×10^{-27} kg $\cong 931.5$ MeV
Electron rest mass		m_e	9.1096×10^{-28} g		9.1096×10^{-31} kg $\cong 5.4859 \times 10^{-4}$ amu
Electron rest energy		$m_e c^2$		0.5110 MeV	
Proton rest mass		m_p	1.6726×10^{-24} g		1.6726×10^{-27} kg $\cong 1.0072766$ amu
Proton rest energy		$m_p c^2$		938.25 MeV	
Neutron rest mass		m_n	1.6749×10^{-24} g		1.6749×10^{-27} kg $\cong 1.0086652$ amu
Neutron rest energy		$m_n c^2$		939.55 MeV	
Mass ratio proton:electron		m_p/m_e		1836.109	
Mass ratio neutron:proton		m_n/m_p		1.0013786	
Specific electron charge		e_0/m_e	5.272759×10^{17} esu/g		1.758803×10^{11} C/kg
Classical elektron radius	$\dfrac{e_0^2}{m_e c^2}$	$r_e \quad \dfrac{1}{4\pi\epsilon_0} \times \dfrac{e_0^2}{m_e c^2}$	2.8179×10^{-13} cm		2.8179×10^{-15} m

Quantity		Symbol		CGS value	SI value
Compton wavelength of electron	$\left\{\begin{array}{l} h/m_ec \\ \hbar/m_ec \end{array}\right.$	$\begin{array}{l}\lambda_c \\ \lambda_c\end{array}$	$\begin{array}{l} h/m_ec \\ \hbar/m_ec \end{array}$	$\begin{array}{l}2.4263 \times 10^{-10}\,\text{cm} \\ 3.8616 \times 10^{-11}\,\text{cm}\end{array}$	$\begin{array}{l}2.4263 \times 10^{-12}\,\text{m} \\ 3.8616 \times 10^{-13}\,\text{m}\end{array}$
Sommerfeld fine-structure constant	$\dfrac{e_0^2}{\hbar c}$	α	$\dfrac{1}{4\pi\varepsilon_0}\times\dfrac{e_0^2}{\hbar c}$	$\dfrac{1}{137.036}$	
Bohr radius of hydrogen ground state	$\dfrac{\hbar^2}{m_e e_0^2}$	a	$4\pi\varepsilon_0 \times \dfrac{\hbar^2}{m_e e_0^2}$	$5.2918 \times 10^{-9}\,\text{cm}$	$5.2918 \times 10^{-11}\,\text{m}$
Rydberg constant (ground state energy of hydrogen)	$\tfrac{1}{2}m_e c^2 \times \alpha^2$	Ry	$\tfrac{1}{2}m_e c^2 \times \alpha^2$	$2.1799 \times 10^{-11}\,\text{erg}$	$2.1799 \times 10^{-18}\,\text{J} \cong 13.6058\,\text{eV}$
Bohr magneton	$\dfrac{e_0\hbar}{2m_e c}$	μ_B	$\dfrac{e_0\hbar}{2m_e}$	$9.2741 \times 10^{-21}\,\text{erg}\,\text{G}^{-1}$	$9.2741 \times 10^{-24}\,\text{J}\,\text{T}^{-1}$
Nuclear magneton	$\dfrac{e_0\hbar}{2m_p c}$	μ_N	$\dfrac{e_0\hbar}{2m_p}$	$5.0509 \times 10^{-24}\,\text{erg}\,\text{G}^{-1}$	$5.0509 \times 10^{-27}\,\text{J}\,\text{T}^{-1}$
Magnetic moment of electron	$\Big\}$	μ_e		$9.2848 \times 10^{-21}\,\text{erg}\,\text{G}^{-1}$	$\begin{array}{l}9.2848 \times 10^{-24}\,\text{J}\,\text{T}^{-1} \\ = 1.00115964\,\mu_B\end{array}$
Magnetic moment of proton	$\Big\}$	μ_p		$1.41062 \times 10^{-23}\,\text{erg}\,\text{G}^{-1}$	$\begin{array}{l}1.41062 \times 10^{-26}\,\text{J}\,\text{T}^{-1} \\ = 2.7928\,\mu_N\end{array}$
Gravitational constant		G		$6.6732 \times 10^{-8}\,\text{dyn}\,\text{cm}^2\,\text{g}^{-2}$	$6.6732 \times 10^{-11}\,\text{N}\,\text{m}^2\,\text{kg}^{-2}$
Standard acceleration of gravity		g		$9.80665 \times 10^2\,\text{cm}\,\text{s}^{-2}$	$9.80665\,\text{m}\,\text{s}^{-2}$
Permeability constant in vacuum		μ_0			$4\pi \times 10^{-7}\,\text{N}\,\text{A}^{-2} = 1.2566 \times 10^{-6}\,\text{N}\,\text{A}^{-2}$
Dielectric constant in vacuum	$\Big\}$	$\begin{array}{l}\varepsilon_0 = 1/(\mu_0 c^2) \\ 1/(4\pi\varepsilon_0)\end{array}$			$\begin{array}{l}8.85418 \times 10^{-12}\,\text{C}^2\,\text{m}^{-2}\,\text{N}^{-1} \\ 8.98755 \times 10^9\,\text{N}\,\text{m}^2\,\text{C}^{-2}\end{array}$
Boltzmann constant		k_B		$1.38062 \times 10^{-16}\,\text{erg}\,\text{K}^{-1}$	

Subject Index